THEORY OF MACHINES

SHIVENDRA NANDAN, RISHIKESH TRIVEDI
& SATYAJEET KANT

Copyright © 2020 The Shivendra Group
All rights reserved.
ISBN: 9798672783222

Dedication

Pt. Devesh Kumar Mishra

CONTENTS

Chapter - 1: Mechanism
Chapter - 2 : Flywheel
Chapter - 3 : Governor
Chapter - 4 : CAM
Chapter - 5 : Balancing of Rigid Rotors and Field Balancing
Chapter - 6: Balancing of single and multi- cylinder engines
Chapter - 7: Linear Vibration Analysis of Mechanical Systems
Chapter - 8: Critical speeds or Whirling of Shafts
Chapter - 9: Gear Train
Chapter - 10: Miscellaneous

Guidelines

(a) How to use this material for preparation of GATE

i. Attend all classes (Coaching) and meticulously read class note OR read book by Ghosh-Mallick

ii. Topic list (which are to be covered for GATE)
   a. *Mechanism (VIMP)*
   b. *Linear Vibration Analysis of Mechanical Systems (VIMP)*
   c. *Gear train (VIMP)*
   d. *Flywheel (Coefficient of Fluctuation of speed, Coefficient of Fluctuation of energy), mass calculation,*
   e. *Critical Speed of Shafts*

iii. Practice all solved examples from the book of ONLY the topics mentioned in the above topic list.

iv. Then solve my question set (GATE +IES+IAS) on your own and cross check with my explanations of ONLY the topics which are to be covered for GATE.(mentioned above)

(b) How to use this material for preparation of IES

v. Attend all classes (Coaching) and meticulously read class note OR read book by Ghosh-Mallick

vi. All Topics should be covered which is given in this booklet (Question set)

vii. Practice all solved examples from the book of all the topics given in this booklet.

viii. Then solve my question set (GATE +IES+IAS) on your own and cross check with my explanations.

Mechanism

Chapter 1

1. Mechanism

Theory at a glance (IES, GATE & PSU)

What is TOM

The subject theory of machine may be defined as that branch of engineering science which deals with the study of relative motion both the various parts of m/c and forces which act on them.

The theory of m/c may be sub divided into the following branches:
1. **Kinemics:** It deals with the relative motion between the various parts of the machine

2. **Dynamics:** It deals with the force and their effects, while acting upon the m/c part in motion.

    Resistance Body: Resistant bodies are those which do not suffer appreciable distortion or change in physical form by the force acting on them e.g., spring, belt.

    Kinematic Link Element: A resistant body which is a part of an m/c and has motion relative to the other connected parts is **term as link.**

    A link may consist of one or more resistant bodies. Thus a link may consist of a number of parts connected in such away that they form one unit and have no relative motion to each other.

- **A link should have the following two characteristics:**
    1. It should have relative motion, and
    2. It must be a resistant body.

Functions of Linkages

The function of a link mechanism is to produce rotating, oscillating, or reciprocating motion from the rotation of a crank or *vice versa*. Stated more specifically linkages may be used to convert:

1. Continuous rotation into continuous rotation, with a constant or variable angular velocity ratio.
2. Continuous rotation into oscillation or reciprocation (or the reverse), with a constant or variable velocity ratio.
3. Oscillation into oscillation, or reciprocation into reciprocation, with a constant or variable velocity ratio.

Linkages have many different functions, which can be classified according on the primary goal of the mechanism:

Mechanism

Chapter 1

- **Function generation**: the relative motion between the links connected to the frame,
- **Path generation**: the path of a tracer point, or
- **Motion generation**: the motion of the coupler link.

Types

1. **Rigid Link:** It is one which does not undergo any deformation while transmitting motion–C.R, etc.

2. **Flexible Link:** Partly deformed while transmitting motion–**spring, belts.**

3. **Fluid Link:** It formed by having the motion which is transmitted through the fluid by pressure. e. g, hydraulic press, hydraulic brakes.

Kinematic Pair

Two element or links which are connected together in such a way that their relative motion is completely or successfully constrained form a kinematic pair. i.e. The term kinematic pairs actually refer to kinematic constraints between rigid bodies.

The kinematic pairs are divided into **lower pairs** and **higher pairs**, depending on how the two bodies are in contact.

- **Lower Pair:** When two elements have **surface contact** while in motion.
- **Higher Pair:** When two elements have **point or line of contact** while in motion.

Lower Pairs

A pair is said to be a lower pair when the connection between two elements is through the area of contact. Its 6 types are:
- **Revolute Pair**
- **Prismatic Pair**
- **Screw Pair**
- **Cylindrical Pair**
- **Spherical Pair**
- **Planar Pair.**

Revolute Pair

A revolute allows only a relative rotation between elements 1 and 2, which can be expressed by a single coordinate angle 'θ'. Thus a revolute pair has a single degree of freedom.

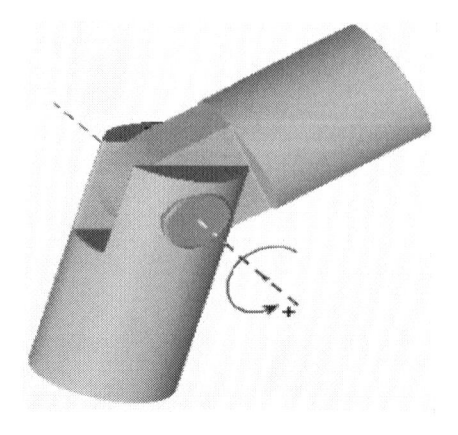

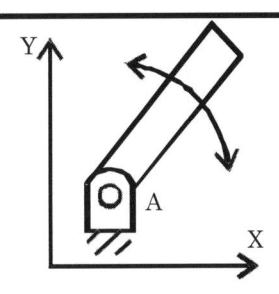

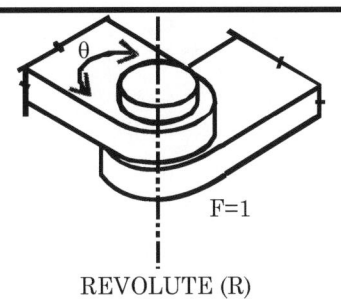

Mechanism

Chapter 1

Prismatic Pair

A prismatic pair allows only a relative translation between elements 1 and 2, which can be expressed by a single coordinate 'x'. Thus a prismatic pair has a single degree of freedom.

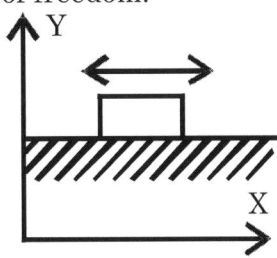

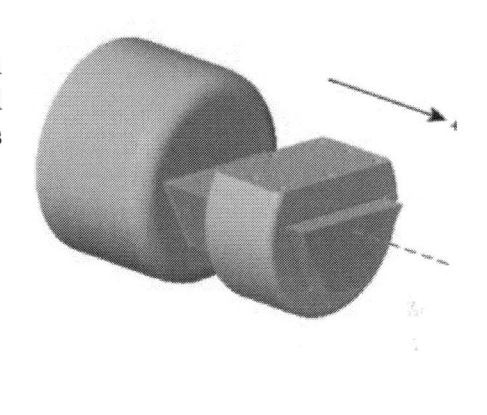

Screw Pair

A screw pair allows only a relative movement between elements 1 and 2, which can be expressed by a single coordinate angle 'θ' or x'. Thus a screw pair has a single degree of freedom. Example-lead screw and nut of lathe, **screw jack.**

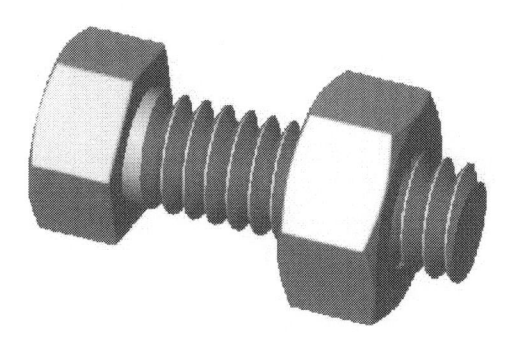

Mechanism

Chapter 1

Cylindrical Pair

A cylindrical pair allows both rotation and translation between elements 1 and 2, which can be expressed as two independent coordinate angle 'θ' and 'x'. Thus a cylindrical pair has two degrees of freedom.

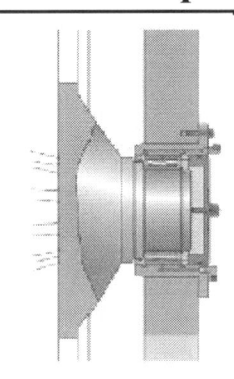

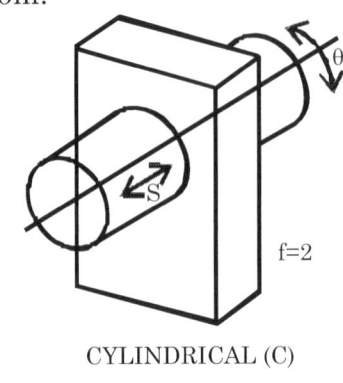

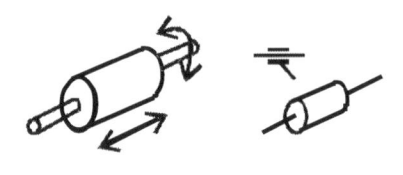

CYLINDRICAL (C)

Spherical Pair

A spherical pair allows three degrees of freedom since the complete description of relative movement between the connected elements needs three independent coordinates. Two of the coordinates 'α' and 'β' are required to specify the position of the axis OA and the third coordinate 'θ' describes the rotation about the axis OA. e.g. – **Mirror attachment of motor cycle.**

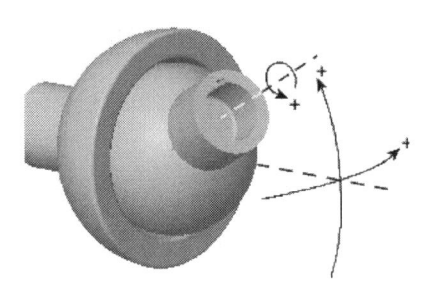

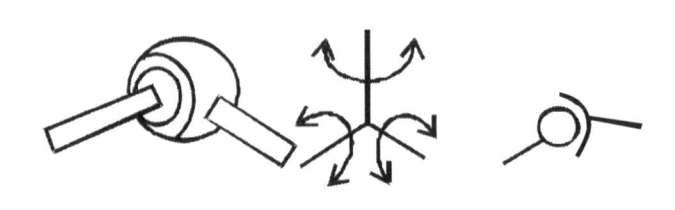

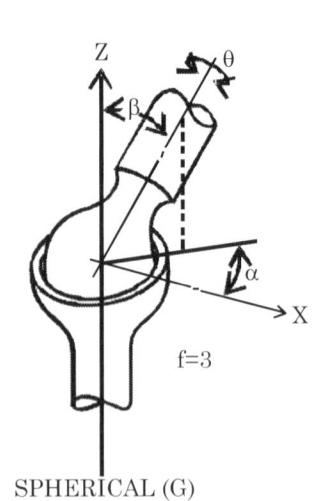

SPHERICAL (G)

Mechanism

Chapter 1

Planar Pair

A planar pair allows three degrees of freedom. Two coordinates x and y describe the relative translation in the xy-plane and the third 'θ' describes the relative rotation about the z-axis.

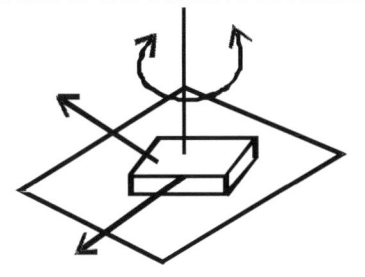

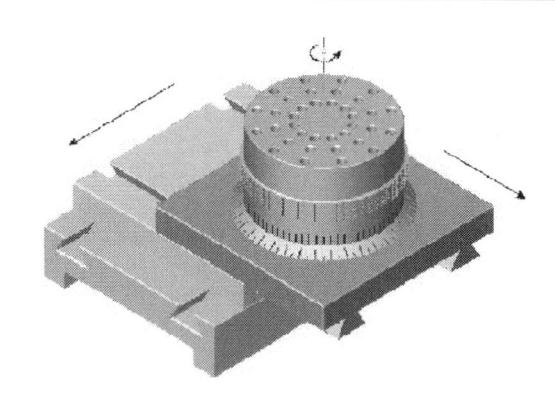

Higher Pairs

A higher pair is defined as one in which the connection between two elements has only **a point or line of contact**. A cylinder and a hole of equal radius and with axis parallel make contact along a surface. Two cylinders with unequal radius and with axis parallel make contact along a line. A point contact takes place when spheres rest on plane or curved surfaces (ball bearings) or between teeth of a skew-helical gears. In roller bearings, between teeth of most of the gears and in cam-follower motion. The degree of freedom of a kinetic pair is given by the number independent coordinates required to completely specify the relative movement.

Wrapping Pairs

Wrapping Pairs comprise belts, chains, and other such devices

Sliding Pair

When the two elements of a pair are connected in such a way that one can only slide relative to the other, the pair is known as a sliding pair. The piston and cylinder, cross-head and guides of a reciprocating steam engine, ram and its guides in shaper, tail stock on the lathe bed etc. are the examples of a sliding pair. A little consideration will show that a sliding pair has a completely constrained motion.

Turning pair

When the two elements of a pair are connected in such a way that one can only turn or revolve about a fixed axis of another link, the pair is known as turning pair. A shaft with collars at both ends fitted into a circular hole, the crankshaft in a journal bearing in an engine, lathe

Mechanism

Chapter 1

spindle supported in head stock, cycle wheels turning over their axles etc. are the examples of a turning pair. A turning pair also has a completely constrained motion.

Rolling pair: When the two elements of a pair are connected in such a way that one rolls over another fixed link, the pair is known as rolling pair. Ball and roller bearings are examples of rolling pair.

According to mechanical constraint between the elements:
1. **Closed Pair:** When two elements of a pair are held together mechanically. e.g., all lower pair and some of higher pair.
2. **Unclosed Pair** (Open Pair): When two elements of a pair are not held together mechanically. e.g., cam and follower.

Kinematic Constraints

Two or more rigid bodies in space are collectively called a *rigid body system*. We can hinder the motion of these independent rigid bodies with **kinematic constraints**. *Kinematic constraints* are constraints between rigid bodies that result in the decrease of the degrees of freedom of rigid body system.

Types of Constrained Motions

Following are the three types of constrained motions:

1. **Completely constrained motion;** When motion between a pair is limited to a definite direction irrespective of the direction of force applied, then the motion is said to be a completely constrained motion. For example, the piston and cylinder (in a steam engine) form a pair and the motion of the piston is limited to a definite direction (i.e. it will only reciprocate) relative to the cylinder irrespective of the direction of motion of the crank, as shown in Fig.5.1

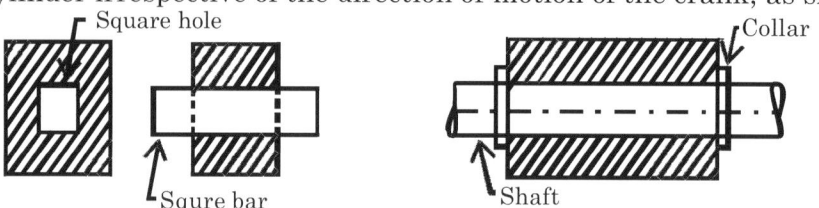

Fig. 5.2 Square bar in a square hole   Fig. 5.3 Shaft with collar in a circular hole.

The motion of a square bar in a square hole, as shown in Fig.5.2, and the motion of a shaft with collars at each in a circular hole, as shown in Fig. 5.3, are also examples of completely constrained motion.

2. **Incompletely constrained motion:** When the motion between a pair can take place in more than one direction, then the motion is called an incomplete constrained motion. The change in the direction of impressed force may alter the direction of relative motion between the pair. A circular bar or shaft in a circular hole, as shown in Fig. 5.4., is an incompletely constrained motion as it may either rotate or slide in a hole . These both motions have no relationship with the other.

Mechanism

Chapter 1

Fig. 5.4 Shaft in a circular hole.

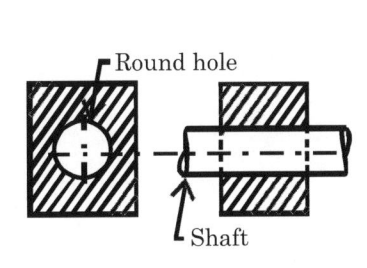

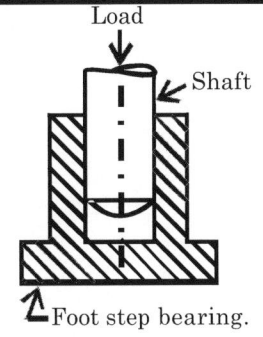

Fig. 5.4 Shaft in a f oot step bearing.

3. Successfully constrained motion: When the motion between the elements, forming a pair, is such that the constrained motion is not completed by itself, but by some other means, then the motion is said to be successfully constrained motion. Consider a shaft in a **foot-step bearing** as shown in Fig 5.5. The shaft may rotate in a bearing or it may rotate in a bearing or it may upwards. This is a case of incompletely constrained motion. But if the load is placed on the shaft to prevent axial upward movement of the shaft, then the motion. But if the pair is said to be successfully constrained motion. The motion of an I.C. engine valve (these are kept on their seat by a spring) and the piston reciprocating inside an engine cylinder are also the examples of successfully constrained motion.

Kinematic chain

A kinematic chain is a **series of links** connected by kinematic pairs. The chain is said to be closed chain if every link is connected to at least two other links, otherwise it is called an open chain. A link which is connected to only one other link is known as singular link. If it is connected to two other links, it is called binary link. If it is connected to three other links, it is called ternary link, and so on. A chain which consists of only binary links is called simple chain.

If each link is assumed to form two pairs with two adjacent links, then relation between the No. of pairs (p) formatting a kinematic chain and the number of links (l) may be expressed in the from of an equation:

$$l = 2p - 4$$

Since in a kinematic chain each link forms a part of two pairs, therefore there will be as many links as the number of pairs.

Another relation between the number of links (l) and the number of joints (j) which constitute a kinematic chain is given by the expression:

$$j = \frac{3}{2}l - 2$$

Where,
l = no. of links

Mechanism

Chapter 1

p = no. of pairs
j = no. of binary joints.

If L.H.S. > R.H.S. $\Rightarrow$ Structure
   L.H.S. = R.H.S. $\Rightarrow$ Constrained chain
   L.H.S. < R.H.S. $\Rightarrow$ Unconstrained chain. e.g.

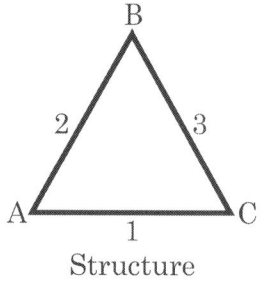

Structure

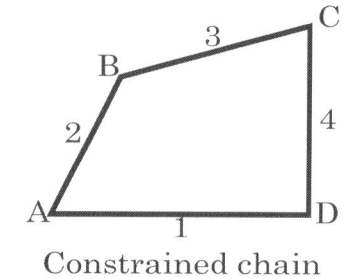

Constrained chain

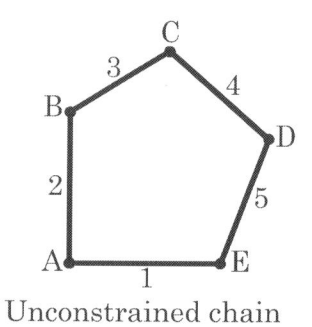

Unconstrained chain

Types of Joins

1. Binary Joint
2. Ternary Joint
3. Quaternary Joint.

Links, Joints and Kinematic Chains

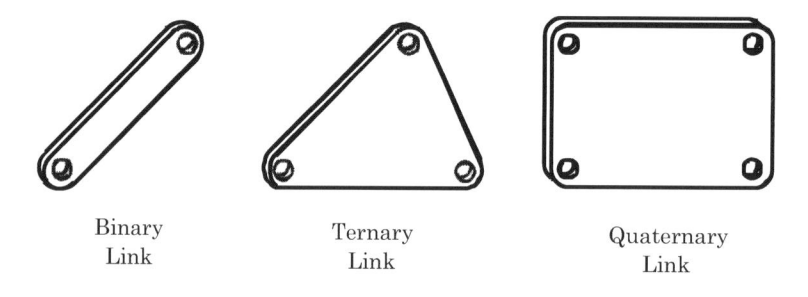

Binary Link

Ternary Link

Quaternary Link

Every link must have nodes. The number of nodes defines the type of link.

- **Binary link** - One link with two nodes
- **Ternary link** - One link with three nodes
- **Quaternary link** - One link with four nodes

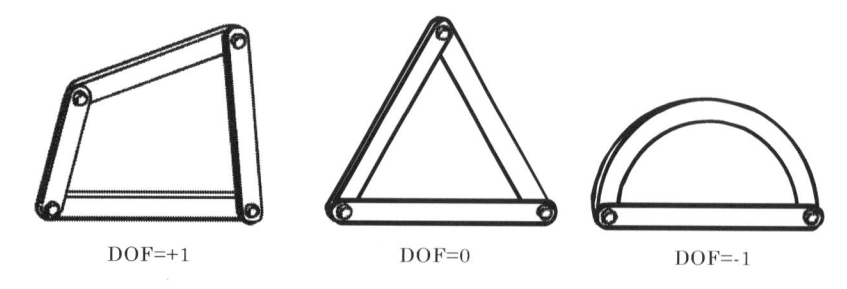

DOF=+1

DOF=0

DOF=-1

Now that we have defined degrees of freedom, We can look at the illustration above and determine the degrees of freedom of each.

Mechanism

Chapter 1

	Joint Type	DOF	Description
A	First order pin joint	1	two binary links joined at a common point
B	Second order pin joint	2	three binary links joined at a common point
C	Half Joint	1 *or* 2	Rolling or sliding *or* both

Mechanism

Mechanism: When one of the link of a kinematic chain is fixed, it will be a mechanism. If the different link of the same kinematic chain is fixed, the result is a different mechanism. The primary function of a mechanism is to transmit or modify motion.

Machine: When a mechanism is required to transmit power or to do some particular kind of work it is known as a machine.

Structure: An assemblage of resistant bodies having no relative motion between them and meant for carrying load having straining action called structure.

Inversions: Mechanism is one in which one of the link of kinematic chain is fixed. Different mechanism are formed by fixing different link of the same kinematic chain are known as inversions of each other.

Mechanisms and Structures

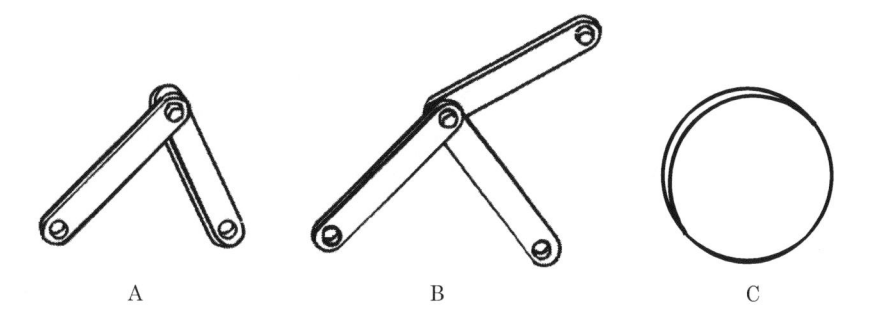

A   B   C

- A **mechanism** is defined by the number of **positive** degrees of freedom. If the assembly has zero or negative degrees of freedom it is a structure.
- A **structure** is an assembly that has **zero** degrees of freedom. An assembly with **negative** degrees of freedom is a structure with residual stresses.

Degrees of freedom

Degree of Freedom: It is the **number of independent variables** that must be specified to define completely the condition of the system.
A kinematic chain is said to be movable when its d.o.f. ≥ 1 otherwise it will be locked. If the d.o.f. is 1 the chain is said to be constrained.

Figure4-1 shows a rigid body in a plane. To determine the DOF this body we must consider how many distinct ways the bar can be moved. In a two dimensional plane such as this

Mechanism

Chapter 1

computer screen, there are 3 DOF. The bar can be translated along the x axis, translated along the y axis, and rotated about its central.

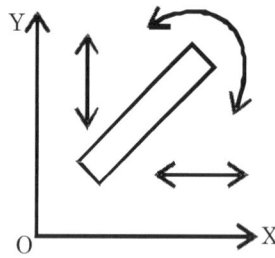

Figure 4-1 Degrees of freedom of a rigid body in a plane

Consider a pencil on a table. If the corner of the table was used as a reference point, two independent variables will be required to fully define its position. Either an X-Y coordinate of an endpoint and an angle or two X coordinates *or* two Y coordinates. No single variable by itself can never fully define its position. Therefore the system has two degrees of freedom.

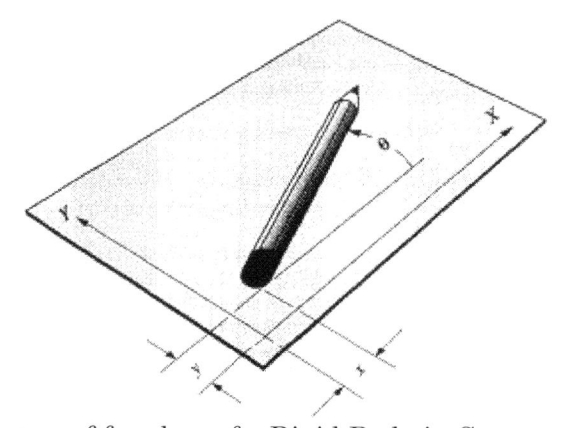

Fig. Degree of freedom of a Rigid Body in Space

An unrestrained rigid body in space has six degrees of freedom: three translating motions Along the x, y and z axes and three rotary motions around the x, y and z axes respectively.

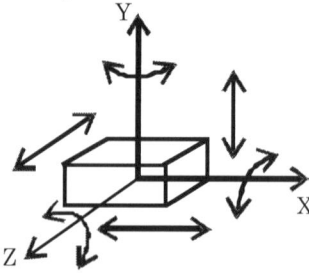

Figure 4-2 Degrees of freedom of a rigid body in space

Unconstrained rigid body in space possesses 6 d.o.f.

Mechanism

Chapter 1

Joint/Pair	D.O.F.	Variable
Pin Joint	1	θ
Sliding Joint	1	S
Screw Pair	1	θ or S
Cost. Pair	2	θ, S
Spherical Pair	3	θ, j, ψ
Planar Pair	3	xy θ

→ Revolute Pair

Kutzbach criterion

The number of degrees of freedom of a mechanism is also called the mobility of the device. The **mobility** is the number of input parameters (usually pair variables) that must be independently controlled to bring the device into a particular position. The **Kutzbach criterion** calculates the mobility.

In order to control a mechanism, the number of independent input motions must equal the number of degrees of freedom of the mechanism.

For example, Figure shows several cases of a rigid body constrained by different kinds of pairs.

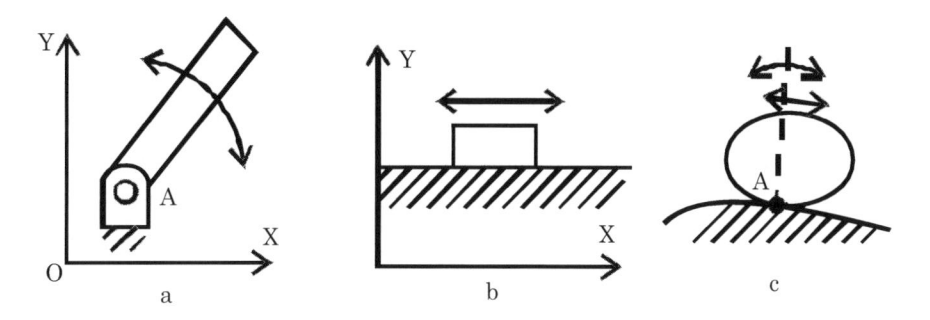

Figure. Rigid bodies constrained by different kinds of planar pairs

In Figure-a, a rigid body is constrained by a **revolute pair** which allows only rotational movement around an axis. It has one degree of freedom, turning around point A. The two lost degrees of freedom are translational movements along the x and y axes. The only way the rigid body can move is to rotate about the fixed point A.

In Figure -b, a rigid body is constrained by a **prismatic pair** which allows only translational motion. In two dimensions, it has one degree of freedom, translating along the x axis. In this example, the body has lost the ability to rotate about any axis, and it cannot move along the y axis.

In Figure-c, a rigid body is constrained by a **higher pair.** It has two degrees of freedom: translating along the curved surface and turning about the instantaneous contact point.

Now let us consider a plane mechanism with *l* number of links. Since in a mechanism, one of the links is to be fixed, therefore the number of movable links will be (l-1) and thus the total

Mechanism

Chapter 1

number of degrees of freedom will be 3 (l-1) before they are connected to any other link. In general, a mechanism with l number of links connected by j number of binary joints or lower pairs (i.e. single degree of freedom pairs) and h number of higher pairs (i.e. two degree of freedom pairs), then the number of degrees of freedom of a mechanism is given by

$$n = 3(l - 1) - 2j - h$$

This equation is called Kutzbach criterion for the movability of a mechanism having plane motion.
If there are no two degree of freedom pairs (i.e. higher pairs), then h = 0. Substituting h= 0 in equation (i), we have

$$n = 3(l - 1) - 2j$$

Where,
    n = degree of freedom
    l = no. of link
    j = no. of joints/no. of lower pair
    h = no. of higher pair.

In general, a rigid body in a plane has three degrees of freedom. Kinematic pairs are constraints on rigid bodies that reduce the degrees of freedom of a mechanism. Figure 4-11 shows the three kinds of pairs in planar mechanisms. These pairs reduce the number of the degrees of freedom. If we create a lower pair (Figure 4-11a, b), the degrees of freedom are reduced to 2. Similarly, if we create a higher pair (Figure 4-11c), the degrees of freedom are reduced to 1.

Turning pair
Two DOF lost
a

Prismatic pair
Two DOF lost
b

Higher pair
One DOF lost
c

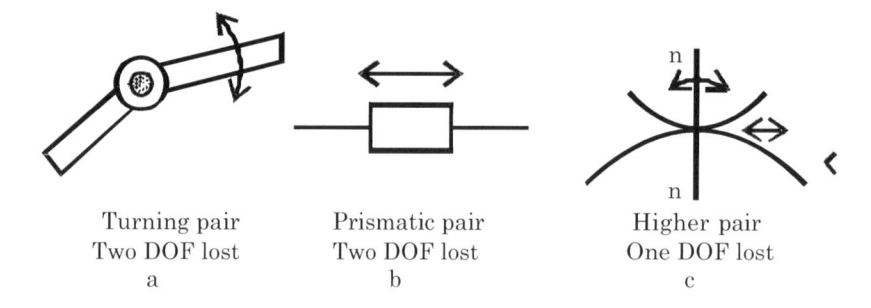

Figure. Kinematic Pairs in Planar Mechanisms

Example 1

Look at the transom above the door in Figure 4-13a. The opening and closing mechanism is shown in Figure 4-13b. Let's calculate its degree of freedom.

Mechanism

Chapter 1

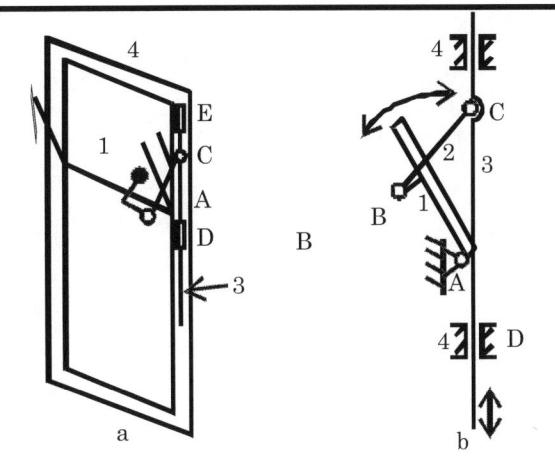

Figure. Transom mechanism

n = 4 (link 1, 3, 3 and frame 4), l = 4 (at A, B, C, D), h = 0

$$F = 3(4-1) - 2 \times 4 - 1 \times 0 = 1$$

(4-2)

Note: D and E function as a same prismatic pair, so they only count as one lower pair.

Example 2

Calculate the degrees of freedom of the mechanisms shown in Figure 4-14b. Figure 4-14a is an application of the mechanism.

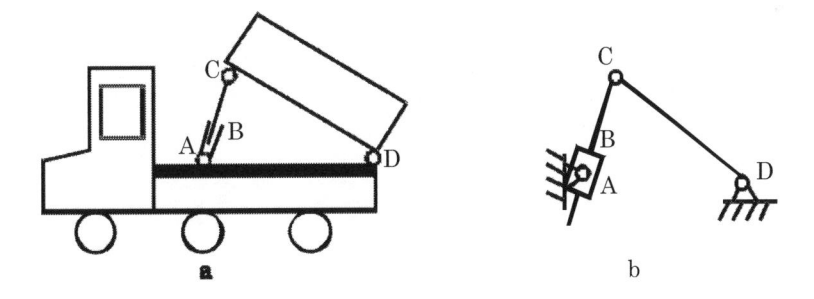

Figure. Dump truck

n = 4, l = 4 (at A, B, C, D), h = 0

$$F = 3(4-1) - 2 \times 4 - 1 \times 0 = 1$$

(4-3)

Example 3

Calculate the degrees of freedom of the mechanisms shown in Figure.

Mechanism

Chapter 1

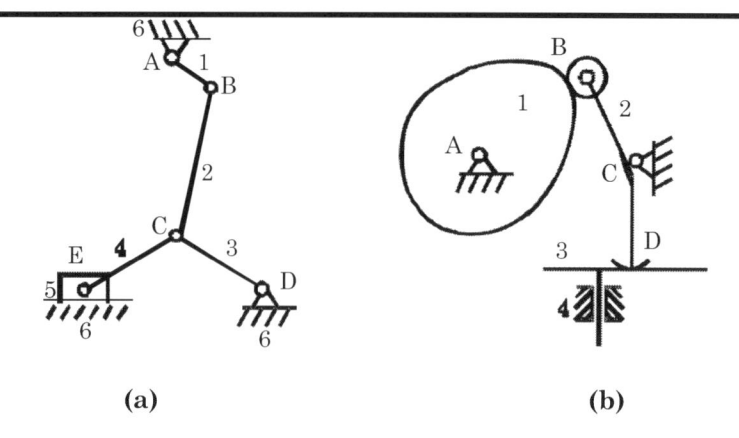

Figure. Degrees of freedom calculation

For the mechanism in Figure-(a)

n = 6, l = 7, h = 0

$$F = 3(6-1) - 2 \times 7 - 1 \times 0 = 1$$

For the mechanism in Figure-(b)

n = 4, l = 3, h = 2

$$F = 3(4-1) - 2 \times 4 - 1 \times 2 = 1$$

Note: The rotation of the roller does not influence the relationship of the input and output motion of the mechanism. Hence, the freedom of the roller will not be considered; it is called a **passive** or **redundant** degree of freedom. Imagine that the roller is welded to link 2 when counting the degrees of freedom for the mechanism.

Redundant link

When a link is move without disturbing other links that links is treated as redundant link.

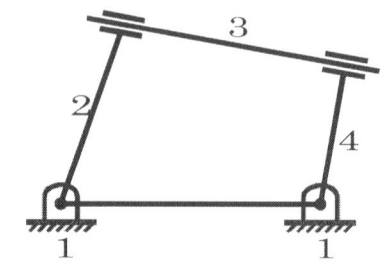

n = 3(l − 1) − 2j − h − R
l = 4, j = 4, R = 1
n = 0

When, n = -1 or less, then there are redundant constraints in the chain and it forms a statically indeterminate structure, as shown in Fig.(e).

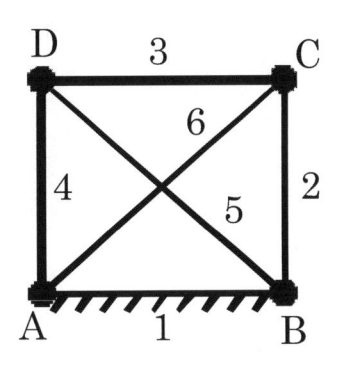

Mechanism

Chapter 1

(e) Six bar mechanism

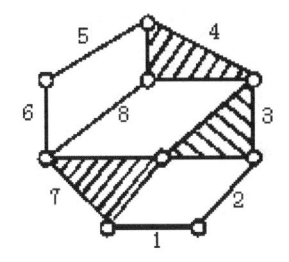

Now, consider the kinematic chain shown in Fig. It has 8 links but only three ternary links. However, the links 6, 7 and 8 constitute a double pair so that the total number of pairs is again 10. The degree of such a linkage will be
$F = 3(8 - 1) - 2 \times 10$
$\quad = 1$

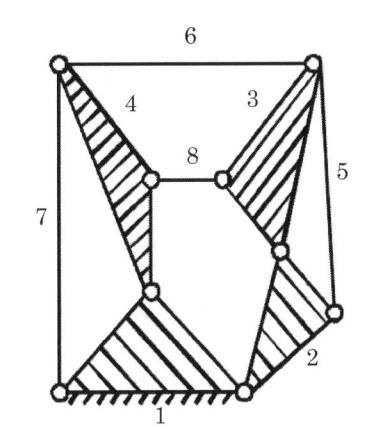

$N_b = 4;\ N_t = 4;\ N_0 = 0;\ N = 8;\ L = 4$
$\quad P_1 = 11$ by counting
or $\quad P_1 = (N + L + 1) = 11$
$\quad F = 3(N - 1) - 2P_1$
$\quad \quad = 3(8-1) - 2 \times 11 = -1$
or $\quad F = N - (2L + 1)$
$\quad \quad = 8 - (2 \times 4 + 1) = -1$

The linkage has negative degree of freedom and thus is a SUPER STRUCTURE.

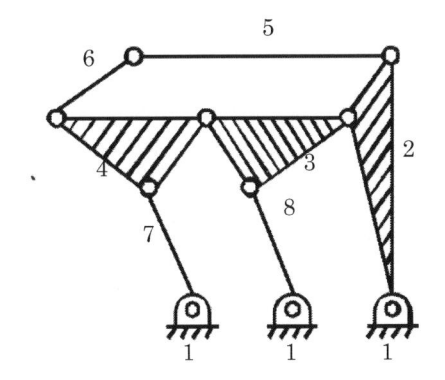

$N_b = 4;\ N_t = 4;\ N_0 = 0;\ N = 8;\ L = 3$
$\quad P_1 = 10$ (by counting)
or $\quad P_1 = (N + L - 1) = 10$
$\quad F = N - (2L + 1) = 8 - (2 \times 3 + 1) = 1$
or $\quad F = 3(N - 1) - 2P_1$
$\quad \quad = 3(8 - 1) - 2 \times 10 = 1$

i.e., the linkage has a constrained motion when one of the seven moving links is driven by an external source.

Mechanism

Chapter 1

$N_b = 7$; $N_t = 2$; $N_0 = 2$; $N = 11$

$L = 5$; $P_1 = 15$

$F = N - (2L + 1) = 11 - (2 \times 5 + 1) = 0$

Therefore, the linkage is a structure.

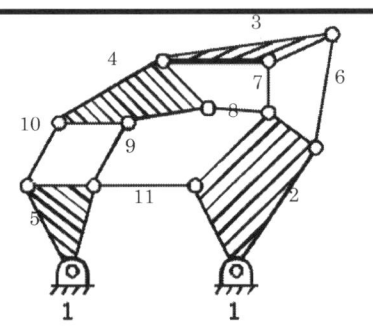

The linkage has 4 loops and 11 links. Referring Table 1.2, it has 2 degrees of freedom. With 4 loops and 1 degree of freedom, the number of joints 13. Three excess joints can be formed by

    6 ternary links or

    4 ternary links and 1 Quaternary link or

    2 ternary links, or

    a combination of ternary and quaternary links with double joints.

Figure 1.20 (b) shows one of the possible solutions.

There are 4 loops and 8 links.

$$F = N - (2L +) = 8 - (4 \times 2 + 1) = -1$$

It is a superstructure. With 4 loops, the number of links must be 10 to obtain one degree of freedom. As the number of links is not to be increased by more than one, the number of loops has to be decreased. With 3 loops, 8 links and 10 joints, the required linkage can be designed. One of the many solutions is shown in Fig.

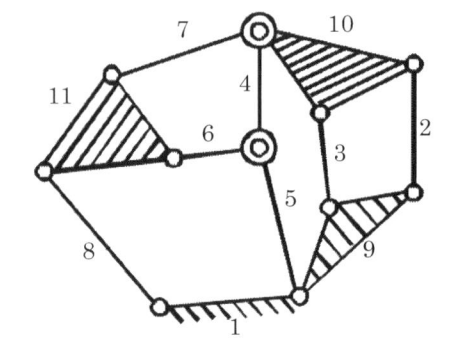

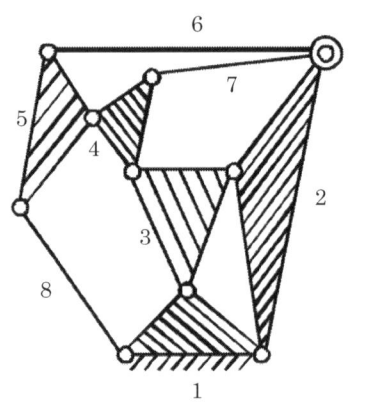

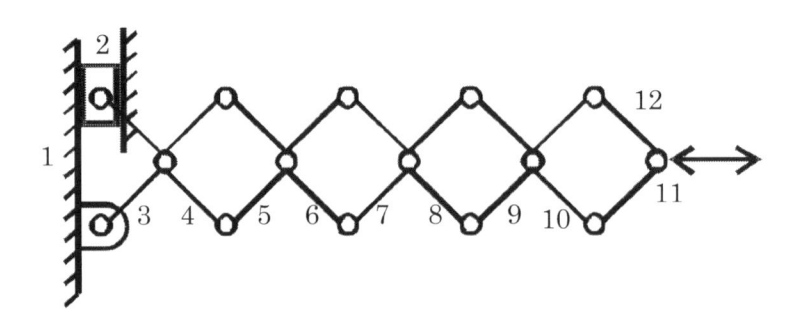

It has 5 loops and 12 links.

It has 1 degree of freedom and thus is a mechanism.

Mechanism

Chapter 1

The mechanism has a sliding pair. Therefore, its degree of freedom must be found from Gruebler's criterion.
Total number of links = 8 (Fig.)
=10
(At the slider, one sliding pair and two turning pairs)

$F = 3(N-1) - 2P_1 - P_2$
$= 3(8-1) - 2 \times 10 - 0 = 1$

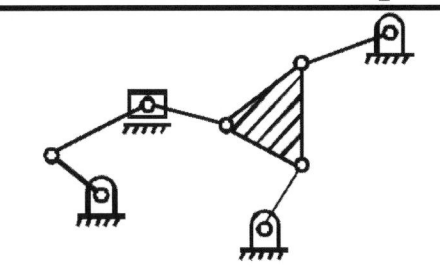

The mechanism has a can pair. Therefore, its degree of freedom must be found from Gruebler's criterion.
Total number of links = 7 (Fig.)
Number of pairs with 1 degree of freedom = 8
Number of pairs with 2 degrees of freedom = 1

$F = 3(N-1) - 2P_1 - P_2$
$= 3(7-1) - 2 \times 8 - 1 = 1$

Thus, it is a mechanism with one degree of freedom.

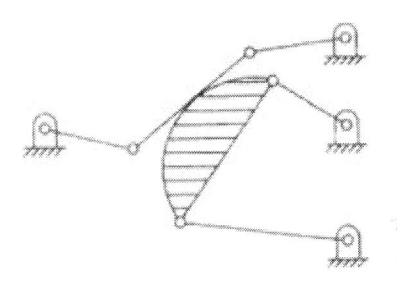

Grubler Criterion

Grubler's Criterion for Plane Mechanisms

The Grubler's criterion applies to mechanisms with only single degree of freedom joints where the overall movability of the mechanism is unity. Substituting n = 1 and h = 0 in Kutzbach equation, we have

$$1 = 3(l-1) - 2j \quad \text{or} \quad 3l - 2j - 4 = 0$$

This equation is known as the Grubler's criterion for plane mechanisms with constrained motion.

A little consideration will show that a plane mechanism with a movability of 1 and only single degree of freedom joints can not have odd number of links. The simplest possible mechanisms of this type are a four bar mechanism and a slider – crank mechanism in which $l = 4$ and $j = 4$.

Grashof's law

In the range of planar mechanisms, the simplest groups of lower pair mechanisms are four bar linkages. A **four bar linkage** comprises four bar-shaped links and four turning pairs as shown in Figure 5-8.

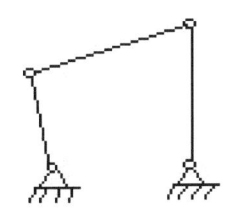

Mechanism

Chapter 1

Fig. four bar linkage

The link opposite the frame is called the **coupler link**, and the links which are hinged to the frame are called **side links**. A link which is free to rotate through 360 degree with respect to a second link will be said to **revolve** relative to the second link (not necessarily a frame). If it is possible for all four bars to become simultaneously aligned, such a state is called a **change point**.

Some important concepts in link mechanisms are:

1. **Crank**: A side link which revolves relative to the frame is called a *crank*.
2. **Rocker**: Any link which does not revolve is called a *rocker*.
3. **Crank-rocker mechanism**: In a four bar linkage, if the shorter side link revolves and the other one rocks (*i.e.*, oscillates), it is called a *crank-rocker mechanism*.
4. **Double-crank mechanism**: In a four bar linkage, if both of the side links revolve, it is called a *double-crank mechanism*.
5. **Double-rocker mechanism**: In a four bar linkage, if both of the side links rock, it is called a *double-rocker mechanism*.

Classification

Before classifying four-bar linkages, we need to introduce some basic nomenclature.

In a four-bar linkage, we refer to the *line segment between hinges* on a given link as a **bar** where:

- s = length of shortest bar
- l = length of longest bar
- p, q = lengths of intermediate bar

Grashof's theorem states that a four-bar mechanism has *at least* one revolving link if

$$s + l \leq p + q$$

and all three mobile links will rock if

$$s + l > p + q$$

The inequality 5-1 is **Grashof's criterion**.

All four-bar mechanisms fall into one of the four categories listed in Table 5-1:

Case	l + s vers. p + q	Shortest Bar	Type
1	<	Frame	Double-crank
2	<	Side	Rocker-crank
3	<	Coupler	Double rocker
4	=	Any	Change point

Mechanism

Chapter 1

| 5 | > | Any | Double-rocker |

Table: Classification of Four-Bar Mechanisms

From Table we can see that for a mechanism to have a crank, the sum of the length of its shortest and longest links must be less than or equal to the sum of the length of the other two links. However, this condition is necessary but not sufficient. Mechanisms satisfying this condition fall into the following three categories:

1. When the shortest link is a side link, the mechanism is a crank-rocker mechanism. The shortest link is the crank in the mechanism.
2. When the shortest link is the frame of the mechanism, the mechanism is a double-crank mechanism.
3. When the shortest link is the coupler link, the mechanism is a double-rocker mechanism.

Inversion of Mechanism

Method of obtaining different mechanisms by fixing different links in a kinematic chain, is known as inversion of the mechanism.

Inversion is a term used in kinematics for a reversal or interchanges of form or function as applied to kinematic chains and mechanisms.

Types of Kinematic Chains

The most important kinematic chains are those which consist of four lower pairs, each pair being a sliding pair or a turning pair. The following three types of kinematic chains with four lower pairs are important from the subject point of view:

1. Four bar chain or quadric cyclic chain,

2. Single slider crank chain, and

3. Double slider crank chain.

For example, taking a different link as the fixed link, the slider-crank mechanism shown in Figure 5-14a can be inverted into the mechanisms shown in Figure 5-14b, c, and d. Different examples can be found in the application of these mechanisms. For example, the mechanism of the pump device in Figure 5-15 is the same as that in Figure 5-14b.

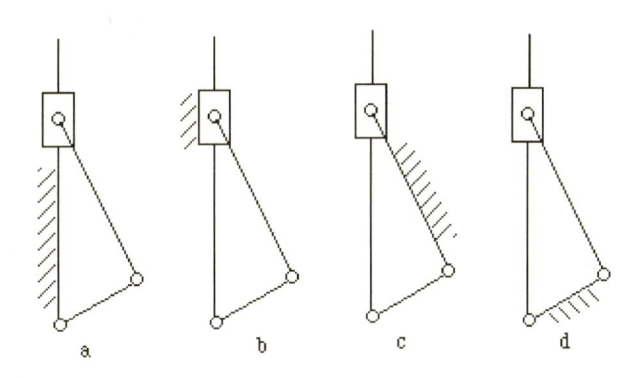

Mechanism

Chapter 1

Figure: Inversions of the crank-slide mechanism

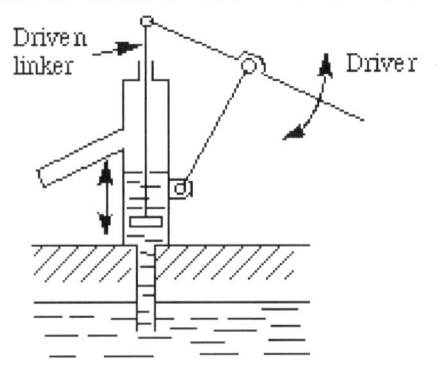

Figure. A pump device

- Keep in mind that the inversion of a mechanism does not change the motions of its links relative to each other but does change their absolute motions.

- Inversion of a kinematic chain has no effect on the relative motion of its links.
- The motion of links in a kinematic chain relative to some other links is a property of the chain and is not that of the mechanism.
- For **L** number of links in a mechanism, the number of possible inversions is equal to L.

1. Inversion of four bar chain

(a) Crank and lever mechanism/Beam engine (1st inversion).

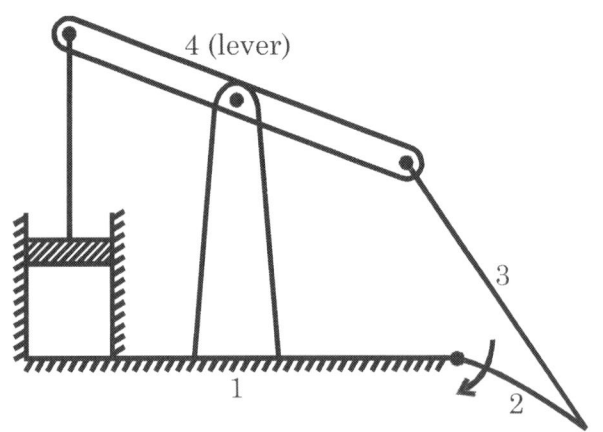

(b) Double crank mechanism (Locomotive mechanism) 2nd inversion.

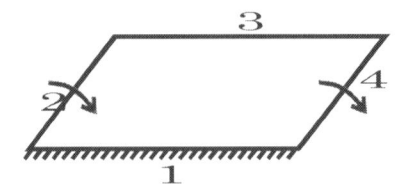

(c) Double lever mechanism (Ackermann steering) 3rd inversion.

2, 4 → Oscillator

3 → Coupler

Mechanism

Chapter 1

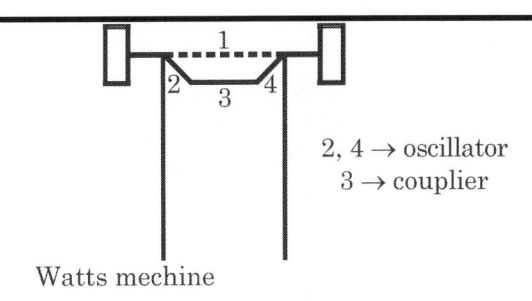

2, 4 → oscillator
3 → couplier

Watts mechine

2. Inversion of the Slider-Crank Mechanism

Inversion is a term used in kinematics for a reversal or interchanges of form or function as applied to kinematic chains and mechanisms. For example, taking a different link as the fixed link, the slider-crank mechanism shown in Figure (a) can be inverted into the mechanisms shown in Figure (b), (c), and d. Different examples can be found in the application of these mechanisms. For example, the mechanism of the pump device in Figure is the same as that in Figure (b).

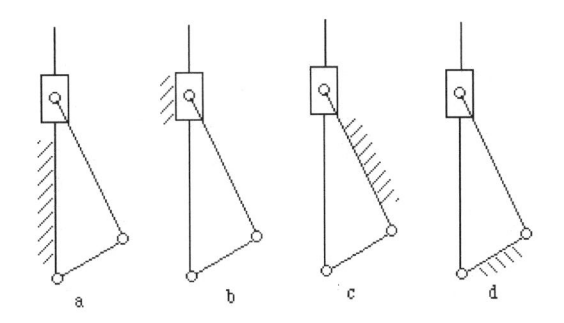

Figure. Inversions of the crank-slide mechanism

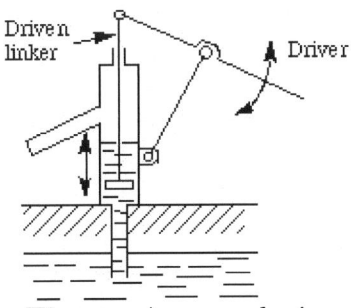

Figure. A pump device

Keep in mind that the inversion of a mechanism does not change the motions of its links relative to each other but does change their absolute motions.

1. Pendulum pump or Bull engine: In this mechanism the inversion is obtained by fixing cylinder or link4 (i.e. sliding pair), as shown in figure below.

Mechanism

Chapter 1

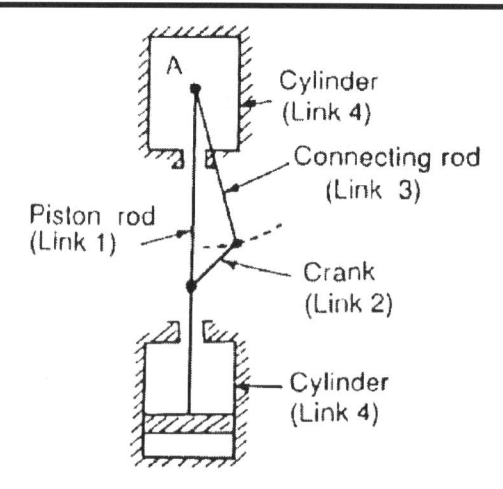

Fig. (Duplex pump mechanism with Piston rod (Link 1), Crank (Link 2), Connecting rod (Link 3), Cylinder (Link 4))

The **duplex** pump which is used to supply feed water to boilers uses this mechanism.

2. Oscillating cylinder engine: It is used to convert reciprocating motion into rotary motion.

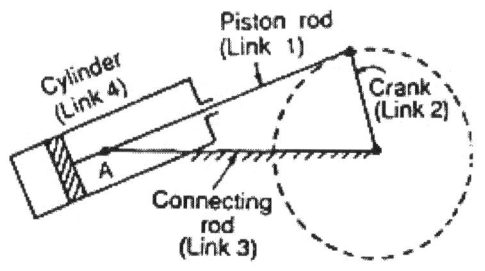

Fig. Oscillating cylinder engine

3. Rotary internal combustion engine or Gnome engine:

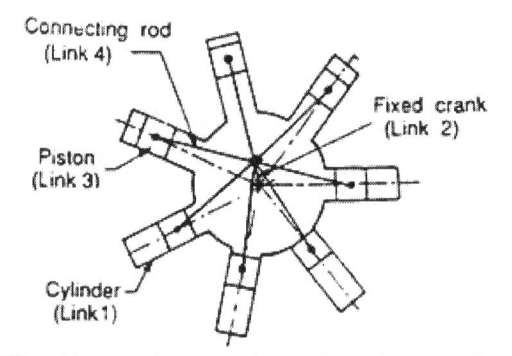

Fig. Rotary internal combustion engine

Sometimes back, rotary internal combustion engines were used in aviation. But now-a-days gas turbines are used in its place.

Quick return motion mechanism

4. Crank and slotted lever quick return motion mechanism: This mechanism is mostly used in shaping machines, slotting machines and in rotary internal combustion engines.

Mechanism

Chapter 1

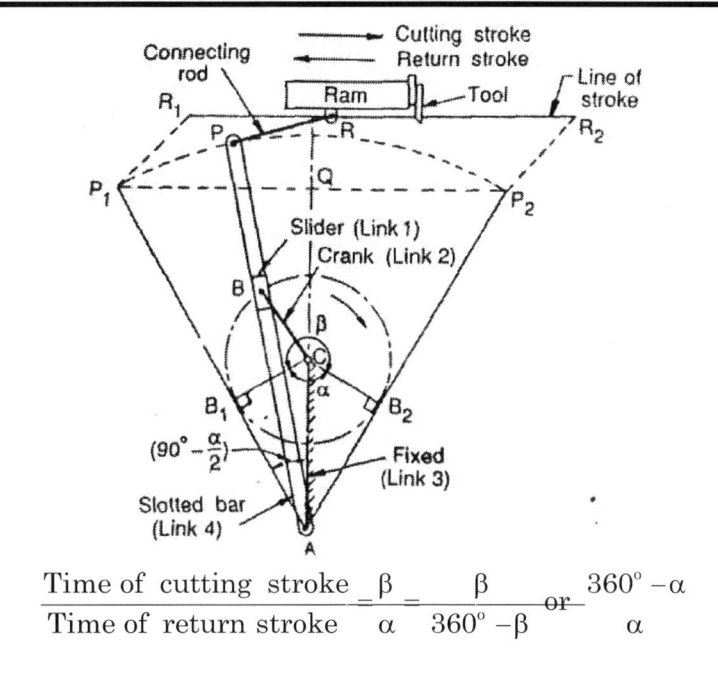

$$\frac{\text{Time of cutting stroke}}{\text{Time of return stroke}} = \frac{\beta}{\alpha} = \frac{\beta}{360° - \beta} \text{ or } \frac{360° - \alpha}{\alpha}$$

Length of stroke

$$= 2\,\text{AP} \times \frac{\text{CB}}{\text{AC}}$$

Note: We see that the angle β made by the forward or cutting stroke is greater than the angle α described by the return stroke. Since the crank rotates with uniform angular speed, therefore the return stroke is completed within shorter time. Thus it is called quick return motion mechanism.

5. Whitworth quick return motion mechanism:

This mechanism is mostly used in shaping and slotting machines. In this mechanism, the link *CD* (link 2) formatting the turning pair is fixed, as shown in Figure. The link 2 corresponds to a crank in a reciprocating steam engine. The driving crank *CA* (link 3) rotates at a uniform angular speed. The slider (link 4) attached to the crank pin at A slides along the slotted bar *PA* (link 1) which oscillates at a pivoted point *D*. The connecting rod *PR* carries the ram at R to which a cutting tool is fixed. The motion of the tool is constrained along the line *RD* produced, i.e. along a line passing through D perpendicular to *CD*.

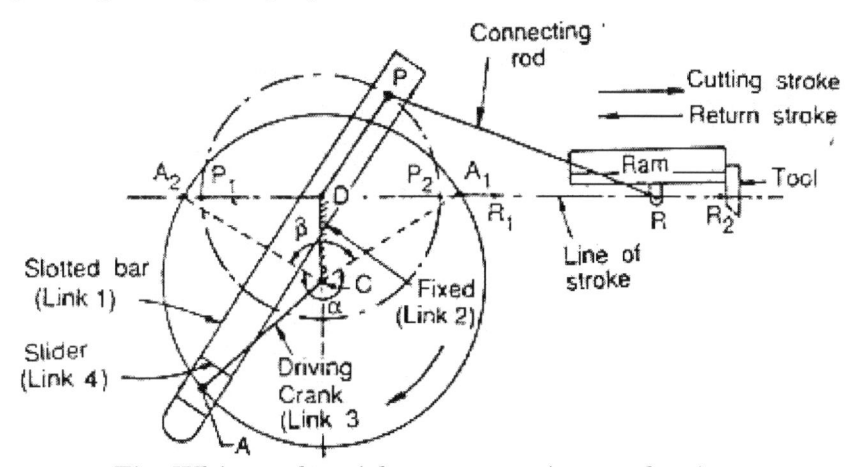

Fig. Whitworth quick return motion mechanism

Mechanism

Chapter 1

$$\frac{\text{Time of cutting stroke}}{\text{Time of return stroke}} = \frac{\alpha}{\beta} = \frac{\alpha}{360° - \alpha} \text{ or } \frac{360° - \beta}{\beta}$$

Note: In order to find the length of effective stroke $R_1 R_2$, mark $P_1 R_1 = P_2 P_2 = PR$. The length of effective stroke is also equal to 2 PD

The Geneva Wheel

An interesting example of intermittent gearing is the **Geneva Wheel** shown in Figure 8-4. In this case the **driven wheel**, B, makes one fourth of a turn for one turn of the **driver**, A, the **pin**, a, working in the **slots**, b, causing the motion of B. The circular portion of the driver, coming in contact with the corresponding hollow circular parts of the driven wheel, retains it in position when the pin or tooth a is out of action. The wheel A is cut away near the pin a as shown, to provide clearance for wheel B in its motion.

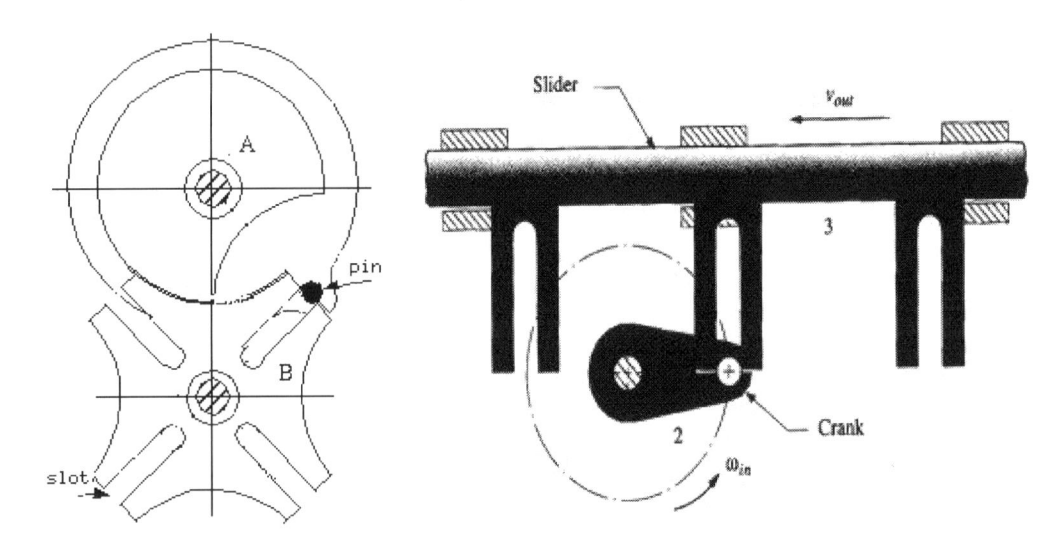

Figure. Geneva wheel

If one of the slots is closed, A can only move through part of the revolution in either direction before pin a strikes the closed slot and thus stop the motion. The device in this modified form was used in watches, music boxes, *etc.*, to prevent over winding. From this application it received the name Geneva stop. Arranged as a stop, wheel A is secured to the spring shaft, and B turns on the axis of the spring barrel. The number of slots or interval units in B depends upon the desired number of turns for the spring shaft.

An example of this mechanism has been made in Sim Design, as in the following picture.

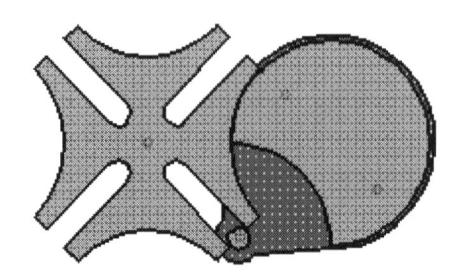

Mechanism

Chapter 1

- Geneva mechanism is used to **transfer components** from one station to the other in a rotary transfer machine
- Geneva mechanism produces intermittent rotary motion from continuous rotary motion.

Hand Pump: Here also the slotted link shape is given to the slider and vice-versa, in order to get the desired motion.

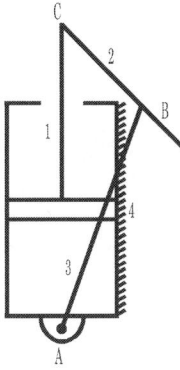

Here the slider (link 4) is fixed and hence, it is possible for link 1 to reciprocate along a vertical straight line. At the same time link 2 will rotate and link 3 will oscillate about the pin.

Inversion of Double slider crank chain

It has four binary links, two revolute pairs, two sliding pairs.
Its various types are:
- Elliptical Trammel
- Scotch Yoke mechanism
- Oldham's coupling.

Elliptical trammels

It is an instrument used for drawing ellipses. This inversion is obtained by fixing the slotted plate (link 4), as shown in Figure. The fixed plate or link 4 has two straight grooves cut in it, at right angles to each other. The link 1 and link 3, are known as sliders and form sliding pairs with link 4. The link AB (link 2) is a bar which forms turning pair with links 1 and 3.

When the links 1 and 3 slide along their respective grooves, any point on the link 2 such as P traces out an ellipse on the surface of link 4, as shown in Figure (a). A little consideration will show that AP and BP are the semi-major axis and semi-major axis of the ellipse respectively. This can be proved as follows:

Mechanism

Chapter 1

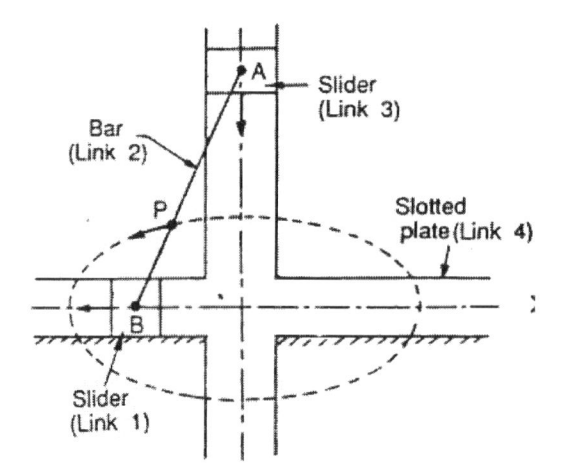

Note: If P is the mid-point of link BA, then $AP = BP$.' Hence if P is the midpoint of link BA, it will trace a circle.

Scotch yoke mechanism

Here the constant rotation of the crank produces harmonic translation of the yoke. Its four binary links are:
1. Fixed Link
2. Crank
3. Sliding Block
4. Yoke

The four kinematic pairs are:
1. revolute pair (between 1 & 2)
2. revolute pair (between 2 & 3)
3. prismatic pair (between 3 & 4)
4. prismatic pair (between 4 & 1)

This mechanism is used for converting rotary motion into a reciprocating motion.

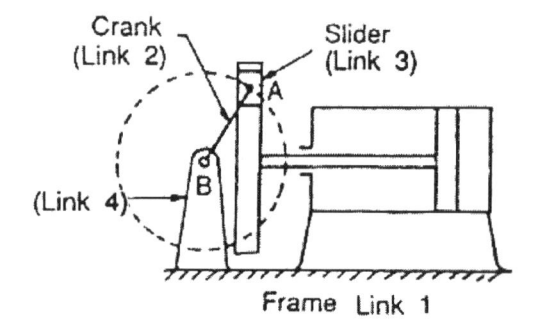

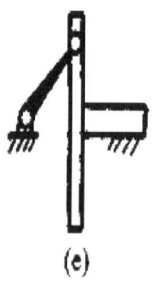

Oldham's coupling
- It is used for transmitting angular velocity between two parallel but eccentric shafts.
- An Oldham's coupling is used for connecting two parallel shafts whose axes are at a small distance apart.
- Oldham's coupling is the inversion of double slider crank mechanism.
- The shafts are coupled in such a way that if one shaft rotates the other shaft also rotates at the same speed.

Mechanism

Chapter 1

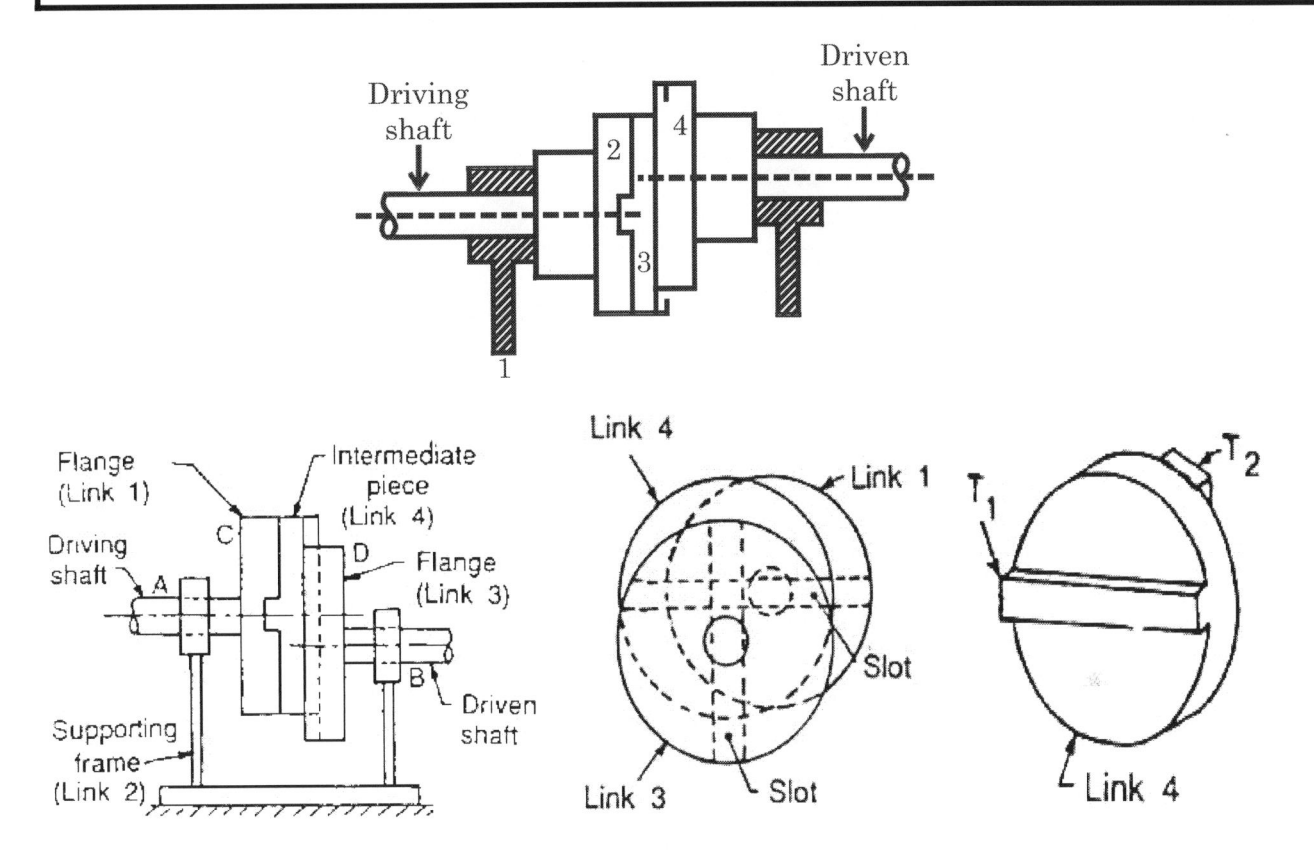

- The link 1 and link 3 form turning pairs with link 2. These flanges have diametrical slots cut in their inner faces.
- The intermediate piece (link 4) which is a circular disc, have two tongues (i.e. diametrical projections) T₁ and T₂ on each face at right angles to each other.
- The tongues on the link 4 closely fit into the slots in the two flanges (link 1 and link 3).
- The link 4 can slide or reciprocate in the slots in the flanges.

Let the distance between the axes of the shafts is constant, the centre of intermediate piece will describe a circle of radius equal to the distance between the axes of the two shafts. Then the maximum sliding speed of each tongue along its slot is equal to the peripheral velocity of the centre of the disc along its circular path.

∴ **Maximum** sliding' speed of each tongue (in m/s),

$$v = \quad . \, r$$

Where ω = Angular velocity of each shaft in rad/s, and
r = Distance between the axes of the shafts in metres.

Hooke's Joint (Universal Coupling)

This joint is used to connect two non-parallel intersecting shafts. It also used for shafts with angular misalignment where flexible coupling does not serve the purpose. Thus Hooke's Joint connecting two rotating shafts whose axes lies in one plane.

Mechanism

Chapter 1

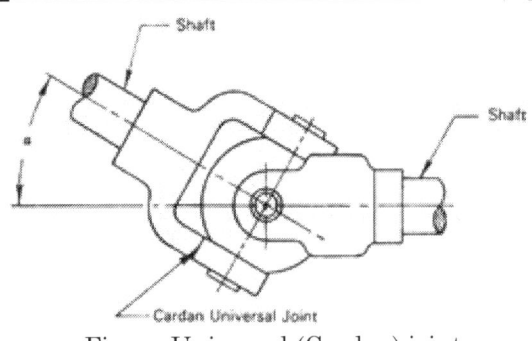

Figure Universal (Cardan) joint

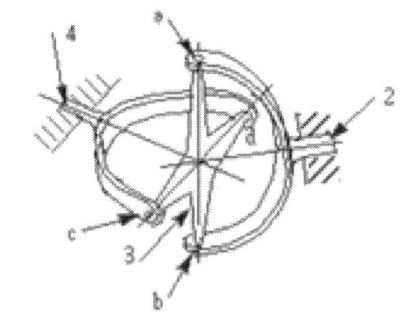

Figure 8-6 General form for a universal joint

There are two types of Hooke's joints in use,
    (a) Single Hooke's Joint
    (b) Double Hooke's Joint

One disadvantage of single Hooke's joint is that the velocity ratio is not constant during rotation. But this can be overcome by using double Hooke's joint.

Let θ = input angular displacement
    ϕ = output angular displacement
    α = shaft angle

$$\boxed{\frac{\tan\theta}{\tan\phi} = \cos\alpha}$$

$\tan\theta = \cos\alpha \tan\phi$

Let ω_1 = angular velocity of driving shaft
    ω_2 = angular velocity of driven shaft

$$\therefore \quad \omega_1 = \frac{d\theta}{dt}$$

$$\omega_2 = \frac{d\phi}{dt}$$

Now different above equation

$$\sec^2\theta \frac{d\theta}{dt} = \cos\alpha \sec^2\phi \frac{d\phi}{dt}$$

$$\sec^2\theta\, \omega_1 = \cos\alpha \sec^2\phi\, \omega_2$$

$$\therefore \quad \frac{\omega_2}{\omega_1} = \frac{\sec^2\theta}{\cos\alpha \sec^2\phi} \qquad \ldots(a)$$

now $\sec^2\phi = 1 + \tan^2\phi$

$$= 1 + \frac{\tan^2\theta}{\cos^2\alpha} \qquad \left[\because \tan\phi = \frac{\tan\theta}{\cos\alpha}\right]$$

$$= 1 + \frac{\sin^2\theta}{\cos^2\theta \cos^2\alpha}$$

$$= \frac{\cos^2\theta \cos^2\alpha + \sin^2\theta}{\cos^2\theta \cos^2\alpha}$$

$\therefore$ Equation (a) will be

$$\frac{\omega_2}{\omega_1} = \frac{\sec^2\theta}{\cos\alpha \left[\dfrac{\cos^2\theta \cos^2\alpha + \sin^2\theta}{\cos^2\theta \cos^2\alpha}\right]}$$

Mechanism

Chapter 1

$$= \frac{\cos\alpha}{\cos^2\theta\cos^2\alpha + \sin^2\theta}$$

$$= \frac{\cos\alpha}{\cos^2\theta[1-\sin^2\alpha] + \sin^2\theta} = \frac{\cos\alpha}{1-\cos^2\theta\sin^2\alpha}$$

$$\therefore \quad \boxed{\omega_2 = \frac{\omega_1 \cos\alpha}{1-\cos^2\theta\sin^2\alpha}}$$

Special Cases:

For a given shaft angle α, the expresses given above is maximum when $\cos\theta \pm 1$

i.e., when $\theta = 0, \pi$, etc.

And will be minimum when $\cos\theta = 0$

i.e., when $\theta = \frac{\pi}{2}, \frac{3\pi}{2}$, etc.

(i) Max. Velocity ratio:

$\frac{\omega_2}{\omega_1}$ is maximum when $\theta = 0, \pi$.

$$\boxed{\left(\frac{\omega_2}{\omega_1}\right)_{max} = \frac{\cos\alpha}{1-\sin^2\alpha} = \frac{1}{\cos\alpha}}$$

Thus $\frac{\omega_2}{\omega_1}$ is maximum at $\theta = 0$ or $180°$, i.e. two times in one revolution of driving shaft.

(ii) Minimum V.R:

$$\boxed{\left(\frac{\omega_2}{\omega_1}\right)_{min} = \cos\alpha}$$

$\frac{\omega_2}{\omega_1}$ is minimum at $\theta = 90°, 270°$. i.e. two times in one revolution.

(iii) For equal speed:

$$\frac{\omega_2}{\omega_1} = 1 = \frac{\cos\alpha}{1-\cos^2\theta\sin^2\alpha}$$

$$\cos\theta = \pm\sqrt{\frac{1}{1+\cos\alpha}}$$

Thus $\frac{\omega_2}{\omega_1}$ is unity at θ given by above equation i.e., four times in one revolution of driving shaft.

Now

$$\cos^2\theta = \frac{1}{1+\cos\alpha}$$

$$\sin^2\theta = 1 - \frac{1}{1+\cos\alpha} = \frac{\cos\alpha}{1+\cos\alpha}$$

$\therefore \quad \tan^2\theta = \cos\alpha$

$$\boxed{\tan^2\theta = \pm\sqrt{\cos\alpha}}$$

Mechanism

Chapter 1

(iv) **Maximum variation of the velocity of driven shaft**

Variation of velocity of driven shaft

$$= \frac{\omega_{2_{max}} - \omega_{2_{min}}}{\omega_{2_{mean}}}$$

But $\omega_{2_{mean}} = \omega_1$, because both shafts complete one revolution during the same interval of time

maximum variation of velocity of $\omega_2 = \dfrac{\dfrac{\omega_1}{\cos\alpha} - \omega_1 \cos\alpha}{\omega_1}$

$\boxed{\text{maximum variation of } \omega_2 = \sin\alpha \tan\alpha}$

$\boxed{\omega_{2_{max}} - \omega_{2_{min}} = \omega_1 \tan\alpha \sin\alpha}$

Maximum fluctuation of speed of driven shaft.

$$\omega_{2max} - \omega_{2min} = \omega_1 \left(\frac{1}{\cos\alpha} - \cos\alpha \right)$$

$$\omega_1 \alpha^2 = \omega_1 \tan\alpha \sin\alpha$$

Since α is a small angle

$\therefore \sin\alpha = \tan\alpha = \alpha$

Average equation of driven shaft:

$$\frac{d\omega_2}{dt} = \frac{d\omega_2}{d\theta} \cdot \frac{d\theta}{dt} = \omega_1 \frac{d\omega_2}{d\theta} = \underbrace{\frac{-\omega_1^2 \cos\alpha \times \sin 2\theta \sin^2\alpha}{(1 - \cos^2\theta \sin^2\alpha)^2}}_{\alpha_2}$$

Condition for maximum acceleration

$$\frac{d\alpha_2}{d\theta} = 0 \Rightarrow \cos 2\theta \approx \frac{2\sin^2\alpha}{2 - \sin^2\alpha}$$

Straight Line Mechanisms

One of the most common forms of the constraint mechanisms is that it permits only relative motion of an oscillatory nature along a straight line. The mechanisms used for this purpose are called straight line mechanisms. These mechanisms are of the following two types:
1. In which only turning pairs are used, and
2. In which one sliding pairs are used,

Approximate Straight Line Motion Mechanisms

The approximate straight line motion mechanisms are the modifications of the four- bar chain mechanisms.

Exact straight line mechanism	Approximate striaght line mechanis
1. Peaculier mechansim	1. Watts mechanism
2. Hart's mechansim	2. Robert's mechanism
3. Scot t Russel mechansim	3. Modified s cot t Russel.
	4. Grass Hooper mech.

Mechanism

Chapter 1

Steering Gear Mechanism

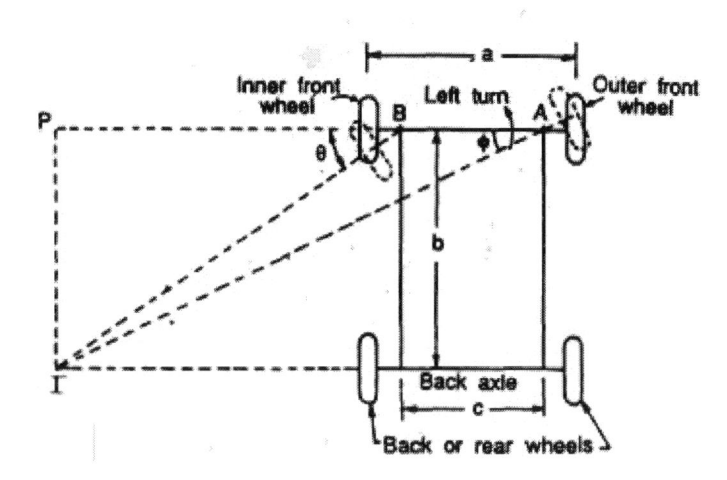

Fig. Steering Gear Mechanism

The steering gear mechanism is used for changing the direction of two or more of the wheel axles with reference to the chassis, so as move the automobile in any desired path. Usually the two back wheels have-a common axis, which is fixed in direction with reference to the chassis and the steering is done by means of the front wheels.

$\cot\phi - \cot\phi = c/b$

Where

a = wheel track,

b = wheel base, and

c = Distance between the pivots A and B of the front axle.

Davis Steering Gear

Ackerman Steering Gear

The Ackerman steering gear mechanism is much simpler than Davis gear. The difference between the Ackerman and Davis steering gears are:

1. The whole mechanism of the Ackerman steering gear is on the back of the front wheels; whereas in Davis steering gear, it is in front of the wheels.

2. The Ackerman steering gear consists of turning pairs, whereas Davis steering gear consists of sliding members.

Velocity and Acceleration

Mechanism

Chapter 1

The concept of velocity and acceleration images is used extensively in the kinematic analysis of mechanisms having ternary, quaternary, and higher- order links. If the velocities and accelerations of any two points on a link are known, then, with the help of images the velocity and acceleration of any other point on the link can be easily determined. An example is

1. Instantaneous Centre Method
2. Relative Velocity Method

Velocity by Instantaneous Centre Method

Instantaneous centre is one point about which the body has pure rotation. Hence for the body which having straight line motion, the radius of curvature of it is at infinity and hence instantaneous centre of this ties at infinite.

Special cases of ICR

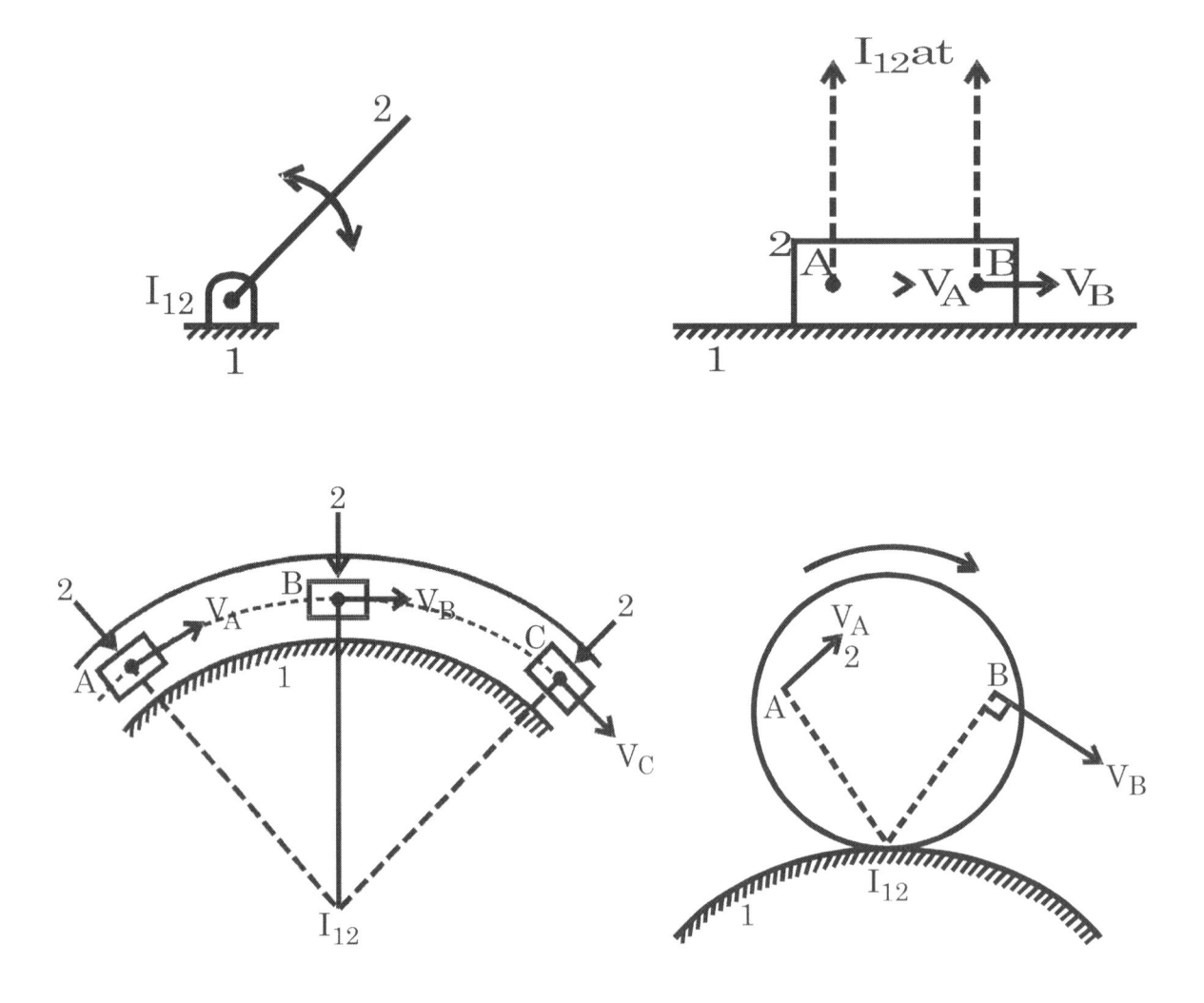

Types of ICR:

Mechanism

Chapter 1

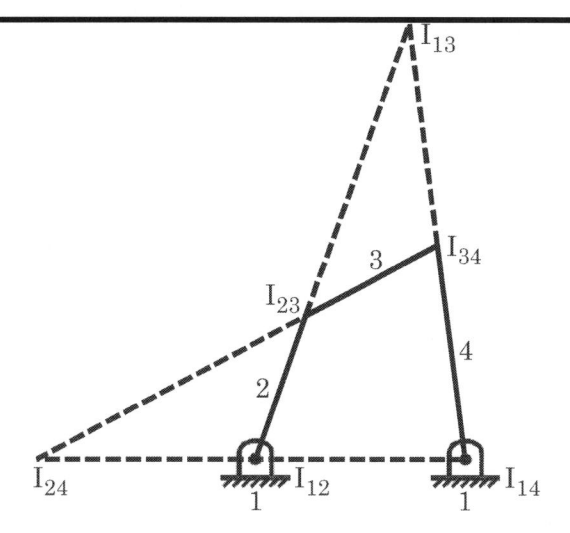

(i) Fixed ICR: I_{12}, I_{14}
(ii) Permanent ICR: I_{23}, I_{34}
(iii) Neither Fixed nor Permanent I.C: I_{13}, I_{24}

Three-Centre-in-line Theorem (Kennedy's Theorem)

Kennedy Theorem states that ―If three links have relative motion with respect to each other, their relative instantaneous centre lies on straight line‖.

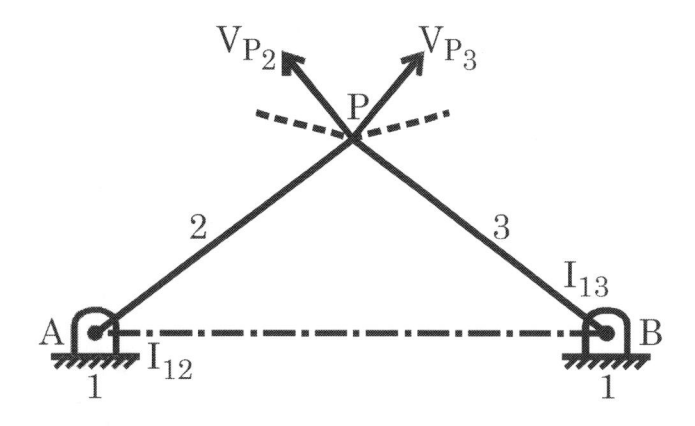

The Theorem can be proved by contradiction.
The Kennedy Theorem states that the three IC I_{12}, I_{13}, I_{23} must all lie on the same straight line on the line connecting two pins.

Let us suppose this is not true and I_{23} is located at the point P. Then the velocity of P as a point on link 2 must have the direction V_{P_2}, $\perp$ to AP. Also the velocity of P as a point on link 3 must have the direction V_{P_3}, $\perp$ to BP. The direction is inconsistent with the definition that an instantaneous centre must have equal absolute velocity as a part of either link. The point P chosen therefore, cannot be the IC I_{23}.

This same contradiction in the direction of V_{P_2} and V_{P_3} occurs for any location chosen for point P, except the position of P chosen on the straight line passing through I_{12} and I_{13}. This justify the Kennedy Theorem.

Mechanism

Chapter 1

Properties of the IC:

1. A rigid link rotates instantaneously relative to another link at the instantaneously centre for the configuration of the mechanism considered.

2. The two rigid links have no linear velocity relative to each other at the instantaneous centre. In other words, the velocity of the IC relative to any third rigid link will be same whether the instantaneous centre is regarded as a point on the first rigid link or on the second rigid link.

Number of I.C in a mechanism:

$$N = \frac{n(n-1)}{2}$$

N = no. of I.C.
n = no. of links.

1. Each configuration of the link has one centre.
   The instantaneous centre changes with alteration of configuration of mechanism.

Method of locating instantaneous centre in mechanism

Consider a pin jointed four bar mechanism as shown in fig. The following procedure is adopted for locating instantaneous centre.

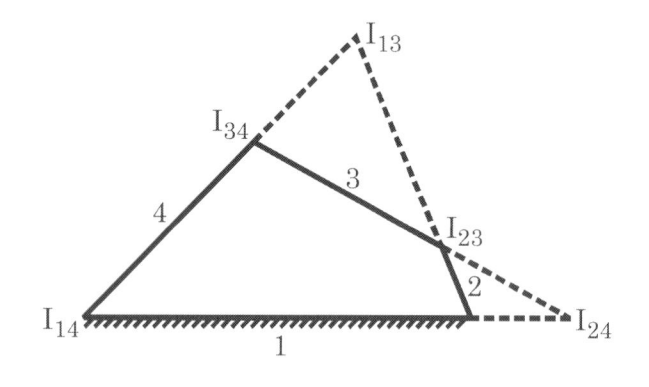

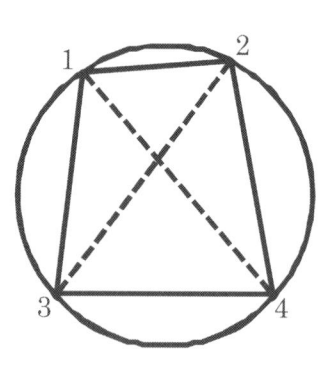

1. First of all, determine the no. of IC.

$$N = \frac{n(4-1)}{3} = \frac{4(4-1)}{2} = 6$$

2. Make a list at all the instantaneous centre in a mechanism.

Links	1	2	3	4
	–	12	23	34	–
IC	13	24		
	14			

3. Locate the fixed and permanent instantaneous centre by inspection. In fig I_{12} and I_{14} are fixed I.Cs and I_{23} and I_{34} are permanent instantaneous centre locate the remaining neither fixed nor permanent IC by Kennedy's Theorem. This is done by circle diagram

Mechanism
Chapter 1

as shown mark the points on a circle equal to the no. of links in mechanism. In present case 4 links.

4. Join the points by solid line to show these centres are already found. In the circle diagram these lines are 12, 23, 34, and 14 to indicate the ICs I_{12}, I_{23}, I_{34} and I_{14},

5. In order to find the other two IC, join two such points that the line joining them forms two adjacent triangles in the circle diagram. The line which is responsible for completing two triangles should be a common side to the two triangles. In fig join 1 and 3 to form triangle 123 and 341 and the instantaneous centre I_{13} will lie on the intersection of I_{12}, I_{23} and $I_{14} I_{34}$. similarly IC I_{24} is located.

Angular Velocity Ratio Theorem

According to this Theorem "the ratio of angular velocity of any two links moving in a constrained system is inversely proportional to the ratio of distance of their common instantaneous centre from their centre of rotation".

$$\frac{\omega_2}{\omega_3} = \frac{I_{13} I_{23}}{I_{12} I_{23}}$$

$$\frac{\omega_2}{\omega_4} = \frac{I_{14} I_{24}}{I_{12} I_{24}}$$

Indices of Merit (Mechanical Advantage)

From previous concept are know that

$$\boxed{\frac{\omega_2}{\omega_4} = \frac{I_{14} I_{24}}{I_{12} I_{24}}}$$ as per angular velocity ratio Theorem.

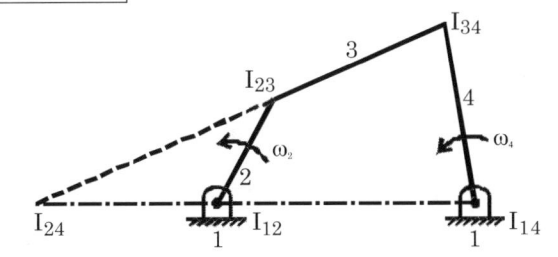

Let T_2 represent the input torque T_4 represent the output torque. Also consider that there is no friction or inertia force.

Then $T_2 \omega_2 = T_4 \omega_4$

$-$ve sign indicates that power is applied to link 2 which is negative of the power applied to link 4 by load.

$$\therefore \boxed{\frac{T_4}{T_2} = \frac{\omega_2}{\omega_4} = \frac{I_{14} I_{24}}{I_{12} I_{24}}}$$

The mechanical advantage of a mechanism is the instantaneous ratio of the output force (torque) to the input force (torque). From above equation we know that mechanical advantage is the reciprocal of the velocity ratio.

Fig shows a typical position of four bar linkage in toggle, where link 2 and 3 are on the same straight line.

Mechanism

Chapter 1

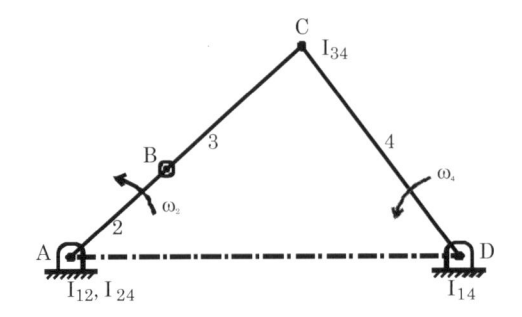

At this position, I_{12} and I_{24} is coincident at A and hence, the distance $I_{24} I_{24}$ is zero,

$$\therefore \frac{\omega_4}{\omega_2} = \frac{I_{12} I_{24}}{I_{14}, I_{24}} = \frac{0}{I_{14} I_{24}} = 0$$

$$\therefore \boxed{\omega_4 = 0}$$

$\therefore$ Mechanical advantage $\dfrac{T_4}{T_2} = \dfrac{\omega_2}{\omega_4} = \infty$

Hence the mechanical advantage for the toggle position is infinity

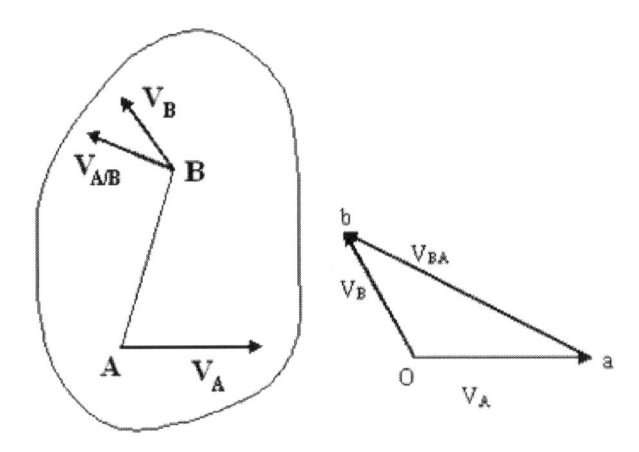

The relative velocity method is based upon the velocity of the various points of the link. Consider two points A and B on a link. Let the absolute velocity of the point A i.e. VA is known in magnitude and direction and the absolute velocity of the point B i.e. VB is known in direction only. Then the velocity of B may be determined by drawing the velocity diagram as shown.
1. Take some convenient point o, Known as the pole.
2. Through o, draw oa parallel and equal to VA, to some convenient scale.
3. Through a, draw a line perpendicular to AB. This line will represent the velocity of B with respect to A, i.e.
4. Through o, draw a line parallel to VB intersecting the line of VBA at b.
5. Measure ob, which gives the required velocity of point B to the scale.

1. Relative Velocity and Acceleration:

Relative velocity of coincident points in two kinematic links:

Mechanism

Chapter 1

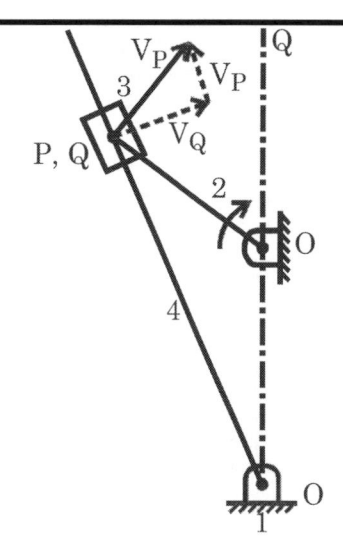

P on link 2 and 3
Q on link 4

Rubbing Velocity:

Let r_b = radius of pin B.

$\omega_{2/3}$ = relative angular velocity between link 2 and 3.

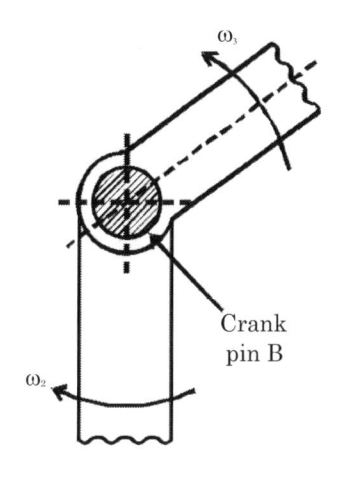

$$v_{rub} = r_b \cdot \omega_{2/3}$$

$\omega_{2/3} = \omega_2 \pm \omega_3$, + for opposite rotation.

Relative Acceleration Method:

$$f = f_c + f_t$$

f = total acceleration
f_c = Centripetal acceleration
f_t = Tangential acceleration

$$f_c = r\omega^2 = \frac{V^2}{r} \qquad f_t = r\alpha$$

Where,
r = radius of rotation of a point on link

Mechanism

Chapter 1

ω = Angular velocity of rotation
V = linear velocity of a point on link
α = Angular acceleration

Direction of f_c is along radius of rotation and towards centre.
Direction of f_t is perpendicular to radius of rotation.

Corioli's Component of Acceleration:

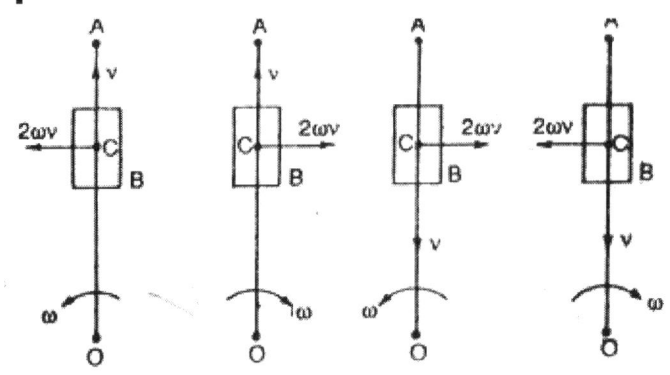

Fig. Direction of coriolis component of acceleration

This tangential component of acceleration of the slider B with respect to the coincident point C on link is known as coriolis component of acceleration and is always perpendicular to the link.

∴ Coriolis component of the acceleration of B with respect to C,
$$a^c_{BC} = a^t_{BC} = 2\omega.v$$

Where
ω = Angular velocity of the link OA, and
V = Velocity of slider B with respect to coincident point C.

Pantograph

A pantograph is an instrument used to reproduce to an enlarged or a reduced scale and as exactly as possible the path described by a given point.

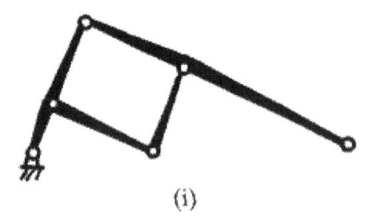

(i)

A pantograph is mostly used for the reproduction of plane areas and figures such as maps, plans etc., on enlarged or reduced scales. It is, sometimes, used as an indicator rig in order to reproduce to a small scale the displacement of the crosshead and therefore of the piston of a reciprocating steam engine. It is also used to guide cutting tools.

Velocity and Acceleration by Analytical method

Kinematic analysis of piston in I.C engine:

Mechanism

Chapter 1

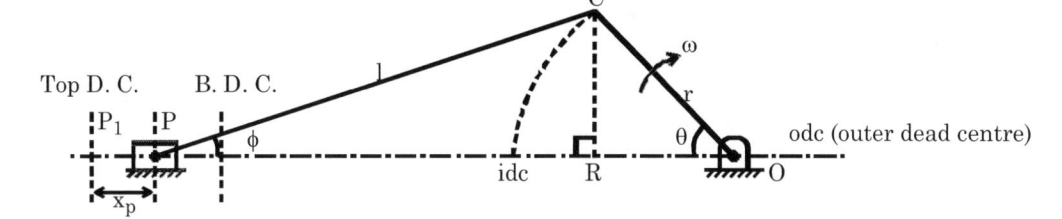

Let, r = length of crank
l = length of connecting rod
n = obliquity ratio
$= \dfrac{l}{r}$
ω = angular velocity of the crank
θ = inclination of the crank to i.d.c.
ϕ = inclination of connecting rod to the line of stroke.
x_P = displacement of piston
V_P = velocity of the piston
f_P = acceleration of the piston.

When the crank rotates through angle θ from its inner dead centre position the piston, receives displacement x_P.

$\therefore$ Displacement $x_P = P_1P$
$= P_1O = PO$
$= (l + r) - (l\cos\phi + r\cos\theta)$
$x_P = r(1 - \cos\theta) + l(1 - \cos\phi)$

Now from figure
$CR = r\sin\theta - l\sin\phi$
$\sin\theta = \dfrac{r}{l}\sin\theta = \dfrac{\sin\theta}{n}$

$\cos\phi = \sqrt{1 - \sin^2\phi} = \sqrt{1 - \left(\dfrac{\sin\theta}{n}\right)^2}$

$= \sqrt{\dfrac{n^2 - \sin^2\theta}{n^2}} = \dfrac{\sqrt{n^2 - \sin^2\theta}}{n}$

$\therefore x_P = r(1 - \cos\theta) + l\left[1 - \dfrac{\sqrt{n^2 - \sin^2\theta}}{n}\right]$

$= r(1 - \cos\theta) + \dfrac{1}{n}(n - \sqrt{n^2 - \sin^2\theta})$

$= r[1 - \cos\theta + n - \sqrt{n^2 - \sin^2\theta}]$

$\therefore \boxed{x_P = r[1 - \cos\theta + n - \sqrt{n^2 - \sin^2\theta}]}$

Now differentiating above equation with respect to „t".

$V_P = \dfrac{d}{dt}x_P = \dfrac{d}{dt}[r(1 - \cos\theta + n - \sqrt{n^2 - \sin^2\theta})]$

$= \dfrac{d}{d\theta}[r(1 - \cos\theta + n - \sqrt{n^2 - \sin^2\theta})] \cdot \dfrac{d\theta}{dt}$

Mechanism

Chapter 1

$$= r\omega \cdot \frac{d}{d\theta}[1 - \cos\theta + n - \sqrt{n^2 - \sin^2\theta}]$$

$$V_P = r\omega \left[\sin\theta + \frac{\sin 2\theta}{2\sqrt{n^2 - \sin^2\theta}}\right] \approx r\omega\left(\sin\theta + \frac{\sin 2\theta}{2n}\right)$$

Again differentiating above equation w.r.to t

$$f_P = r\omega^2\left[\cos\theta + \frac{\cos 2\theta}{n}\right]$$

Analysis of connecting rod in I.C. engine:

Let, ω_r = angular velocity of the C.R.
α_r = angular velocity of the C.R

now $CR = r\sin\theta = l\sin\phi$

$$\sin\phi = \frac{\sin\theta}{n}$$

now Differentiating above w.r. to 't'

$$\frac{d}{dt}\sin\phi = \frac{1}{n}\frac{d}{dt}\sin\theta$$

$$\frac{d}{d\phi}\sin\phi \cdot \frac{d\phi}{dt} = \frac{1}{n}\frac{d}{d\theta}\sin\theta \cdot \frac{d\theta}{dt}$$

$$\cos\phi \cdot \omega_r = \frac{\cos\theta}{n} \cdot \omega$$

$$\omega_r = \frac{\omega\cos\theta}{n\cos\phi}$$

Also from previous section

$$\cos\phi = \sqrt{\frac{n^2 - \sin^2\theta}{n^2}}$$

$$\therefore \omega_r = \frac{\omega\cos\theta}{n \cdot \sqrt{\frac{n^2 - \sin^2\theta}{n^2}}} \approx \frac{\omega\cos\theta}{n}$$

$$\alpha_r = \frac{-\omega^2 \sin\theta}{n}$$

Q. In a slider crank mechanism the stroke of the slider is 200 mm and the obliquity ratio is 4.5. The crank rotates uniformly at 400 rpm clockwise. While the crank is approaching the inner dead center and the connecting rod is normally to the crank. Find
(i) Velocity of piston and angular velocity of the C.R.
(ii) Acceleration of the piston and angular acceleration of the C.R.

Solution:
Stroke length = 200 mm = 2r
n = 4.5, N = 400 rpm, clockwise.
There find $V_P, f_P, \omega_r, \alpha_r$

Mechanism

Chapter 1

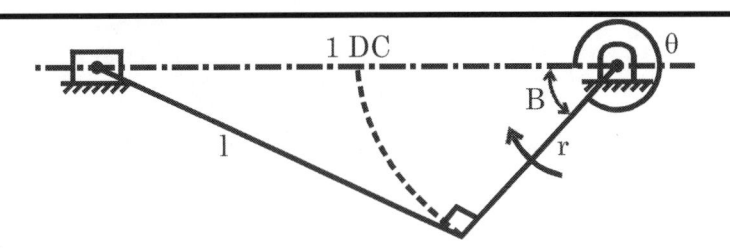

Now r = 100 mm
n = 4.5
∴ l = 450 mm
ω = 41.88 rad/sec. clockwise
$\beta = \tan^{-1}\dfrac{l}{r} = 77.47°$
∴ $\theta = 360 - 77.47 = 282.53°$

(i) $V_P = r\omega\left[\sin\theta + \dfrac{\sin 2\theta}{2n}\right]$

$= 100 \times 41.88\left[\sin 282.53 + \dfrac{\sin 565.05}{9}\right]$

$= -4286.17$ mm/sec

$= -4.29$ m/sec (direction away from the crank)

(ii) Acceleration
$f_P = r\omega^2\left[\cos\theta + \dfrac{\cos 2\theta}{n}\right]$

$= 100 \times 41.88^2\left[\cos 282.53 + \dfrac{\cos 565.05}{4.5}\right]$

$= 2.75$ m/sec² (direction towards the crank)

(iii) $\omega_r = \dfrac{\omega\cos\theta}{n} = 2.02$ rad/sec.

(Direction of ω_r is opposite to that of ω)

(iv) $\alpha_r = \dfrac{-\omega^2\sin\theta}{n} = -380.63$ rad/sec².

(Direction of α_r is same as that of ω).

Q. In an I.C engine mechanism having obliquity ratio n, show that for uniform engine speed the ratio for piston acceleration at the beginning of stroke and end of the stroke is given by $\dfrac{1+n}{1-n}$.

Solution:

Mechanism

Chapter 1

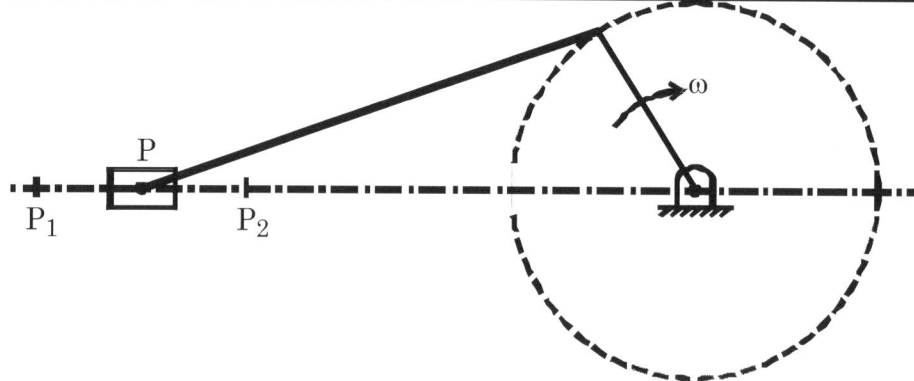

P_1 is beginning of stroke
P_2 is end of stroke
at P_1, $\theta = 0$
and at P_2 $\theta = 180°$

$$f_P = r\omega^2 \left(\cos\theta + \frac{\cos 2\theta}{n} \right)$$

$$f_{P_1} = r\omega^2 \left(1 + \frac{1}{n}\right) = r\omega^2 \left(\frac{1+n}{n}\right)$$

$$f_{P_2} = r\omega^2 \left(-1 + \frac{1}{n}\right) = r\omega^2 \left(\frac{1-n}{n}\right)$$

$$\therefore \boxed{\frac{f_{P_1}}{f_{P_2}} = \frac{1+n}{1-n}}.$$

Velocity and Acceleration by Klein's Construction:

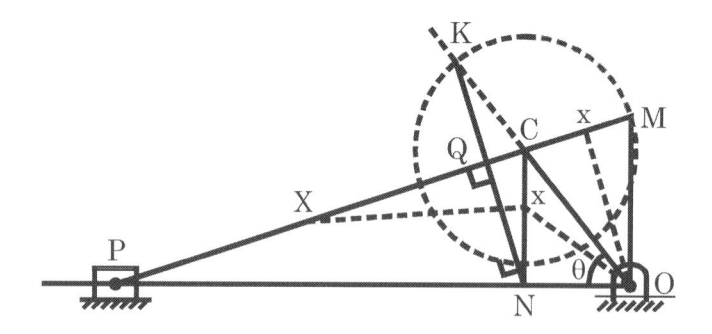

OC — crank
CP — connecting rod
θ — Angle made by crank with i.d.c
ω — Angular velocity of crank.

Procedure to draw velocity diagram:

1. Firstly draw the configuration diagram of slider crank mechanism to the scale 1: K.
2. After getting configuration diagram OCP, now draw a line through ‗O' ⊥ to the line of stroke OP.

Mechanism

Chapter 1

3. Extend the connecting rod length PC, to meet this ⊥ name the intersection point as M.
4. Δ OCM Represent the velocity polygon of slider crank mechanism to the scale ‗KW'.

In Δ OCM,

OC represents V_C,

CM represents $V_{P/C}$

And OM represents V_P

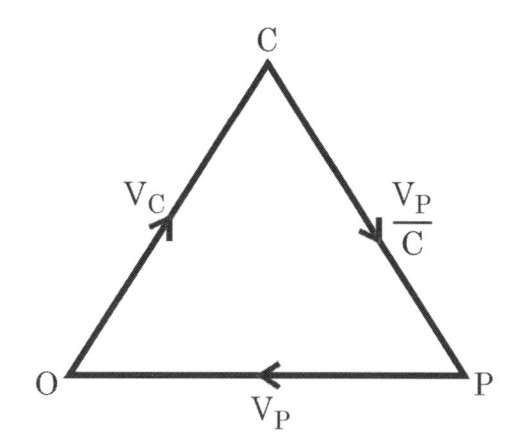

∴ $\boxed{V_C = OC.K\omega}$

$\boxed{V_{P/C} = OM.K\omega}$

$\boxed{V_P = OM.K\omega}$

Velocity of any point x lying on connecting rod:

$$\frac{Cx}{CP} = \frac{Cx}{CM} \Rightarrow Cx = \frac{Cx}{CP} \times CM$$

∴ $\boxed{V_X = Ox . K\omega}$

Acceleration Analysis:

1. As discussed earlier Δ OCM is velocity polygon of slider crank mechanism to scale ‗kW'.
2. With C as centre and with CM as radius draw a circle.
3. Draw another circle with diameter as length PC.
4. Then KL represents the common chord of these two circles.
5. Extend KL to meet the line of stroke at N. Also KL is intersecting to PC at Q.
6. Then Δ OCM represent the acceleration polygon.

Special Cases of Klein's Construction:

When the crank is at i.d.c.

Mechanism

Chapter 1

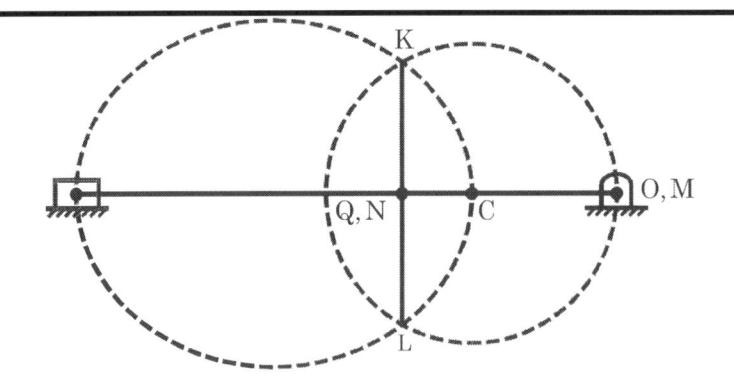

$OM = 0$, $\therefore V_P = OM \times k\omega = 0$

$V_C = OC \times k\omega$, $V_{P/C} = CM \times k\omega$ [$\because CM = OC$]

$\therefore$ $\boxed{V_C = V_{P/C}}$

Acceleration polygon is OCQN.
Here $QN = 0$

$\therefore$ $f_{P/C}^t = NQ \times k\omega^2 = 0$

i.e., $\alpha_{P/C} = \dfrac{f_{P/C}^t}{PC} = 0$

it means angular velocity of connecting rod is maximum.

$f_C = CO \times k\omega^2$

$f_P = NO \times k\omega^2$

$f_{P/C} = QC \times k\omega^2$

(ii) When the crank is at 90° from idc.

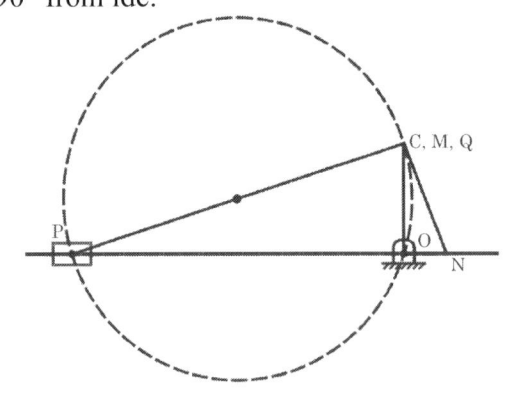

$OCM \rightarrow$ Velocity triangle
$OCQN \rightarrow$ Acceleration polygon.

$V_C = OC \times k\omega$

$V_{P/C} = CM \times k\omega = 0$

$\therefore$ $W_{P/C} = \dfrac{V_{P/C}}{PC} = 0$

$\therefore$ at $\theta = 90°$ $W_{P/C} = 0$

$V_P = PM \times k\omega$

$\therefore$ $\boxed{V_P = V_C}$

$f_{P/C}^t = QC \cdot k\omega^2 = 0$

$f_C = CO \times k\omega^2$

Mechanism

Chapter 1

$f_P = -NO \times k\omega^2 \rightarrow$ case of retardation

$f^t_{P/C} = NQ \cdot k\omega^2$

When the crank is at 180° from i.d.c.

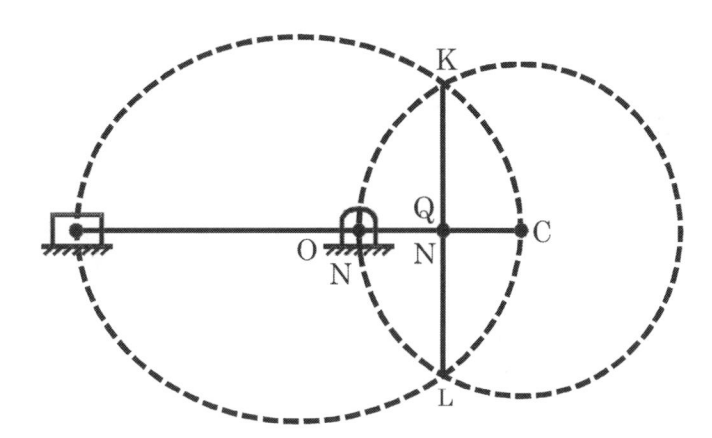

OCM → Velocity triangle
OCQN → Acceleration polygon.
$V_P = (OM) k\omega = 0$
$V_C = (OC) \cdot k\omega$
$V_{P/C} = CM \cdot k\omega$ $\qquad V_C = V_{P/C}$
$f^t_{P/C} = NQ \cdot k\omega^2 = 0$ $\qquad \therefore \alpha_{P/C} = 0$
i.e., it means angular velocity of C.R. is maximum
$f_C = (CO) k\omega^2$
$f_P = -(NO) k\omega^2 \rightarrow$ Retardation
$f^c_{P/C} = (QC) k\omega^2$

(v) When crank and connecting rod are mutual perpendicular.

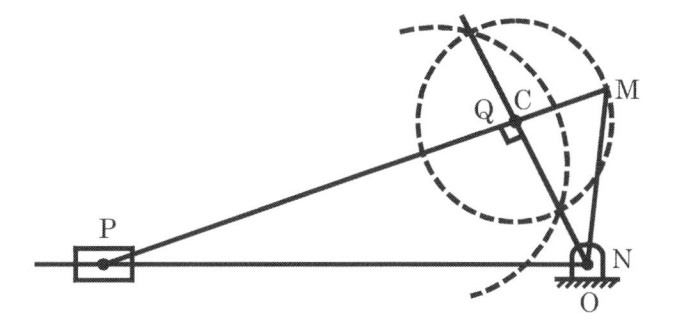

OCM → velocity triangle
OCQN → Acceleration polygon.
In polygon,
CO represents $f^C_C = f_C$, No represents $f^t_P = f_{,P}$
NQ represents $f^t_{P/C}$ and NC represents $f_{P/C}$

$\therefore \quad f_C = CO \cdot k\omega^2$ direction C to O
$\qquad f_P = No \cdot k\omega^2$ direction N to O
$\qquad f^t_{P/C} = NQ \cdot K\omega^2$ direction Q to C
$\qquad f^C_{P/C} = QC \cdot K\omega^2$ direction Q to C

Mechanism

Chapter

$f_{P/C}$ = NC. K ω^2 direction N to C.

Acceleration of any point X lying on connecting rod:

$$\frac{Cx}{CP} = \frac{Cx_1}{CN} \Rightarrow Cx_1 = \frac{Cx}{CP}.CN$$

$\therefore \quad \boxed{f_X = Ox_1 .k\omega^2}$

Force analysis in I.C engine Mechanism:

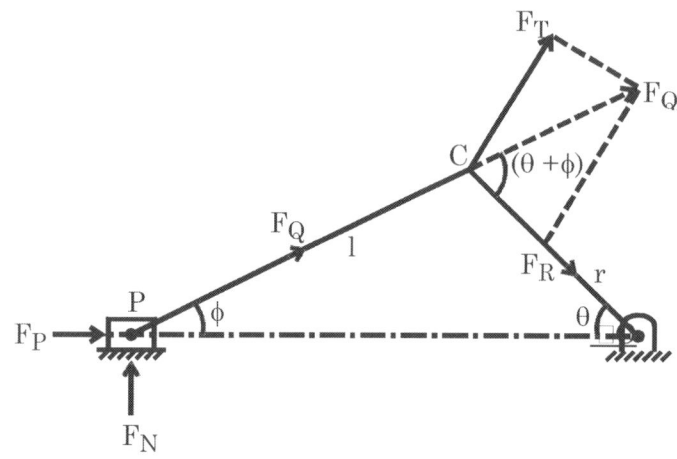

F_P = Net axial force on the piston or piston effort.
F_Q = Force acting along connecting rod
F_N = Normal reaction on the side of the cylinder or piston side thrust.
F_T = Tangential force at crank pin or force perpendicular to the crank.
F_R = Radial load on the crank shaft bearing.
T = Turning moment or Torque on the crank

$$\boxed{T = F_T . r}$$

*Net axial force (F_P) : During acceleration

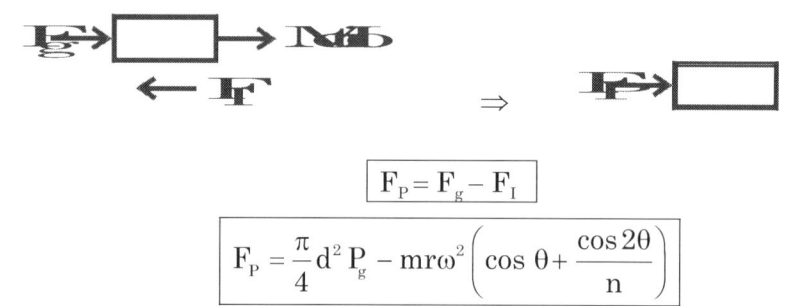

$\therefore \quad \boxed{F_P = F_g - F_I}$

$$\boxed{F_P = \frac{\pi}{4}d^2 P_g - mr\omega^2\left(\cos\theta + \frac{\cos 2\theta}{n}\right)}$$

During retardation:

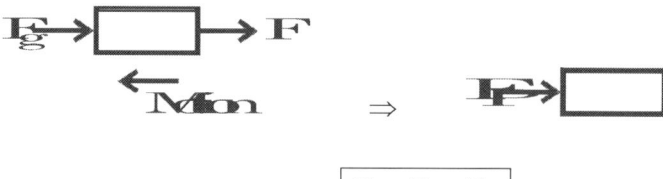

$\therefore \quad \boxed{F_P = F_g + F_I}$

d = Diameter of the piston (m)
P_g = intensity of pressure in the cylinder (N/m²)

Mechanism

Chapter 1

or Different in the intensity of pressure on two sides of the piston.

F_I = inertia force = mf_P

*In case of vertical engine, the wt of the reciprocating parts (W_R) and friction force (F_f).

$$F_P = F_g \pm F_I - F_f + W_R$$

Q. The mass of reciprocating parts of a steam engine is 225 kg, diagram of the cylinder is 400 mm, length of the stroke is 500 mm and the ratio of length of connecting rod to crank is 4.2. When the crank is at inner dead center, the difference in the pressure of the two sides of the piston is 5 bar. At what speed must the engine run so that the thrust in the connecting rod in this piston is equal to 5200 N?

Solution: Data

   m = 225 kg.

   d = 400 mm = 0.4 m

   Stroke length = 500 mm = 2r

   ∴ r = 250 mm = 0.25 m

   n = 4.2

   θ = 0°

   p_g = 5 bar

   F_Q = 5200 N

   ω = ?

   Here $F_Q = F_P$

   Now $F_P = F_g - F_I$

   $= \dfrac{\pi}{4} d^2 P_g - mr\omega^2 \left(\cos\theta - \dfrac{\cos 2\theta}{n} \right)$

   $5200 = \dfrac{\pi}{4}(0.4)^2 \times 5 \times 10^5 - 225 \times 0.25 \times \omega^2 \left(1 + \dfrac{1}{4.2}\right)$

   ∴ ω = 28.767 rad/sec.

   N = 274.7 rpm.

Q. A single cylinder two stroke vertical engine a bore of 30 cm and a stroke of 40 cm with a connecting rod of 80 cm long. The mass of the reciprocating parts is 120 kg. When the piston is at quarter stroke and moving down, the pressure on it is 70 N/cm². If the speed of the engine crank shaft is 250 rpm clockwise find the turning moment on the crank shaft. Neglect the mass and inertia effects on connecting rods and crank.

Solution: Data

   d = 30 cm

   Stroke = 40 cm = 2r

   ∴ r = 20 cm   n = 4

   l = 80 cm

   m = 120 kg

   x_P = 0.25 × stroke length

   P_g = 70 N / cm²

   N = 250 rpm (clockwise), ω = 26.166

   T = ?

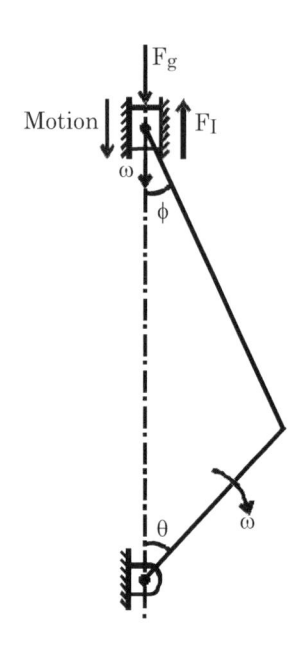

Mechanism

Chapter 1

Now $x_P = r(1 - \cos\theta + n - \sqrt{n^2 - \sin^2\theta})$

$0.25 \times 2r = r(1 - \cos\theta + 4 - \sqrt{16 - \sin^2\theta})$

$\dfrac{1}{2} = 5 - \cos\theta - \sqrt{16 - \sin^2\theta}$

$\sqrt{16 - \sin^2\theta} = 4.5 - \cos\theta$

$16 - \sin^2\theta = 20.25 + \cos^2\theta - 9\cos\theta$

$9\cos\theta = 20.25 - 16 + \sin^2\theta + \cos^2\theta$

$\qquad = 5.25$

$\cos\theta = \dfrac{5.25}{9}$

$\therefore \theta = 54.31°$

Now $\sin\phi = \dfrac{\sin\theta}{n}$

$\phi = 11.71°$

Now $F_P = F_G - F_I + \omega$

$F_G = \dfrac{\pi}{4} d^2 \cdot P_g$

$\qquad = \dfrac{\pi}{4} 30^2 \times 70$

$\qquad = 49455 \text{ N}$

$F_I = mr\omega^2 \left(\cos\theta + \dfrac{\cos 2\theta}{n}\right)$

$\qquad = 120 \times 0.20 \times 26.166^2 \left(\cos 54.31 + \dfrac{\cos 108.62}{4}\right)$

$\qquad = 8274.7 \text{ N}$

$W = mg = 1177.2 \text{ N}$

$\therefore F_P = 49455 - 8274.7 + 1177.2$

$\qquad = 42357.7 \text{ N}$

Mechanism

Chapter 1

Now $F_Q = \dfrac{F_P}{\cos\phi} = 43257.8$ N

Now $F_T = F_Q \sin(\theta + \phi)$
           $= 39524.1$ N

$\therefore T = F_T \times r$
            $= 7904.8$ Nm

$\therefore \boxed{T = 7.904 \text{ kNm}}$

Objective Questions (IES, IAS, GATE)

Previous 20-Years GATE Questions

Kinematic pair

GATE-1. Match the items in columns I and II [GATE-2006]

Column I	Column II
P. Higher kinematic pair	1. Grubler's equation
Q. Lower kinematic pair	2. Line contact
R. Quick return mechanism	3. Euler's equation
S. Mobility of a linkage	4. Planer
	5. Shaper
	6. Surface contact

(a) P-2, Q-6, R-4, S-3 (b) P-6, Q-2, R-4, S-1
(c) P-6, Q-2, R-5, S-3 (d) P-2, Q-6, R-5, S-1

GATE-2. The minimum number of links in a single degree-of-freedom planar mechanism with both higher and lower kinematic pairs is [GATE-2002]
(a) 2 (b) 3 (c) 4 (d) 5

Degrees of freedom

GATE-3. The number degrees of freedom of a planar linkage with 8 links and 9 simple revolute joints is [GATE-2005]
(a) 1 (b) 2 (c) 3 (d) 4

GATE-4. When a cylinder is located in a Vee-block, then number of degrees of freedom which are arrested is [GATE-2003]
(a) 2 (b) 4 (c) 7 (d) 8

GATE-5. The number of degrees of freedom of a five link plane mechanism with five revolute pairs as shown in the figure is [GATE-1993]
(a) 3 (b) 4 (c) 2 (d) 1

Mechanism

Chapter 1

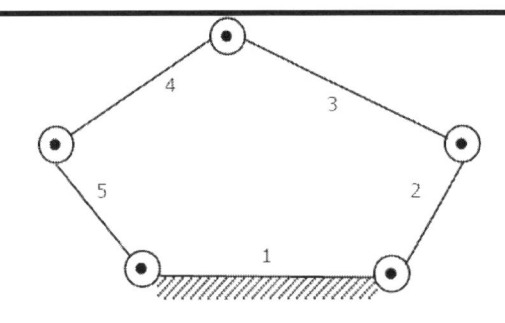

GATE-6. Match the following with respect to spatial mechanisms. [GATE-2004]

Type of Joint	Degrees of constraint
P. Revolute	1. Three
Q. Cylindrical	2. Five
R. Spherical	3. Four
	4. Two
	5. Zero

(a) P-1 Q-3 R-3   (b) P-5 Q-4 R-3   (c) P-2 Q-3 R-1   (d) P-4 Q-5 R-3

Grubler Criterion

GATE-7. A planar mechanism has 8 links and 10 rotary joints. The number of degrees of freedom of the mechanism, using Grubler's criterion, is

[GATE-2008]

(a) 0   (b) 1   (c) 2   (d) 3

GATE-8. Match the approaches given below to perform stated kinematics/dynamics analysis of machine. [GATE -2009]

Analysis	Approach
P. Continuous relative rotation	1. D'Alembert's principle
Q. Velocity and acceleration	2. Grubler's criterion
R. Mobility	3. Grashoff's law
S. Dynamic-static analysis	4. Kennedy's theorem

(a) P-1, Q-2, R-3, S-4   (b) P-3, Q-4, R-2, S-1
(c) P-2, Q-3, R-4, S-1   (d) P-4, Q-2, R-1, S-3

Grashof's law

GATE-9. Which of the following statements is incorrect [GATE-2010]
(a) Grashof's rule states that for a planar crank-rocker four bar mechanism, the sum of the shortest and longest link lengths cannot be less than the sum of the remaining two link lengths.
(b) Inversions of a mechanism are created by fixing different links one at a time.
(c) Geneva mechanism is an intermittent motion device.
(d) Gruebler's criterion assumes mobility of a planar mechanism to be one.

GATE-10. In a four-bar linkage, S denotes the shortest link length, L is the longest link length, P and Q are the lengths of other two links. At least one of the three moving links will rotate by 360° if [GATE-2006]
(a) S + L ≤ P + Q   (b) S + L > P + Q
(c) S + P ≤ L + Q   (d) S + P > L + Q

Mechanism

Chapter 1

Inversion of Mechanism

GATE-11. The number of inversions for a slider crank mechanism is [GATE-2006]
(a) 6   (b) 5   (c) 4   (d) 3

Inversion of Single Slider crank chain

GATE-12. The mechanism used in a shaping machine is [GATE-2003]
(a) A closed 4-bar chain having 4 revolute pairs
(b) A closed 6-bar chain having 6 revolute pairs
(c) A closed 4-bar chain having 2 revolute and 2 sliding pairs
(d) An inversion of the single slider-crank chain

Quick return motion mechanism

GATE-13. A simple quick return mechanism is shown in the figure. The forward to return ratio of the quick return mechanism is 2: 1. If the radius of the crank O_1P is 125 mm, then the distance 'd' (in mm) between the crank centre to lever pivot centre point should be
(a) 144.3   (b) 216.5
(c) 240.0   (d) 250.0

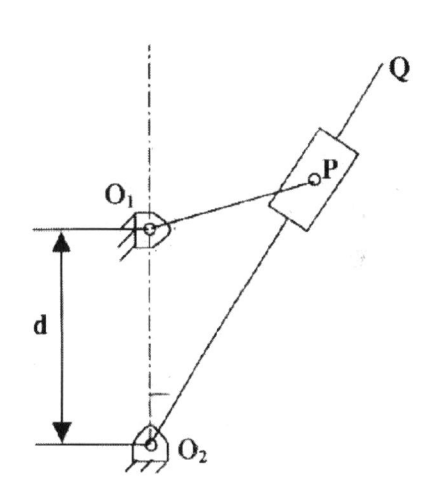

[GATE-2009]

GATE-14. Match the following [GATE-2004]

Type of Mechanism	Motion achieved
P. Scott - Russel mechanism	1. Intermittent motion
Q. Geneva mechanism	2. Quick return motion
R. Off-set slider-crank mechanism	3. Simple harmonic motion
S. Scotch Yoke mechanism	4. Straight line motion

(a) P-2 Q-3 R-1 S-4   (b) P-3 Q-2 R-4 S-1
(c) P-4 Q-1 R-2 S-3   (d) P-4 Q-3 R-1 S-2

Mechanism

Chapter 1

GATE 15. Figure shows a quick return mechanism. The crank OA rotates clockwise uniformly.
OA = 2 cm.
OO = 4 cm.
(a) 0.5
(b) 2.0
(c) $\sqrt{2}$
(d) 1

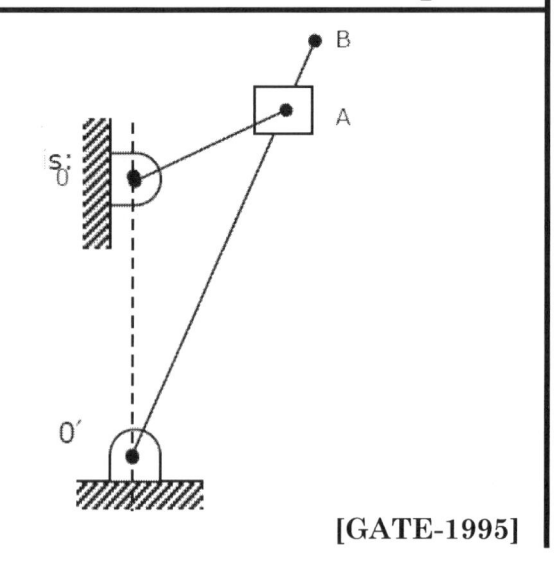

[GATE-1995]

Inversion of Double slider crank chain

GATR-16. The lengths of the links of a 4-bar linkage with revolute pairs only are p, q, r, and s units. Given that p < q < r < s. Which of these links should be the fixed one, for obtaining a "double crank" mechanism? [GATE-2003]
(a) Link of length p   (b) link of length q
(c) Link of length r   (d) link of length s

Velocity of a point on a link

GATE-17. There are two points P and Q on a planar rigid body. The relative velocity between the two points [GATE-2010]
(a) Should always be along PQ
(b) Can be oriented along any direction
(c) Should always be perpendicular to PQ
(d) Should be along QP when the body undergoes pure translation

GATE-18. The input link O_2P of a four bar linkage is rotated at 2 rad/s in counter clockwise direction as shown below. The angular velocity of the coupler PQ in rad/s, at an instant when $\angle O_4O_2P = 180°$, is

$PQ = O_4Q = \sqrt{2}\,a$ and $O_2P = O_2O_4 = a$.

(a) 4   (b) $2\sqrt{2}$ (c) 1   (d) $1/\sqrt{2}$

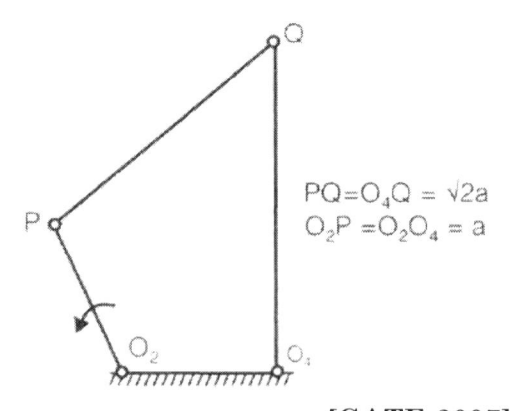

$PQ = O_4Q = \sqrt{2}a$
$O_2P = O_2O_4 = a$

[GATE-2007]

Common Data Questions

Common Data for Questions 19, 20, 21:

Mechanism

Chapter 1

An instantaneous configuration of a four-bar mechanism, whose plane is horizontal, is shown in the figure below. At this instant, the angular velocity and angular acceleration of link $O_2 A$ are ($\omega = 8$ rad/s and $\alpha = 0$, respectively, and the driving torque (τ) is zero. The link $O_2 A$ is balanced so that its centre of mass falls at O_2

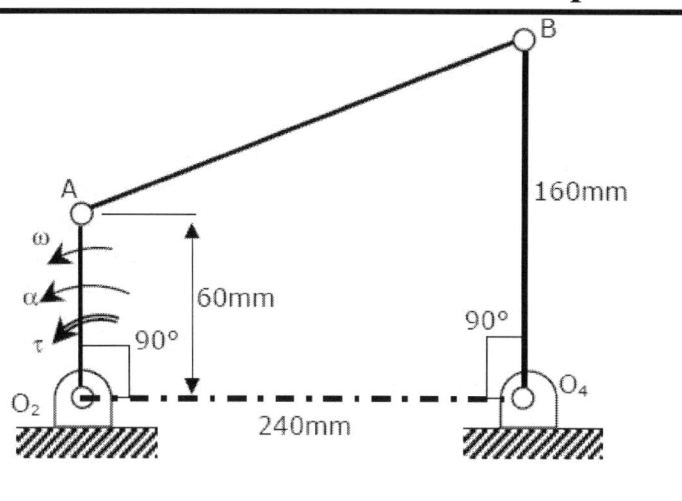

GATE-19. Which kind of 4-bar mechanism is O_2ABO_4? [GATE-2005]
(a) Double-crank mechanism (b) Crank-rocker mechanism
(c) Double-rocker mechanism (d) Parallelogram mechanism

GATE-20. At the instant considered, what is the magnitude of the angular velocity of Q4B? [GATE-2005]
(a) 1 rad/s (b) 3 rad/s (c) 8 rad/s (d) $\frac{64}{3}$ rad/s

GATE-21. At the same instant, if the component of the force at joint A along AB is 30 N, then the magnitude of the joint raction at O_2 [GATE-2005]
(a) is zero (b) is 30 N
(c) is 78 N (d) cannot be determined from the given data

GATE-22. For the planar mechanism shown in figure select the most appropriate choice for the motion of link 2 when link 4 is moved upwards.
(a) Link 2 rotates clockwise
(b) Link 2 rotates counter – clockwise
(c) Link 2 does not move
(d) Link 2 motion cannot be determined

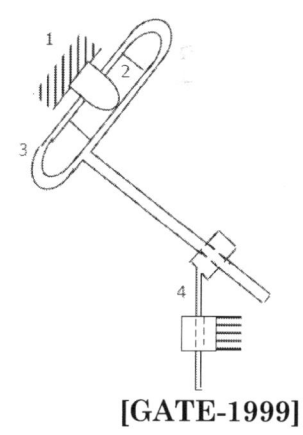

[GATE-1999]

Mechanism

Chapter 1

Location of Instantaneous centres

GATE-23. The figure below shows a planar mechanism with single degree of freedom. The instant centre 24 for the given configuration is located at a position
(a) L   (b) M
(c) N   (d) ∞

[GATE-2004]

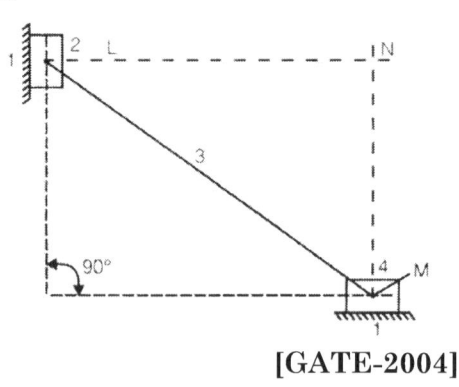

GATE-24. For the audio cassette mechanism shown in Figure given below where is the instantaneous centre of rotation (point) of the two spools?

[GATE-1999]

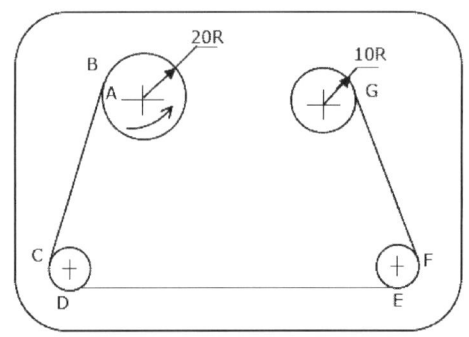

(a) Point P lies to the left of both the spools but at infinity along the line joining A and H

(b) Point P lies in between the two spools on the line joining A and H, such that PH = 2 AP

(c) Point P lies to the right of both the spools on the line joining A and H, such that AH = HP

(d) Point P lies at the intersection of the line joining B and C and the line joining G and F

GATE-25. Instantaneous centre of a body rolling with sliding on a stationary curved surface lies
(a) at the point of contact   [GATE-1992]
(b) on the common normal at the point of contact
(c) on the common tangent at the point of contact
(d) at the centre of curvature of the stationary surface

Mechanism

Chapter 1

Number of Instantaneous centres in Mechanism and Kennedy Theorem

GATE-26. In the figure shown, the relative velocity of link 1 with respect to link 2 is 12 m/sec. Link 2 rotates at a constant speed of 120 rpm. The magnitude of Carioles component of acceleration of link 1 is
(a) 302 m/s^2 (b) 604 m/s^2
(c) 906 m/s^2 (d) 1208 m/s^2

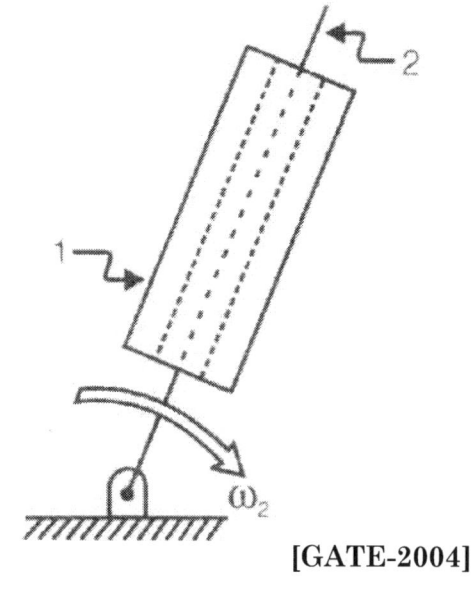

[GATE-2004]

GATE-27. The Carioles component of acceleration is present [GATE-2002]
(a) 4-bar mechanisms with 4 turning pairs (b) shaper mechanism
(c) slider-crank mechanism (d) Scotch Yoke mechanism

Hooke's Joint (Universal Joint)

GATE-28. The coupling used to connect two shafts with large angular misalignment is
(a) a Flange coupling (b) an Oldham's coupling [GATE-2002]
(c) a Flexible bush coupling (d) a Hooker's joint

Previous 20-Years IES Questions

Kinematic pair

IES-1. Match List I with List II and select the correct answer [IES-2002]
List I (Kinematic pairs) List II (Practical example)
A. Sliding pair 1. A road roller rolling over the ground
B. Revolute pair 2. Crank shaft in a journal bearing in an engine
C. Rolling pair 3. Ball and socket joint
D. Spherical pair 4. Piston and cylinder
5. Nut and screw

	A	B	C	D		A	B	C	D
(a)	5	2	4	3	(b)	4	3	1	2
(c)	5	3	4	2	(d)	4	2	1	3

Mechanism

Chapter 1

IES-2. A round bar A passes through the cylindrical hole in B as shown in the given figure. Which one of the following statements is correct in this regard?
(a) The two links shown form a kinematic pair.
(b) The pair is completely constrained.
(c) The pair has incomplete constraint.
(d) The pair is successfully constrained.
[IES-1995]

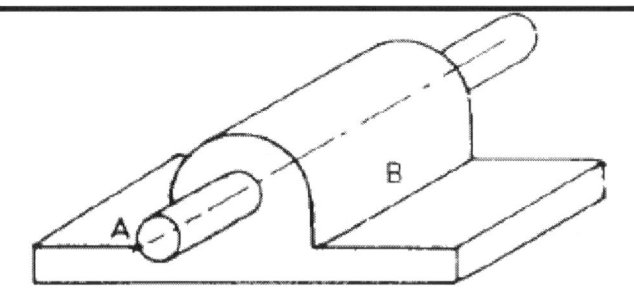

IES-3. Consider the following statements [IES-2000]
1. A round bar in a round hole form a turning pair.
2. A square bar in a square hole forms a sliding pair.
3. A vertical shaft in a footstep bearing forms a successful constraint.
Of these statements
(a) 1 and 2 are correct (b) 1 and 3 are correct
(c) 2 and 3 are correct (d) 1, 2 and 3 are correct

IES-4. Match List-I with List-II and select the correct answer using the codes given below the Lists: [IES-1999]

List-I
A. 4 links, 4 turning pairs
B. 3 links, 3 turning pairs
C. 5 links, 5 turning pairs
D. Footstep bearing

List-II
1. Complete constraint
2. Successful constraint
3. Rigid frame
4. Incomplete constraint

Code: A   B   C   D   A   B   C   D
(a)   3   1   4   2   (b) 1   3   2   4
(c)   3   1   2   4   (d) 1   3   4   2

IES-5. Consider the following statements: [IES-2005]
1. The degree of freedom for lower kinematic pairs is always equal to one.
2. A ball-and-socket joint has 3 degrees of freedom and is a higher kinematic pair
3. Oldham's coupling mechanism has two prismatic pairs and two revolute pairs.
Which of the statements given above is/are correct?
(a) 1, 2 and 3   (b) 1 only   (c) 2 and 3   (d) 3 only

IES-7. Which of the following are examples of forced closed kinematic pairs?
1. Cam and roller mechanism   2. Door closing mechanism [IES-2003]
3. Slider-crank mechanism   4. Automotive clutch operating mechanism
Select the correct answer using the codes given below:
Codes:
(a) 1, 2 and 4   (b) 1 and 3   (c) 2, 3 and 4   (d) 1, 2, 3 and 4

Mechanism

Chapter 1

IES-8. Assertion (A): Hydraulic fluid is one form a link. [IES-1996]
Reason (R): A link need not necessarily be a rigid body but it must be a resistant body.
(a) Both A and R are individually true and R is the correct explanation of A
(b) Both A and R are individually true but R is **not** the correct explanation of A
(c) A is true but R is false
(d) A is false but R is true

IES-9 Assertion (A): When a link has pure translation, the resultant force must pass through the centre of gravity. [IES-1994]
Reason (R): The direction of the resultant force would be in the direction of acceleration of the body.
(a) Both A and R are individually true and R is the correct explanation of A
(b) Both A and R are individually true but R is **not** the correct explanation of A
(c) A is true but R is false
(d) A is false but R is true

Lower pair

IES-10. Consider the following statements: [IES-2006]
1. Lower pairs are more resistant than the higher pairs in a plane mechanism.
2. In a 4-bar mechanism (with 4 turning pairs), when the link opposite to the shortest link is fixed, a double rocker mechanism results.
Which of the statements given above is/are correct?
(a) Only 1   (b) Only 2   (c) Both 1 and 2   (d) Neither 1 nor 2

Higher pair

IES-11. Consider the following pairs of parts: [IES-2000]
1. Pair of gear in mesh   2. Belt and pulley
3. Cylinder and piston   4. Cam and follower
Among these, the higher pairs are
(a) 1 and 4   (b) 2 and 4   (c) 1, 2 and 3   (d) 1, 2 and 4

IES-12. Assertion (A): The elements of higher pairs must be force closed.
Reason (R): This is required in order to provide completely constrained motion.
(a) Both A and R are individually true and R is the correct explanation of A
(b) Both A and R are individually true but R is not the correct explanation of A
(c) A is true but R is false
(d) A is false but R is true [IES-1995]

Kinematic chain

IES-13. In a Kinematic chain, a quaternary joint is equivalent to: [IES-2005]
(a) One binary joint   (b) Two binary joints
(c) Three binary joints   (d) Four binary joints

Mechanism

Chapter 1

IES-14. The kinematic chain shown in the above figure is a
(a) structure
(b) mechanism with one degree of freedom
(c) mechanism with two degree of freedom
(d) mechanism with more than two degrees of freedom

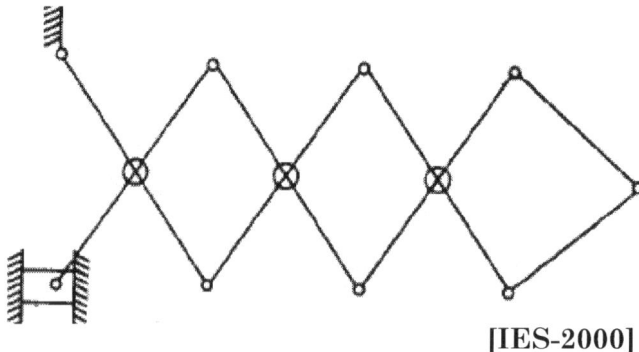

[IES-2000]

IES-15. Which of the following are examples of a kinematic chain? [IES-1998]

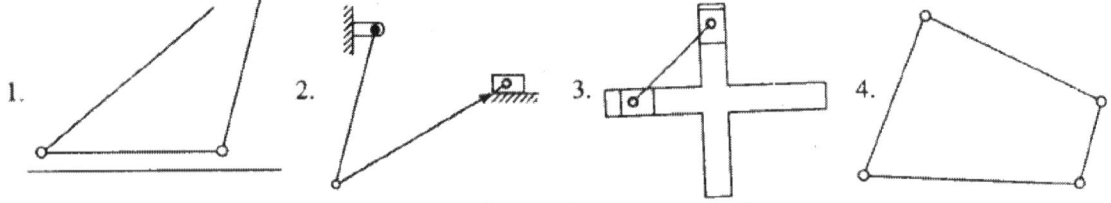

Select the correct answer using the codes given below:
Codes: (a) 1, 3 and 4   (b) 2 and 4   (c) 1, 2 and 3   (d) 1, 2, 3 and 4

IES-16. A linkage is shown below in the figure in which links ABC and DEF are ternary Jinks whereas AF, BE and CD are binary links.
The degrees of freedom of the linkage when link ABC is fixed are
(a) 0   (b) 1
(c) 2   (d) 3

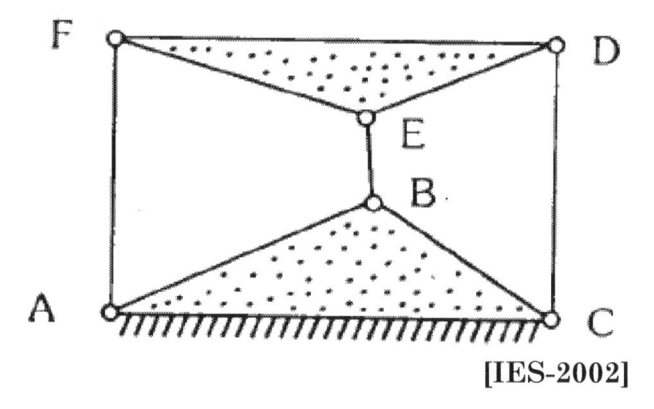

[IES-2002]

Degrees of freedom

IES-18. Match List-I with List-II and select the correct answer using the codes given below the lists: [IES-2001]

List-I	List-II
A. 6 d.o.f. system	1. Vibrating beam
B. 1 d.o.f. system	2. Vibration absorber
C. 2 d.o.f. system	3. A rigid body in space
D. Multi d.o.f. system	4. Pure rolling of a cylinder

Codes: A   B   C   D   A   B   C   D
(a)   1   2   4   3   (b)   1   4   2   3
(c)   3   2   4   1   (d)   3   4   2   1

Mechanism

Chapter 1

IES-19. The two-link system, shown in the given figure, is constrained to move with planar motion. It possesses
(a) 2-degrees of freedom
(b) 3-degrees of freedom
(c) 4-degrees of freedom
(d) 6-degrees of freedom

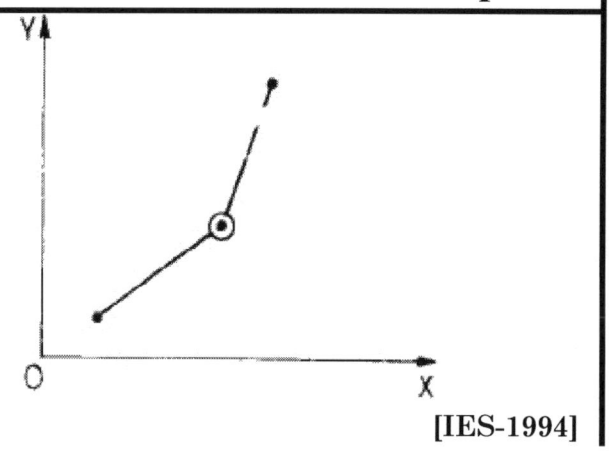

[IES-1994]

IES-20. When supported on three points, out of the 12 degrees of freedom the number of degrees of reedom arrested in a body is [IES-1993]
(a) 3　　　　(b) 4　　　　(c) 5　　　　(d) 6

Grubler criterion

IES-21. $f = 3(n - 1) - 2j$. In the Grubler's equation for planar mechanisms given, j is the [IES-2003]
(a) Number of mobile links　　(b) Number of links
(c) Number of lower pairs　　(d) Length of the longest link

IES-22. Match List-I with List-II and select the correct answer using the codes given below the lists:

List-I	List-II
A. Cam and follower	1. Grubler's rule
B. Screw pair	2. Grashof's linkage
C. 4-bar mechanism	3. Pressure angle
D. Degree of freedom of planar mechanism	4. Single degree of freedom

Codes:　　A　B　C　D　　　　A　B　C　D
(a)　　　　3　4　2　1　(b)　1　2　4　3
(c)　　　　1　4　2　3　(d)　3　2　4　1

Grashof's law

IES-24. Inversion of a mechanism is [IES-1992]
(a) changing of a higher pair to lower pair
(b) obtained by fixing different links in a kinematic chain
(b) turning it upside down
(d) obtained by reversing the input and output motion

IES-26. Match List I (Kinematic inversions) with List II (Applications) and select the correct answer using the codes given below the Lists: [IES-2000]

Mechanism

Chapter 1

	List I		List II
A.	(crank fixed, slider horizontal)	1.	Hand pump
B.	(crank pivot at top-left fixed)	2.	Compressor
C.	(slider along top fixed)	3.	Whitworth quick return mechanism
D.	(slider at bottom fixed)	4.	Oscillating Cylinder Engine

Code: A B C D   A B C D
(a) 1 3 4 2   (b) 2 4 3 1
(c) 2 3 4 1   (d) 1 4 3 2

Inversion of four bar chain

IES-27. Which of the following pairs are correctly matched? Select the correct answer using the codes given below the pairs. [IES-1998]
Mechanism   Chain from which derived
1. Whitworth quick return motion..... Single slider crank chain
2. Oldham's coupling ....Four bar chain
3. Scotch Yoke... ....Double slider crank chain
Codes: (a) 1 and 2   (b) 1, 2 and 3   (c) 1 and 3   (d) 2 and 3

IES-28. Which one of the following conversions is used by a lawn-sprinkler which is a four bar mechanisms? [IES-2004]
(a) Reciprocating motion to rotary motion
(b) Reciprocating motion to oscillatory motion
(c) Rotary motion to oscillatory motion
(d) Oscillatory motion to rotary motion

IES-29. A four-bar chain has [IES-2000]
(a) All turning pairs
(b) One turning pair and the others are sliding pairs
(c) One sliding pair and the others are turning pairs
(d) All sliding pairs

Mechanism

Chapter 1

IES-30. Assertion (A): The given line diagram of Watt's indicator mechanism is a type of crank and lever mechanism.
Reason (R): BCD acts as a lever.
(a) Both A and R are individually true and R is the correct explanation of A
(b) Both A and R are individually true but R is **not** the correct explanation of A
(c) A is true but R is false
(d) A is false but R is true

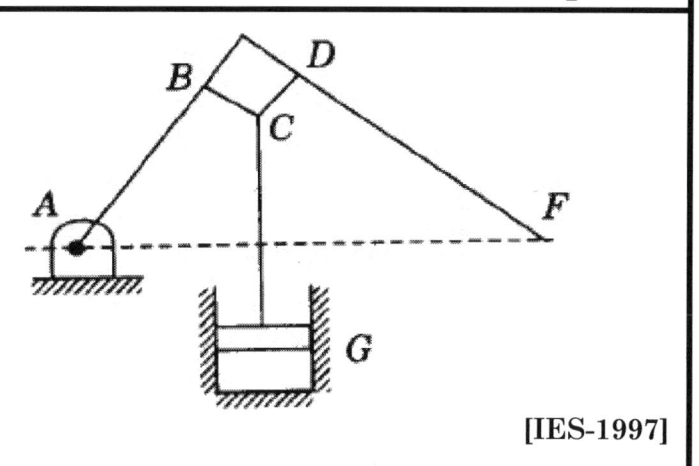

[IES-1997]

IES-31. The centre of gravity of the coupler link in a 4-bar mechanism would experience
(a) No acceleration (b) only linear acceleration [IES-1996]
(c) Only angular acceleration (d) both linear and angular accelerations.

IES-32. In the given figure, ABCD is a four-bar mechanism. At the instant shown, AB and CD are vertical and BC is horizontal AB is shorter than CD by 30 cm. AB is rotating at 5 radius and CD is rotating at 2 rad/s. The length of AB is
(a) 10cm (b) 20 cm
(c) 30 cm (d) 50 cm.

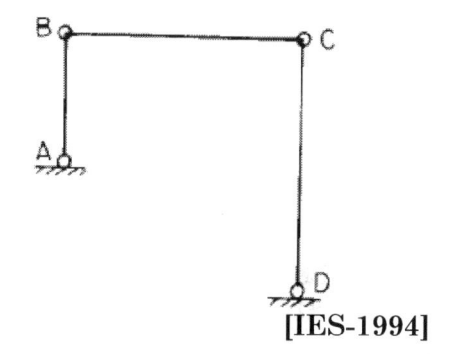

[IES-1994]

Inversion of Single Slider crank chain

IES-33. In a single slider four-bar linkage, when the slider is fixed, it forms a mechanism of [IES-1999]
(a) hand pump (sb) reciprocating engine (c) quick return (d) oscillating cylinder

IES-34. Match List-I with List-II and select the correct answer using the codes given below the Lists: [IES-1997]

List-I
A. Quadric cycle chain
B. Single slider crank chain
C. Double slider crank chain
D. Crossed slider crank chain

List-II
1. Rapson's slide
2. Oscillating cylinder engine mechanism
3. Ackermann steering mechanism
4. Oldham coupling

Codes: A B C D   A B C D
(a) 1 2 4 3   (b) 4 3 2 1
(c) 3 4 1 2   (d) 3 2 4 1

Mechanism

Chapter 1

IES-35. Which one of the following mechanisms represents an inversion of the single slider crank chain? [IES-2008]
(a) Elliptical trammel
(b) Oldham's coupling
(c) Whitworth quick return mechanism
(d) Pantograph mechanism

IES-36. Match List I with List II and select the correct answer using the codes given below the lists: [IES-1993]

List II
A. Quadric cycle chain
B. Single slider crank chain
C. Double slider crank chain
D. Crossed slider crank chain

List II
1. Elliptical trammel
2. Rapsons slide
3. Ackerman steering
4. Eccentric mechanism
5. Pendulum pump

Codes:	A	B	C	D		A	B	C	D
(a)	5	4	2	1	(b)	3	1	5	4
(c)	5	3	4	2	(d)	3	5	1	2

Quick return motion mechanism

IES-37. Match List I with List II and select the correct answer: [IES-2002]

List I (Mechanism)
A. Hart mechanism
B. Pantograph
C. Whitworth mechanism
D. Scotch yoke

List II (Motion)
1. Quick return motion
2. Copying mechanism
3. Exact straight line motion
4. Simple harmonic motion
5. Approximate straight line motion

	A	B	C	D		A	B	C	D
(a)	5	1	2	3	(b)	3	2	1	4
(c)	5	2	1	3	(d)	3	1	2	4

IES-38. The crank and slotted lever quick-return motion mechanism is shown in figure. The length of links O_1O_2, O_1C and O_2A are 10 cm, 20 cm and 5 cm respectively.
The quick return ratio of the mechanism is
(a) 3.0   (b) 2.75
(c) 2.5   (d) 2.0

[IES-2002]

IES-39. Match List I with List II and select the correct answer using the codes given below the Lists: [IES-2000]

Mechanism

Chapter 1

List-I | List-II
(a) Quick return mechanism | 1. Lathe
(b) Apron mechanism | 2. Milling machine
(a) (c) Indexing mechanism | 3. Shaper
(d) Regulating wheel | 3. Shaper
 | 4. Centreless grinding

Codes: A B C D   A B C D
(a) 3 2 1 4   (b) 2 3 4 1
(c) 4 2 3 1   (d) 3 1 2 4

IES-40. Match List I with List II and select the correct answer using the codes given below the Lists: [IES-2000]

List I
A. Compound train
B. Quick return mechanism
C. Exact straight line motion bends and corners
D. Approximate straight line motion

List II
1. Hart mechanism
2. Corioli's force
3. Transmission of motion around
4. Watt mechanism

Code: A B C D   A B C D
(a) 1 2 3 4   (b) 3 2 1 4
(c) 3 4 1 2   (d) 1 4 3 2

IES-41. The type of quick return mechanism employed mostly in shaping machines is: [IES-1997]
(a) DC reversible motor (b) Fast and loose pulleys
(c) Whitworth motion (d) Slotted link mechanism

IES-43. In order to draw the acceleration diagram, it is necessary to determine the Corioli's component of acceleration in the case of [IES-1997]
(a) crank and slotted lever quick return mechanism
(b) slider-crank mechanism (c) four bar mechanism (d) pantograph

IES-44. Which mechanism produces intermittent rotary motion from continuous rotary motion? [IES-2008]
(a) Whitworth mechanism (b) Scotch Yoke mechanism
(c) Geneva mechanism (d) Elliptical trammel

Inversion of Double slider crank chain

IES-45 ABCD is a mechanism with link lengths AB = 200, BC = 300, CD = 400 and DA = 350. Which one of the following links should be fixed for the resulting mechanism to be a double crank mechanism? (All lengths are in mm)

[IES-2004]

(a) A B (b) BC (c) CD (d) DA

Mechanism

Chapter 1

Elliptical trammels

IES-46. A point on a link connecting a double slider crank chain will trace a
[IES-2000]

(a) straight line (b) circle (c) parabola (d) ellipse

IES-47. An elliptic trammel is shown in the given figure. Associated with the motion of the mechanism are fixed and moving centrodes. It can be established analytically or graphically that the moving centrode is a circle with the radius and centre respectively of
(a) l and 0 (b) l/2 and B
(c) l/2 and C (d) l/2 and D

AB = l
BD = DA
Elliptic trammel

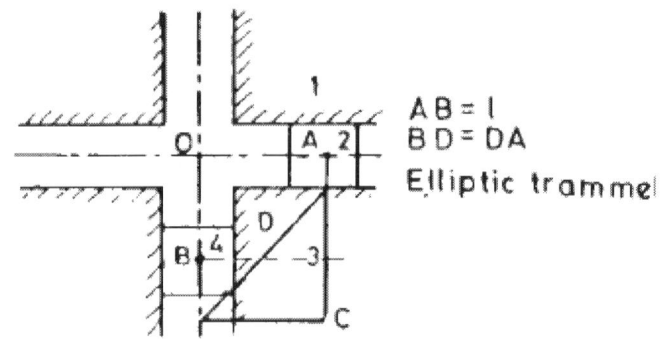

[IES-1994]

Scotch yoke mechanism

IES-49. Scotch yoke mechanism is used to generate [IES-1992]
(a) Sine functions (b) Square roots (c) Logarithms (d) Inversions

IES-51. Which of the following are inversions of a double slider crank chain?
[IES-1993]

1. Whitworth return motion 2. Scotch Yoke
3. Oldham's Coupling 4. Rotary engine
Select correct answer using the codes given below:
Codes:
(a) 1 and 2 (b) 1, 3 and 4 (c) 2 and 3 (d) 2, 3 and 4

Oldham's coupling

IES-52. When two shafts are neither parallel nor intersecting, power can be transmitted by using [IES-1998]
(a) a pair of spur gears (b) a pair of helical gears
(c) an Oldham's coupling (d) a pair of spiral gears

IES-53 Match List I (Coupling) with List II (Purpose) and select the correct answer using the codes given below the lists: [IES-2004]

List I
Muff coupling
B. Flange coupling with
C. Oldham's coupling for power
D. Hook's joint some

List II
1. To transmit power between two parallel shafts
2. To transmit power between two intersecting shafts flexibility
3. For rigid connection between two aligned shafts flexibility
4. For flexible connection between two shafts with misalignment for transmitting power

	A	B	C	D		A	B	C	D
(a)	1	4	3	2	(b)	3	4	2	1

Mechanism

Chapter 1

(c) 3 2 1 4 (d) 1 2 3 4

IES-54. The double slider-crank chain is shown below in the diagram in its three possible inversions. The link shown hatched is the fixed link: [IES-2004]

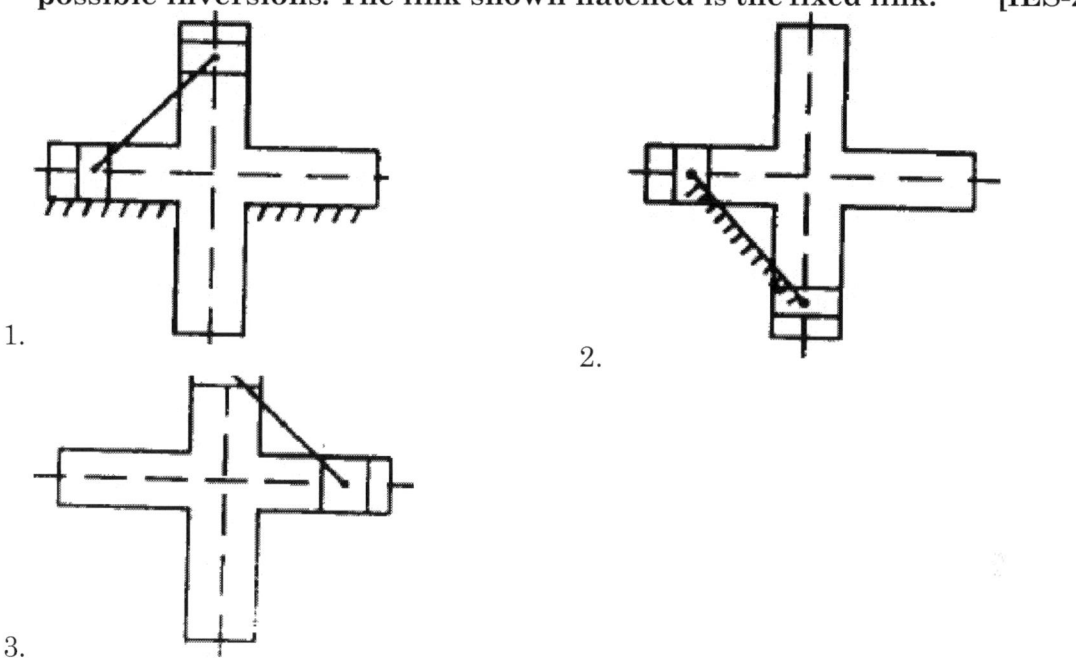

1.

2.

3.

Which one of the following statements is correct?
 (a) Inversion (1) is for ellipse trammel and inversion (2) is for Oldham coupling
 (b) Inversion (1) is for ellipse trammel and inversion (3) is for Oldham coupling
 (c) Inversion (2) is for ellipse trammel and inversion (3) is for Oldham coupling
 (d) Inversion (3) is for ellipse trammel and inversion (2) is for Oldham coupling

IES-55. Match List I with List II and select the correct answer: [IES-2002]

List I (Connecting shaft)	List II (Couplings)
A. In perfect alignment	1. Oldham coupling
B. With angular misalignment of 10°	2. Rigid coupling
C. Shafts with parallel misalignment	3. Universal joint
D. Where one of the shafts may undergo more coupling·	4. Pin type flexible deflection with respect to the other

     A   B   C   D   A   B   C   D
(a) 2   1   3   4   (b) 4   3   1   2
(c) 2   3   1   4   (d) 4   1   3   2

IES-56. Match List-I (Positioning of two shafts) with List-II (Possible connection) and select the correct answer using the codes given below the Lists:
[IES-1997]

List-I	List-II
A. Parallel shafts with slight offset	1. Hooks joint
B. Parallel shafts at a reasonable distance	2. Worm and wheel
C. Perpendicular shafts	3. Oldham coupling
D. Intersecting shafts	4. Belt and pulley

Code: A   B   C   D   A   B   C   D

Mechanism

Chapter 1

(a)	4	3	2	1	(b)	4	3	1	2
(c)	3	4	1	2	(d)	3	4	2	1

IES-57. Match List I with List II and select the correct answer using the codes given below the lists: [IES-1995]

List I (Name)
A. Oldham coupling
B. Flange coupling
C. Universal coupling
D. Friction coupling

List II (Type)
1. Joins collinear shafts and is of rigid type.
2. Joins non-collinear shafts and is adjustable.
3. Joins collinear shafts and engages and Disengages them during motion.
4. Compensates peripheral shafts, longitudinal and angular shifts of shafts

	A	B	C	D		A	B	C	D
Codes: (a)	2	1	4	3	(b)	3	2	1	4
(c)	1	4	2	3	(d)	3	4	2	1

IES-59. Assertion (A): Oldham coupling is used to transmit power between two parallel shafts which are slightly offset. [IES-1994]
Reason (R): There is no sliding member to reduce power in Oldham coupling.
(a) Both A and R are individually true and R is the correct explanation of A
(b) Both A and R are individually true but R is **not** the correct explanation of A
(c) A is true but R is false
(d) A is false but R is true

IES-60. In Oldham's coupling' the condition for maximum speed ratio is. [IES-1992]
(a) $\dfrac{w_1}{W}\cos\alpha$
(b) $\dfrac{w_1}{W}\sin\alpha$
(c) $\dfrac{w_1}{W} = \dfrac{1}{\cos\alpha}$
(d) $\dfrac{w_1}{W} = \dfrac{1}{\sin\alpha}$

Velocity of a point on a link

IES-61. Which one of the following statements is correct? [IES-2004]
In a petrol engine mechanism the velocity of the piston is maximum when the crank is
(a) at the dead centers
(b) at right angles to the line of stroke
(c) slightly less than 90° to line of stroke
(d) slightly above 90° to line of stroke

IES-62. A wheel is rolling on a straight level track with a uniform velocity 'v'. The instantaneous velocity of a point on the wheel lying at the mid-point of a radius
(a) varies between 3 v/2 and - v/2
(b) varies between v/2 and - v/2 **[IES-2000]**
(c) varies between 3 v/2 and - v/2
(d) does not vary and is equal to v

Mechanism

Chapter 1

IES-63. Two points, A and B located along the radius of a wheel, as shown in the figure above, have velocities of 80 and 140 m/s, respectively. The distance between points A and B is 300 mm. The radius of wheel is
(a) 400 mm (b) 500 mm
(c) 600 mm (d) 700 mm

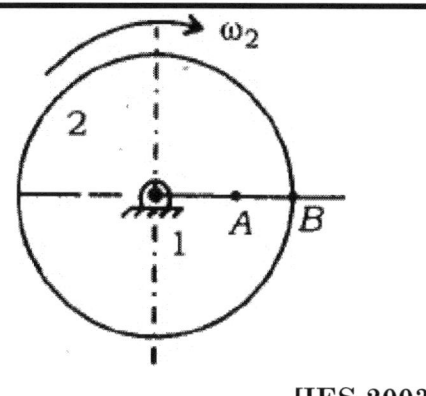

[IES-2003]

IES-64. The crank of the mechanism shown in the side the diagram rotates at a uniform angular velocity θ:

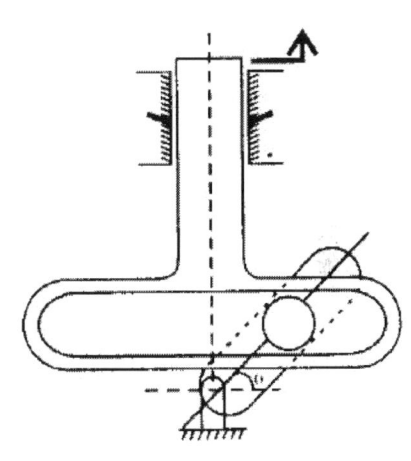

Which one of the following diagrams shows the velocity of slider $\dot{x}$ with respect to the crank angle?

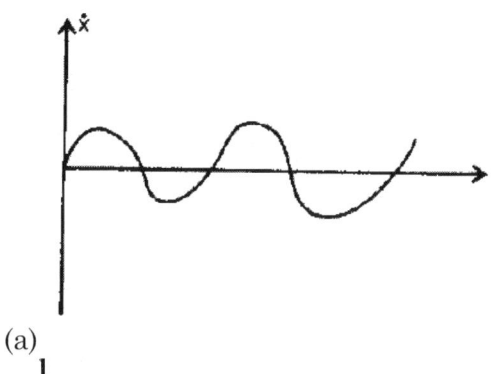

(a) 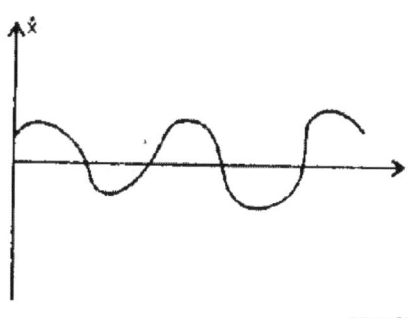 (b) [IES-2004]

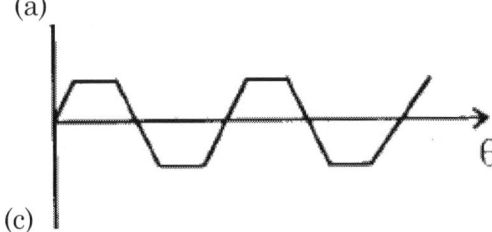

(c) 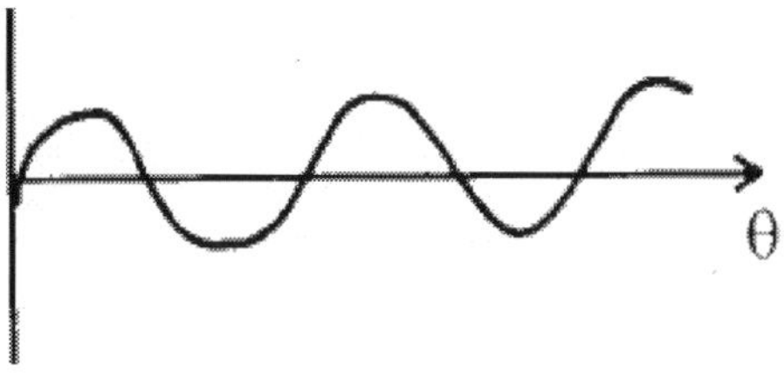 (d)

IES-65. In a slider-crank mechanism, the velocity of piston becomes maximum when
(a) Crank and connecting rod are in line with each other [IES-2003]
(b) Crank is perpendicular to the line of stroke of the piston
(c) Crank and connecting rod are mutually perpendicular
(d) Crank is 120° with the line of stroke

IES-65(i)

Mechanism

Chapter 1

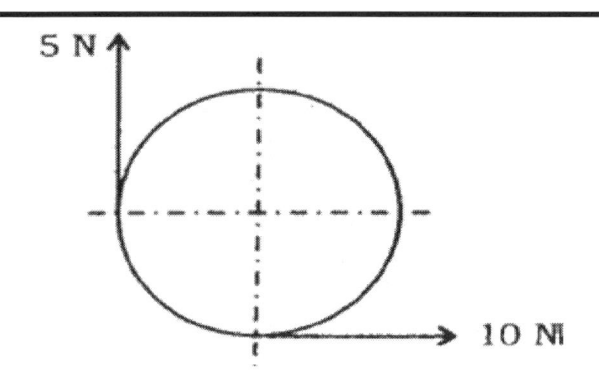

The above figure shows a circular disc of 1kg mass and 0.2 m radius undergoing unconstrained planar motion under the action of two forces as shown. The magnitude of angular acceleration a of the disc is [IES-2003]
(a) 50 rad/s^2 (b) 100 rad/s^2 (c) 25 rad/s^2 (d) 20 rad/s^2

IES-66. Consider the following statements regarding motions in machines:
[IES-2001]
1. Tangential acceleration is a function of angular velocity and the radial acceleration is a function of angular acceleration.
2. The resultant acceleration of a point A with respect to a point B on a rotating link is perpendicular to AB.
3. The direction of the relative velocity of a point A with respect to a point B on a rotating link is perpendicular to AB.
Which of these statements is/are correct?
(a) 1 alone   (b) 2 and 3   (c) 1 and 2   (d) 3 alone

IES-67. Consider a four-bar mechanism shown in the given figure.
The driving link DA is rotating uniformly at a speed of 100 r.p.m. clockwise.
The velocity of A will be
(a) 300 cm/s
(b) 314 cm/s
(c) 325 cm/s
(d) 400 cm/s

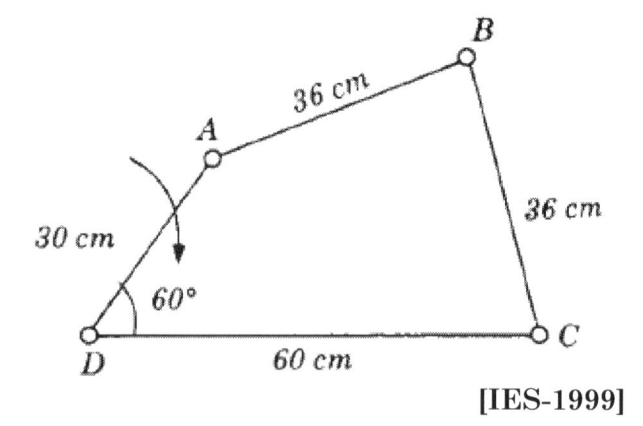

[IES-1999]

IES-68. ABCD is a four-bar mechanism in which AD = 30 cm and CD = 45 cm. AD and CD are both perpendicular to fixed link AD, as shown in the figure. If velocity of B at this condition is V, then velocity of C is

[IES-1993]

Mechanism

Chapter

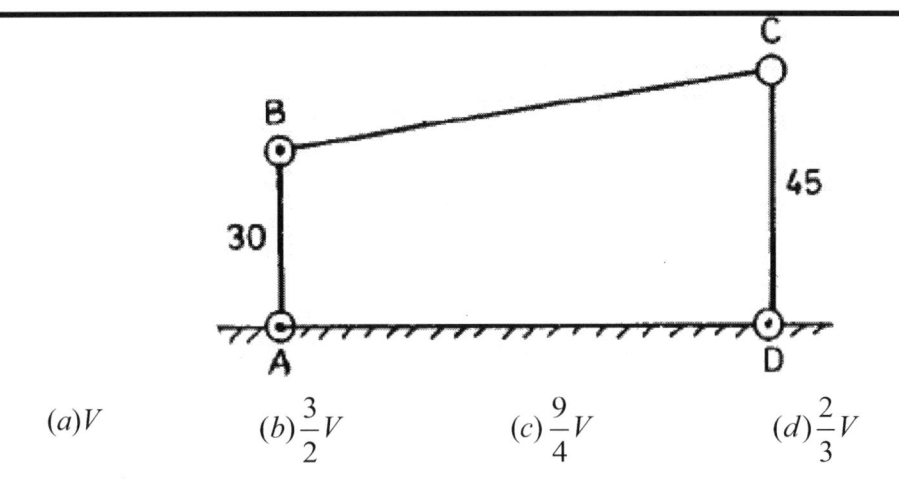

(a) V (b) $\dfrac{3}{2}V$ (c) $\dfrac{9}{4}V$ (d) $\dfrac{2}{3}V$

IES-69 A rod of length 1 m is sliding in a corner as shown in the figure below. At an instant when the rod makes an angle of $60°$ with the horizontal plane, the downward velocity of point A is 1 m/s. What is the angular velocity of the rod at this instant? [IES-2009]

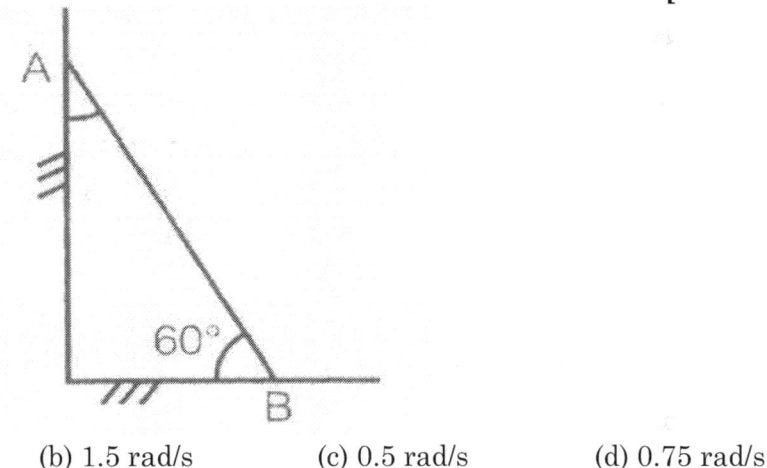

(a) 2.0 rad/s (b) 1.5 rad/s (c) 0.5 rad/s (d) 0.75 rad/s

IES-70. Maximum angular velocity of the connecting rod with a crank to connecting rod ratio 1: for a crank speed of 3000 rpm is around: [IES-2008]
(a) 300 rad/s (b) 60 rad/s (c) 30 rad/s (d) 3000 rad/s

IES-71. The figure as shown below is a rigid body undergoing planar motion. The absolute tangential accelerations of the points R and S on the body are 150 mm/sec² and 300 mm/sec² respectively in the directions shown. What is the angular acceleration of the rigid body? [IES-2009]

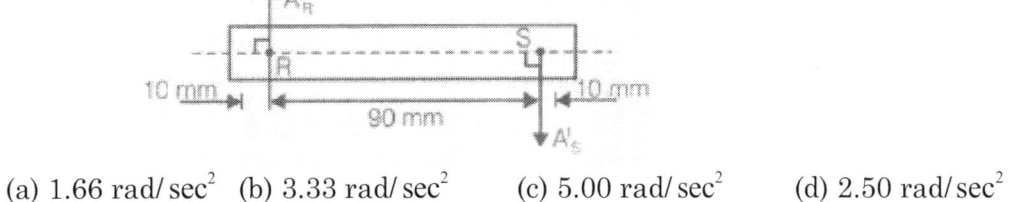

(a) 1.66 rad/sec² (b) 3.33 rad/sec² (c) 5.00 rad/sec² (d) 2.50 rad/sec²

Location of Instantaneous centres

Mechanism

Chapter 1

IES-72. ABCD is a bar mechanism, in which AD is the fixed link, and link BC, is in the form of a circular disc with centre P. In which one of the following cases P will be the instantaneous centre of the disc?
(a) If it lies on the perpendicular bisector of line BC
(b) If it lies on the intersection of the perpendicular bisectors of BC & AD
(c) If it lies on the intersection of the perpendicular bisectors of AB & CD
(d) If it lies on the intersection of the extensions of AB and CD

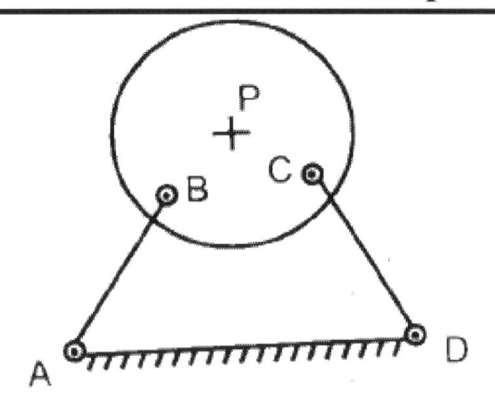

[IES-2004]

IES-73. The instantaneous centre of rotation of a rigid thin disc rolling without slip on a plane rigid surface is located at [IES-1995, 2002]
(a) the centre of the disc (b) an infinite distance perpendicular to the plane surface
(c) the point of contact
(d) the point on the circumference situated vertically opposite to the contact point

IES-74. The relative acceleration of two points which are at variable distance apart on a moving link can be determined by using the [IES-2002]
(a) three centers in line theorem (b) instantaneous centre of rotation method
(C) Corioli's component of acceleration method (d) Klein's construction

IES-75. In the mechanism ABCD shown in the given figure, the fixed link is denoted as (1), Crank AB as (2), rocker BD as (3), Swivel trunnion at C as (4). The instantaneous centre I_{41} is at
(a) the centre of swivel trunnion.
(b) the intersection of line AB and a perpendicular to BD to
(c) Infinity along AC
(d) Infinity perpendicular to BD.

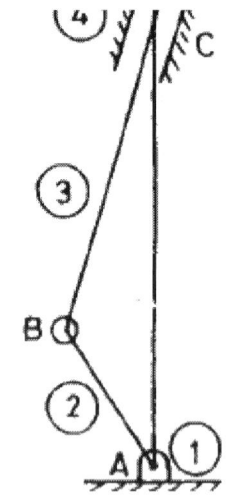

[IES-1996]

Mechanism

Chapter 1

IES-76. The instantaneous centre of motion of a rigid-thin-disc-wheel rolling on plane rigid surface shown in the figure is located at the point.
(a) A (b) B (c) C (d) D.

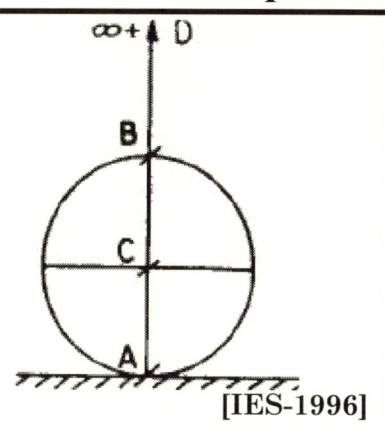

[IES-1996]

Number of Instantaneous centres in Mechanism and Kennedy Theorem

IES-77. What is the number of instantaneous centres of rotation for a 6-link mechanism? **[IES-2006]**
(a) 4 (b) 6 (c) 12 (d) 15

IES-78. The total number of instantaneous centers for a mechanism consisting of 'n' links is

(a) n/2 (b) n (c) $\frac{n-1}{2}$ (d) $\frac{n(n-1)}{2}$ **[IES-1998]**

Force acting in a mechanism

IES-79. A link AB is subjected to a force F (→) at a point P perpendicular to the link at a distance a from the CG as shown in the figure.
This will result in
(a) an inertia force F (→) through the CG and no inertia torque
(b) all inertia force F.a (clockwise) and no inertia force
(c) both inertia force F (→) through the CG and inertia torque Fa (clockwise)
(d) both inertia force F (→) through the CG and inertia torque Fa (anti-clockwise)

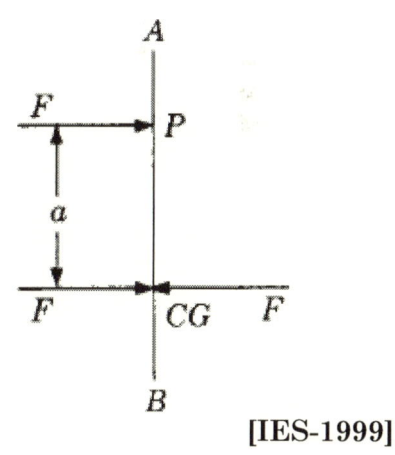

[IES-1999]

Acceleration of a link in a mechanism

IES-80. In the diagram given below, the magnitude of absolute angular velocity of link 2 is 10 radians per second while that of link 3 is 6 radians per second. What is

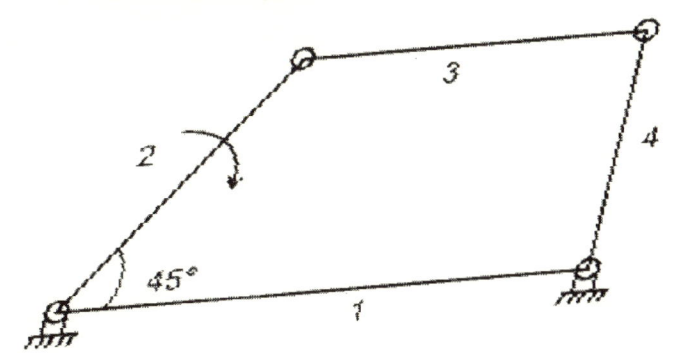

Mechanism

Chapter 1

the angular velocity of link 3 relative to 2?
(a) 6 radians per second
(b) 16 radians per second
(c) 4 radians per second
(d) 14 radians per second   **[IES-2004]**

Coriolis component of Acceleration

IES-81. When a slider moves with a velocity 'V' on a link rotating at an angular speed of ω, the Corioli's component of acceleration is given by   **[IES-1998]**
(a) $\sqrt{2}V\omega$
(b) $V\omega$
(c) $\dfrac{V\omega}{2}$
(d) $2V\omega$

IES-82.

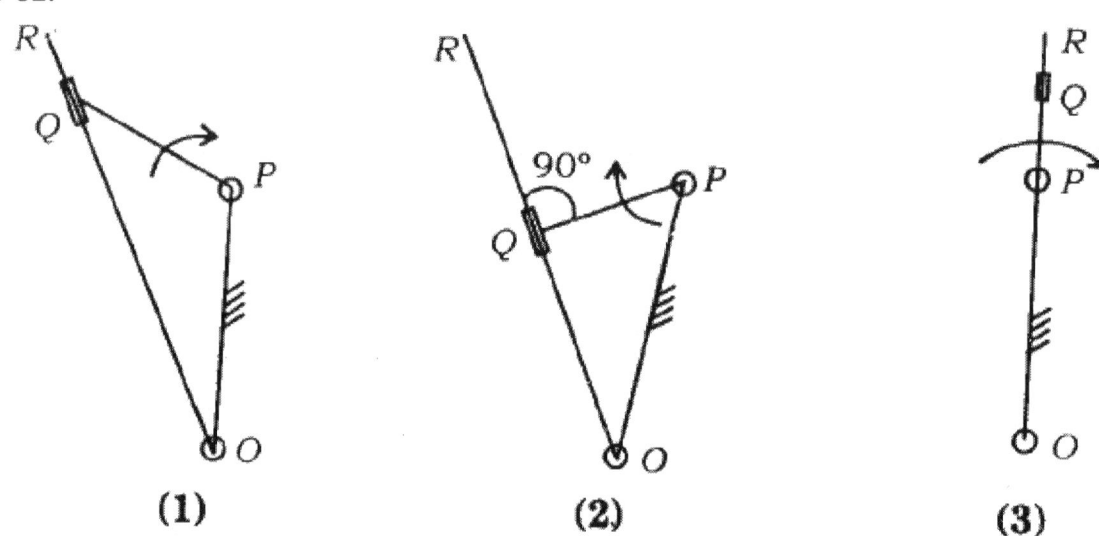

(1)   (2)   (3)

Three positions of the quick-return mechanism are shown above. In which of the cases does the Corioli's component of acceleration exist? **[IES-2003]**
Select the correct answer using the codes given below:
Codes: (a) 1 only   (b) 1 and 2   (c) 1, 2 and 3   (d) 2 and 3

IES-83. **Assertion (A):** The direction of Corioli's acceleration shown in the given figure is correct.
Reason (R): The direction of Corioli's acceleration is such that it will rotate at a velocity v about its origin in the direction opposite to ω.
(a) Both A and R are individually true and R is the correct explanation of A
(b) Both A and R are individually true but R is **not** the correct explanation of A
(c) A is true but R is false
(d) A is false but R is true

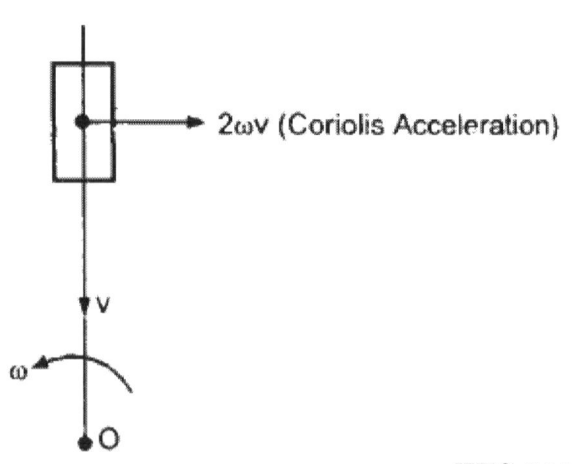

[IES-2000]

Mechanism

Chapter 1

IES-84 The directions of Coriolis component of acceleration, 2ωV, of the slider A with respect to the coincident point B is shown in figures 1, 2, 3 and 4. Directions shown by figures
(a) 2 and 4 are wrong
(b) 1 and 2 are wrong
(c) 1 and 3 are wrong
(d) 2 and 3 are wrong.

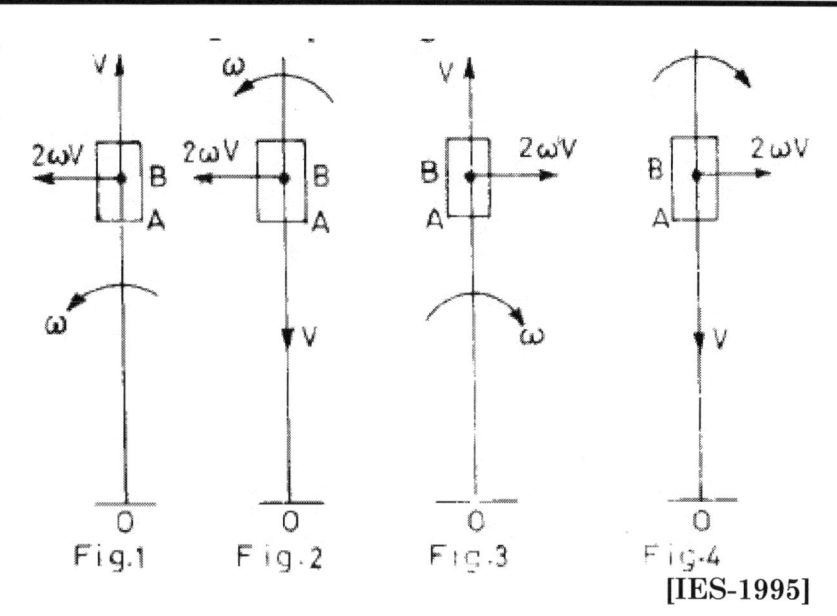

[IES-1995]

IES-85. Consider the following statements: [IES-1993]
Coriolis component of acceleration depends on
1. velocity of slider   2. angular velocity of the link
3. acceleration of slider   4. angular acceleration of link
Of these statements
(a) 1 and 2 are correct   (b) 1 and 3 are correct
(c) 2 and 4 are correct   (d) 1 and 4 are correct

IES-86. The sense of Coriolis component $2\omega V$ is the same as that of the relative velocity vector V rotated.
(a) 45° in the direction of rotation of the link containing the path [IES-1992]
(b) 45° in the direction opposite to the rotation of the link containing the path
(c) 90° in the direction of rotation of the link containing the path
(d) 180° in the direction opposite to the rotation of the link containing the path

IES-87. What is the direction of the Coriolis component of acceleration in a slotted lever-crank mechanism? [IES 2007]
(a) Along the sliding velocity vector
(b) Along the direction of the crank
(c) Along a line rotated 90° from the sliding velocity vector in a direction opposite to the angular velocity of the slotted lever
(d) Along a line rotated 90° from the sliding velocity vector in a direction same as that of the angular velocity of the slotted lever

Mechanism

Chapter 1

IES-88. Assertion (A): Link A experiences Corioli's acceleration relative to the fixed link.

Reason (R): Slotted link A is rotating with angular velocity ω and the Block B slides in the slot of A.

(a) Both A and R are individually true and R is the correct explanation of A
(b) Both A and R are individually true but R is not the correct explanation of A
(c) A is true but R is false
(d) A is false but R is true

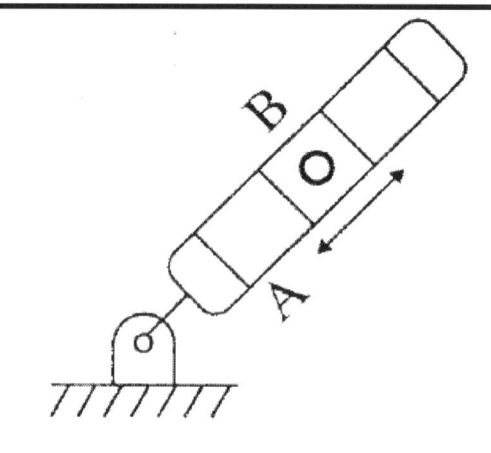

[IES-2006]

IES-89. Consider the following statements: [IES-2005]
1. Corioli's acceleration component in a slotted bar mechanism is always perpendicular to the direction of the slotted bar.
2. In a 4-link mechanism, the instantaneous centre of rotation of the input link and output link always lies on a straight line along the coupler.
Which of the statements given above is/are correct?
(a) 1 only   (b) 2 only   (c) Both 1 and 2   (d) Neither 1 nor 2

IES-90. In the figure given above, the link 2 rotates at an angular velocity of 2 rad/s. What is the magnitude of Corioli's, acceleration experienced by the link 4?
(a) 0
(b) 0.8 m/s²
(c) 0.24 m/s²
(d) 0.32 m/s²

[IES-2005]

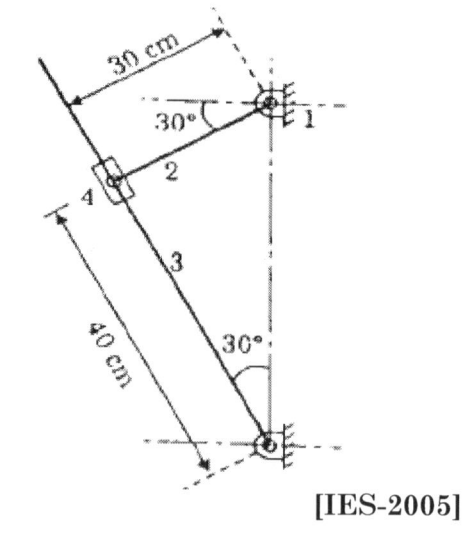

IES-91. At a given instant, a disc is spinning with angular velocity ω in a plane at right angles to the paper, (see the figure) and after a short interval of time δt, it is spinning with angular velocity ω+δω and the axis of spin has changed direction by the amount δθ.

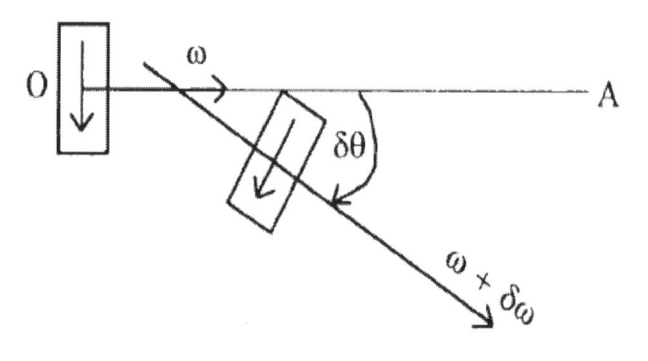

[IES-2008]

Mechanism

Chapter 1

In this situation what is the component of acceleration parallel to OA?
(a) dθ/dt (b) ω(dθ/dt) (c) dω/dt (d) dθ/dω

IES-92. Which one of the following sets of accelerations is involved in the motion of the piston inside the cylinder of a uniformly rotating cylinder mechanism? [IES-2000]
(a) Coriolis's and radial acceleration
(b) Radial and tangential acceleration
(c) Coriolis's and gyroscopic acceleration
(d) Gyroscopic and tangential acceleration

Pantograph

IES-93. Match List I with List II and select the correct answer using the codes given below the lists [IES-1993]

List I	List II
A. Governor	1. Pantograph device
B. Automobile differential	2. Feed-back control
C. Dynamic Absorber	3. Epicyclic train
D. Engine Indicator	4. Two-mass oscillator

Codes:
	A	B	C	D		A	B	C	D
(a)	1	2	3	4	(b)	4	1	2	3
(c)	2	3	4	1	(d)	4	3	2	1

Steering gear mechanism

IES-94. Assertion (A): The Ackermann steering gear is commonly used in all automobiles. [IES-1996]
Reason (R): It has the correct inner turning angle for all positions.
(a) Both A and R are individually true and R is the correct explanation of A
(b) Both A and R are individually true but R is **not** the correct explanation of A
(c) A is true but R is false
(d) A is false but R is true

IES-95. Match List-I with List-II and select the correct answer using the codes given below the Lists.(Notations have their usual meanings) : [IES-2001]

List I
Law of correct steering
B. Displacement relation of Hook's joint
C. Relation between kinematic pairs and links
D. Displacement equation of reciprocating engine piston

List II
1. $f = 3(n-1) - 2j$
2. $x = R\left[(1-\cos\theta) + \dfrac{\sin^2\theta}{2n}\right]$
3. $\cot\phi - \cot\theta = c/b$
4. $\tan\theta = \tan\phi \cos\alpha$

Codes:
	A	B	C	D		A	B	C	D
(a)	1	4	3	2	(b)	1	2	3	4
(c)	3	4	1	2	(d)	3	2	1	4

Mechanism

Chapter

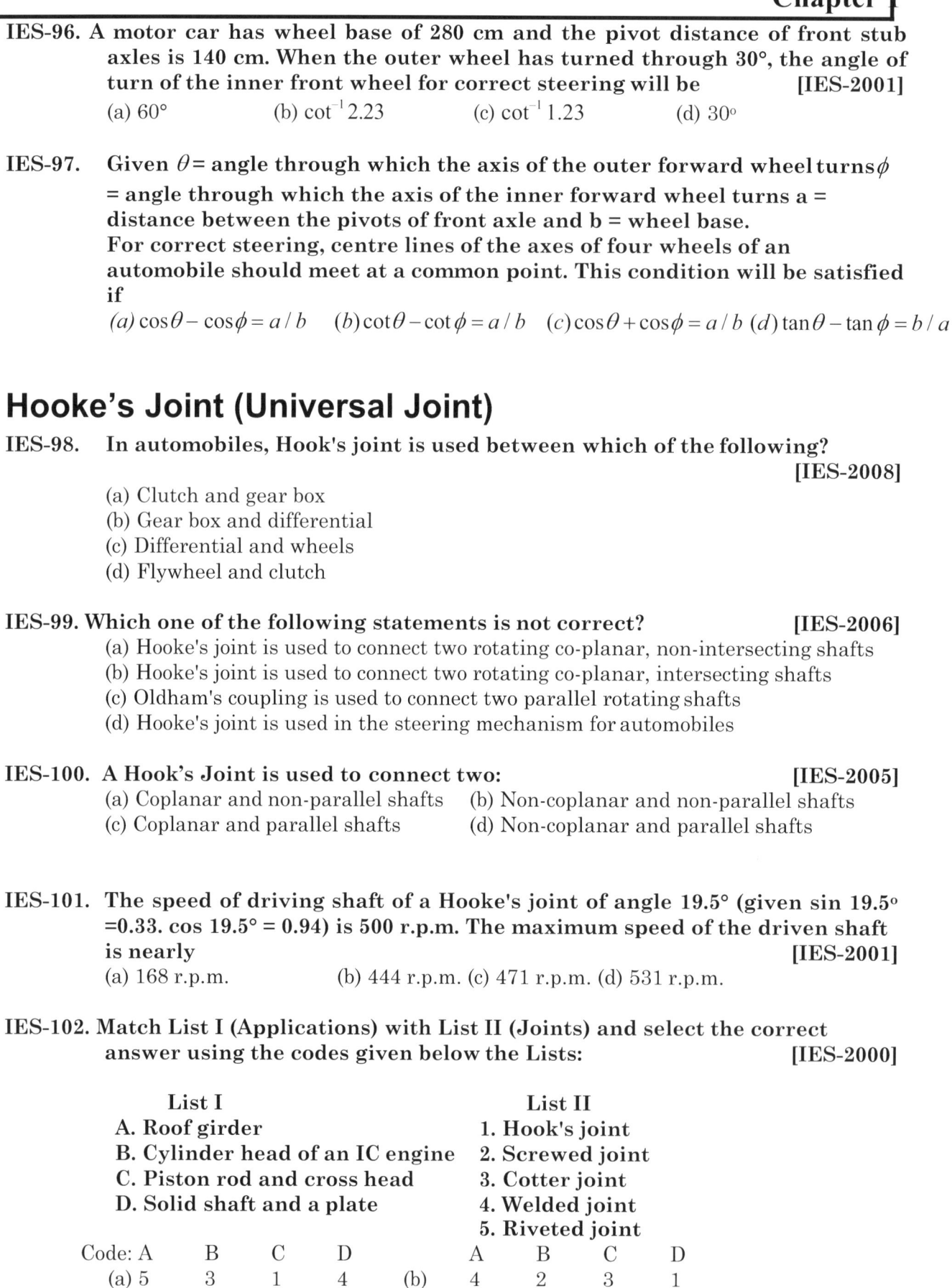

IES-96. A motor car has wheel base of 280 cm and the pivot distance of front stub axles is 140 cm. When the outer wheel has turned through 30°, the angle of turn of the inner front wheel for correct steering will be [IES-2001]
 (a) 60° (b) $\cot^{-1} 2.23$ (c) $\cot^{-1} 1.23$ (d) 30°

IES-97. Given θ = angle through which the axis of the outer forward wheel turns ϕ = angle through which the axis of the inner forward wheel turns a = distance between the pivots of front axle and b = wheel base.
For correct steering, centre lines of the axes of four wheels of an automobile should meet at a common point. This condition will be satisfied if
 (a) $\cos\theta - \cos\phi = a/b$ (b) $\cot\theta - \cot\phi = a/b$ (c) $\cos\theta + \cos\phi = a/b$ (d) $\tan\theta - \tan\phi = b/a$

Hooke's Joint (Universal Joint)

IES-98. In automobiles, Hook's joint is used between which of the following?
[IES-2008]
 (a) Clutch and gear box
 (b) Gear box and differential
 (c) Differential and wheels
 (d) Flywheel and clutch

IES-99. Which one of the following statements is not correct? [IES-2006]
 (a) Hooke's joint is used to connect two rotating co-planar, non-intersecting shafts
 (b) Hooke's joint is used to connect two rotating co-planar, intersecting shafts
 (c) Oldham's coupling is used to connect two parallel rotating shafts
 (d) Hooke's joint is used in the steering mechanism for automobiles

IES-100. A Hook's Joint is used to connect two: [IES-2005]
 (a) Coplanar and non-parallel shafts (b) Non-coplanar and non-parallel shafts
 (c) Coplanar and parallel shafts (d) Non-coplanar and parallel shafts

IES-101. The speed of driving shaft of a Hooke's joint of angle 19.5° (given sin 19.5° =0.33. cos 19.5° = 0.94) is 500 r.p.m. The maximum speed of the driven shaft is nearly [IES-2001]
 (a) 168 r.p.m. (b) 444 r.p.m. (c) 471 r.p.m. (d) 531 r.p.m.

IES-102. Match List I (Applications) with List II (Joints) and select the correct answer using the codes given below the Lists: [IES-2000]

List I	List II
A. Roof girder	1. Hook's joint
B. Cylinder head of an IC engine	2. Screwed joint
C. Piston rod and cross head	3. Cotter joint
D. Solid shaft and a plate	4. Welded joint
	5. Riveted joint

Code:	A	B	C	D		A	B	C	D
(a)	5	3	1	4	(b)	4	2	3	1
(c)	5	2	3	4	(d)	4	3	1	5

Mechanism

Chapter 1

IES-103. Which one of the following figures representing Hooke's jointed inclined shaft system will result in a velocity ratio of unity? [IES-1998]

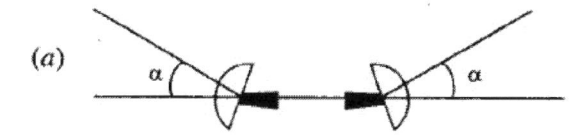

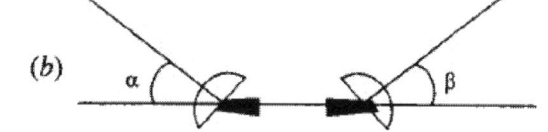

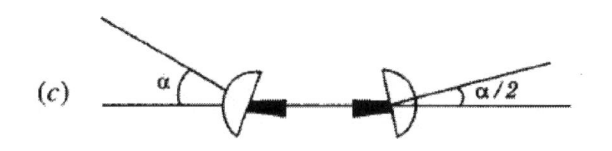

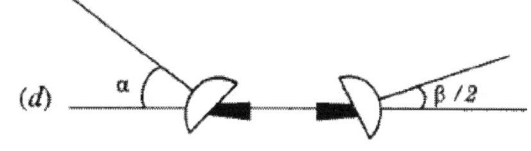

(a) [figure with angles α and α]
(b) [figure with angles α and β]
(c) [figure with angles α and α/2]
(d) [figure with angles α and β/2]

Previous 20-Years IAS Questions

Kinematic pair

IAS-1. Consider the following statements [IAS 1994]
1. A round bar in a round hole form a turning pair.
2. A square bar in a square hole forms a sliding pair.
3. A vertical shaft in a footstep bearing forms a successful constraint.
Of these statements
(a) 1 and 2 are correct
(b) 1 and 3 are correct
(c) 2 and 3 are correct
(d) 1, 2 and 3 are correct

IAS-2. The connection between the piston and cylinder in a reciprocating engine corresponding to [IAS 1994]
(a) completely constrained kinematic pair
(b) incompletely constrained kinematic pair
(c) successfully constrained kinematic pair
(d) single link

IAS-3. Which one of the following "Kinematic pairs" has 3 degrees of freedom between the pairing elements? [IAS-2002]

(a) [figure]
(b) [figure]

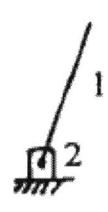

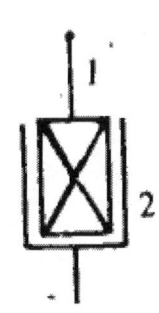

Mechanism

Chapter 1

(c) (d)

Higher pair

IAS-4. Which of the following is a higher pair?
(a) Belt and pulley (b) Turning pair (c) Screw pair (d) Sliding pair

IAS-5. **Assertion (A):** A cam and follower is an example of a higher pair. [IAS 1994]
Reason (R): The two elements have surface contact when the relative motion takes place.
(a) Both A and R are individually true and R is the correct explanation of A
(b) Both A and R are individually true but R is not the correct explanation of A
(c) A is true but R is false
(d) A is false but R is true

Kinematic chain

IAS-6. The given figure shows a / an
(a) locked chain
(b) constrained kinematic chain
(c) unconstrained kinematic chain
(d) mechanism

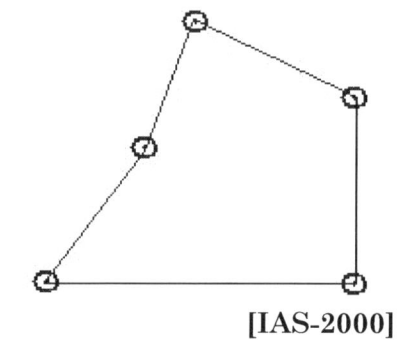

[IAS-2000]

IAS-6a. In a four-link kinematic chain, the relation between the number of links (L) and number of pairs (j) is [IAS-2000]
(a) L=2j+4 (b) L=2j-4 (c) L =4j+ 2 (d) L =4j-2

IAS-7. **Assertion (A):** The kinematic mechanisms shown in Fig. 1 and Fig. 2 above are the kinematic inversion of the same kinematic chain. [IAS-2002]
Reason (R): Both the kinematic mechanisms have equal number of links and revolute joints, but different fixed links.
(a) Both A and R are individually true and R is the correct explanation of A
(b) Both A and R are individually true but R is not the correct explanation of A
(c) A is true but R is false
(d) A is false but R is true

Mechanism

Chapter 1

Degrees of freedom

IAS-8. **Assertion (A)**: The mechanical system shown in the above figure is an example of a 'two degrees of freedom' system undergoing vibrations.
Reason (R): The system consists of two distinct moving elements in the form of a pulley undergoing rotary oscillations and a mass undergoing linear
(a) Both A and R are individually true and R is the correct explanation of A
(b) Both A and R are individually true but R is not the correct explanation of A
[IAS-2002]
(c) A is true but R is false
(d) A is false but R is true

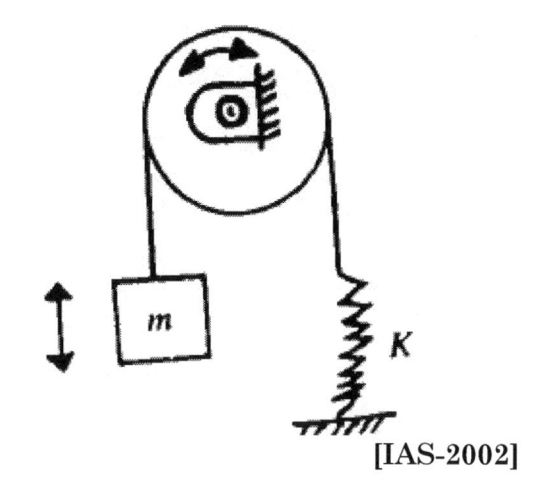

Grubler criterion

IAS-9. For one degree of freedom planar mechanism having 6 links, which one of the following is the possible combination? [IAS-2007]
(a) Four binary links and two ternary links
(b) Four ternary links and two binary links
(c) Three ternary links and three binary links
(d) One ternary link and five binary links

Grashof's law

IAS-10. Consider the following statements in respect of four bar mechanism:
1. It is possible to have the length of one link greater than the sum of lengths of the other three links.
2. If the sum of the lengths of the shortest and the longest links is less than the sum of lengths of the other two, it is known as Grashof linkage.
3. It is possible to have the sum of the lengths of the shortest and the longest links greater than that of the remaining two links. [IAS-2003]
Which of these statements is/are correct?
(a) 1, 2 and 3   (b) 2 and 3   (c) 2 only   (d) 3 only

Inversion of Mechanism

IAS-11. **Assertion (A):** Inversion of a kinematic chain has no effect on the relative motion of its links.
Reason(R): The motion of links in a kinematic chain relative to some other links is a property of the chain and is not that of the mechanism. [IAS-2000]
(a) Both A and R are individually true and R is the correct explanation of A
(b) Both A and R are individually true but R is **not** the correct explanation of A
(c) A is true but R is false
(d) A is false but R is true

IAS-12. **Assertion (A):** An inversion is obtained by fixing in turn different links in a kinematic chain.

Mechanism

Chapter 1

Reason (R): Quick return mechanism is derived from single slider crank chain by fixing the ram of a shaper with the slotted lever through a link. [IAS-1997]
(a) Both A and R are individually true and R is the correct explanation of A
(b) Both A and R are individually true but R is **not** the correct explanation of A
(c) A is true but R is false
(d) A is false but R is true

IAS-13. For L number of links in a mechanism, the number of possible inversions is equal to [IAS-1996]
(a) L - 2   (b) L – 1   (c) L   (d) L + 1

Inversion of four bar chain

IAS-14. The four bar mechanism shown in the figure
(Given: OA = 3 cm, AB = 5 cm
BC = 6 cm, OC = 7 cm) is a
(a) Double crank mechanism
(b) Double rocker mechanism
(c) Crank rocker mechanism
(d) Single slider mechanism

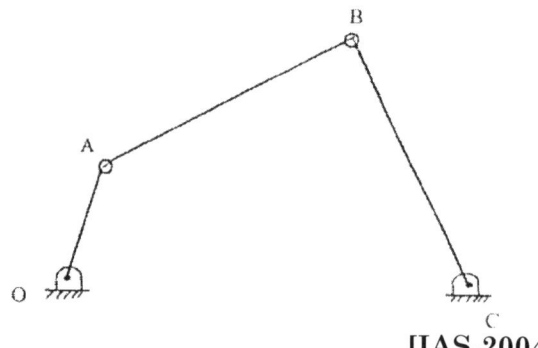

[IAS-2004]

IAS-15. In the four bar mechanism shown in the given figure, link 2 and 4 have equal length. The point P on the coupler 3 will generate a/an
(a) ellipse
(b) parabola
(c) approximately straight line
(d) circle

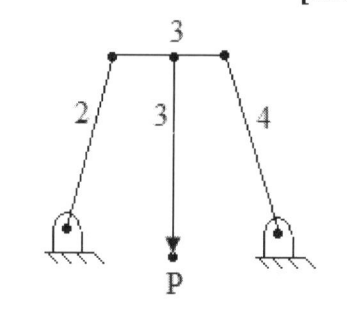

[IAS-1995]

IAS-16. The mechanism shown in the given figure represents
(a) Hart's mechanism
(b) Toggle mechanism
(c) Watts's mechanism
(d) Beam Engine mechanism

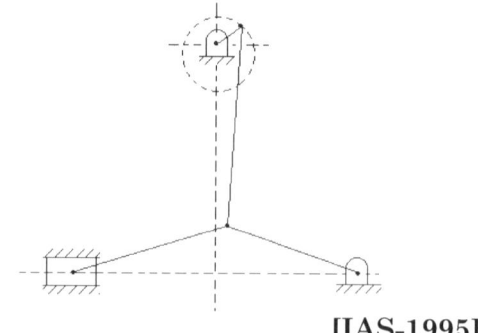

[IAS-1995]

Inversion of Single Slider crank chain

IAS-17. Match List-I with List -II and select the correct answer using the codes given below the List [IAS-1997]

List – I   List-II

Mechanism

Chapter 1

	A. Pantograph	1. Scotch yoke mechanism
	B. Single slider crank chain	2. Double lever mechanism
	C. Double slider crank chain	3. Tchebicheff mechanism
	D. Straight line motion	4. Double crank mechanism
		5. Hand pump

Codes: A   B   C   D   A   B   C   D
   (a)   4   3   5   1   (b)   2   5   1   3
   (c)   2   1   5   3   (d)   4   5   2   1

Quick return motion mechanism

IAS-18. Consider the following mechanisms: [IAS-2002]
1. Oscillating cylinder engine mechanism
2. Toggle mechanism
3. Radial cylinder engine mechanism
4. Quick Return Mechanism
Which of the above are inversions of Slider-crank mechanism?
(a) 1, 2 and 4   (b) 2, 3 and 4   (c) 1, 2 and 3   (d) 1, 3 and 4

IAS-19. In a shaping operation, the average cutting speed is (Stroke length S, Number of strokes per minute N, Quick return ratio R)   [IAS-2000]
(a) NSR   (b) NSR/2   (c) NS (1+ R)   (d) NS (1 +R)/2

IAS-20. Match List-I (Mechanism) with List-II (Associated function) and select the correct answer using the codes given below the List: [IAS-1997]

List-I	List-II
A. Geneva gearing	1. Feed motion in shaper
B. Rachet and Pawl	2. Feed motion in drilling machine
C. Whitworth	3. Indexing of turret
D. Rack and pinion	4. Quick return motion in shaper

Codes: A   B   C   D   A   B   C   D
   (a)   3   1   2   4   (b)   1   3   2   4
   (c)   1   3   4   2   (d)   3   1   4   2

IAS-21. A standard gear has outside diameter of 96mm and module 3 mm. The number of teeth on the gear is   [IAS-1997]
(a) 32   (b) 30   (c) 16   (d) 15

IES-23. Which of the following are the inversions of double slider crank mechanism?   [IAS-1995]
1. Oldham coupling   2. Whitworth quick return mechanism
3. Beam engine mechanism   4. Elliptic trammel mechanism   [IAS-1995]
Select the correct answer from the codes given below.-
Codes: (a) 1 and 2   (b) 1 and 4   (c) 1, 2 and 3   (d) 2, 3 and 4

IAS-24. The Whitworth quick return mechanism is formed in a slider-crank chain when the
(a) coupler link is fixed   (b) longest link is a fixed link
(c) slider is a fixed link   (d) smallest link is a fixed link

IAS-25. Geneva mechanism is used to transfer components from one station to the other in   [IAS-1996]

Mechanism

Chapter 1

(a) an inline transfer machine (b) a rotary transfer machine
(c) a linked line (d) an unlinked flow line

Elliptical trammels

IAS-26. Consider the following statements: [IAS-2007]
1. In a kinematic inversion, the relative motions between links of the mechanism change as different links are made the frame by turns.
2. An elliptical trammel is a mechanism with three prismatic pairs and one revolute pair.
Which of the statements given above is/are correct?
(a) 1 only (b) 2 only (c) Both 1 and 2 (d) Neither 1 nor 2

IAS-27. A point on a connecting line (excluding end points) of a double 'slider crank mechanism traces a [IAS-1995]
(a) straight line path (b) hyperbolic path (c) parabolic path (d) elliptical path

Oldham's coupling

IAS-28. It two parallel shafts are to be connected and the distance between the axes of shafts is small and variable, then one would need to use [IAS-1998]
(a) a clutch (b) a universal joint
(c) an Oldham's coupling (d) a knuckle joint

IAS-29. Oldham's coupling is the inversion of [IAS-1996]
(a) four bar mechanism (b) crank and lever mechanism
(c) single slider crank mechanism (d) double slider crank mechanism

Velocity of a point on a link

IAS-30. A four-bar mechani8m ABCD is shown in the given figure. If the linear velocity 'V_B' of the point 'B' is 0.5 m/s, then the linear velocity 'V_c' of point 'c' will be
(a) 1.25 m/s
(b) 0.5 m/s
(c) 0.4 m/s
(d) 0.2 m/s

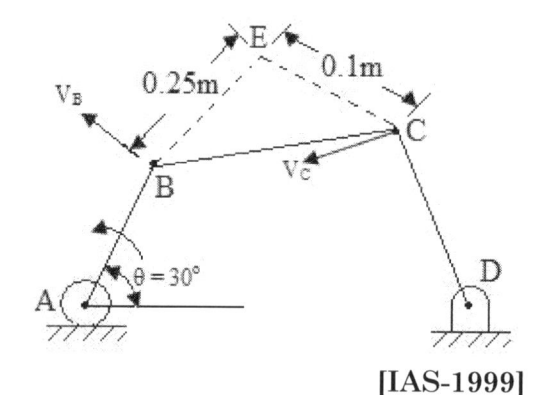

[IAS-1999]

Number of Instantaneous centres in Mechanism and Kennedy Theorem

Mechanism

Chapter 1

IAS-31. How many nstantaneous centers of rotation are there for the mechanism shown in the figure given above?
(a) 6
(b) 10
(c) 15
(d) 21

[IAS-2007]

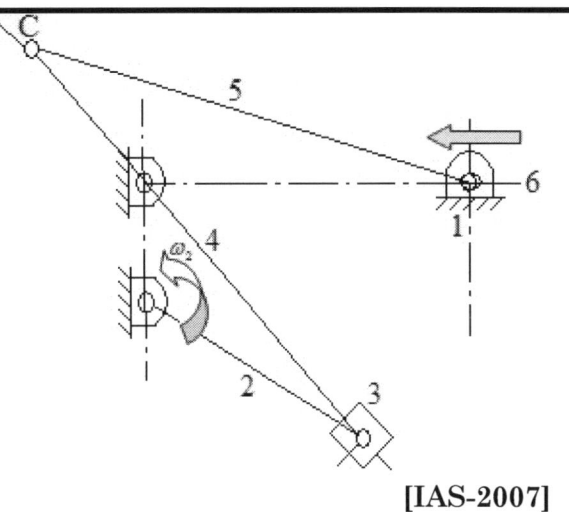

IAS-32. What is the number of instantaneous centers for an eight link mechanism?
(a) 15   (b) 28   (c) 30   (d) 8   [IAS-2004]

IAS-33. The given figure shows a slider crank mechanism in which link 1 is fixed. The number of instantaneous centers would be
(a) 4   (b) 5
(c) 6   (d) 12

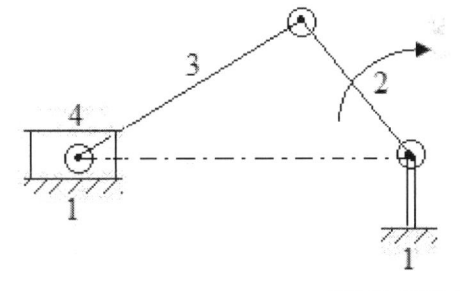

[IAS-1998]

Force acting in a mechanism

Acceleration of a link in a mechanism

IAS-34. Consider the following statements: [IAS-2007]
1. Corioli's component of acceleration is a component of translatory acceleration.
2. If the relative motion between two links of a mechanism is pure sliding, then the relative instantaneous centre for these- two links does not exist.
Which of the statements given above is/are correct?
(a) 1 only   (b) 2 only   (c) Both 1 and 2   (d) Neither 1 nor 2

IAS-35. Consider the following statements:
Corioli's acceleration component appears in the acceleration analysis of the following planar mechanisms: [IAS-2003]
1. Whitworth quick-return mechanism. Slider-crank mechanism.
2. Scotch-Yoke mechanism.
Which of these statements is/are correct?
(a) 1, 2 and 3   (b) 1 and 2   (c) 2 and 3   (d) 1 only

Mechanism

Chapter 1

IAS-36. The above figure shows a four bar mechanism. If the radial acceleration of the point C is 5 cm/s², the length of the link CD is
(a) 2 cm
(b) 10 cm
(c) 20 cm
(d) 100 cm

[IAS-2002]

IAS-37. A slider sliding at 10 cm/s on a link which is rotating at 60 r.p.m. is subjected to Corioli's acceleration of magnitude [IAS-2002]
(a) $40\pi^2 \, cm/s^2$
(b) $0.4\pi \, cm/s^2$
(c) $40\pi \, cm/s^2$
(d) $4\pi \, cm/s^2$

IAS-38. A body in motion will be subjected to Corioli's acceleration when that body is [IAS 1994]
(a) in plane rotation with variable velocity
(b) in plane translation with variable velocity
(c) in plane motion which is a resultant of plane translation and rotation
(d) restrained to rotate while sliding over another body

IAS-39. Match List I (Mechanism) with List II [IAS-2002]
(Name) and select the correct answer using the codes given below the Lists:

List I (Mechanism)	List II (Name)
A. Mechanism used to reproduce a diagram to an enlarged or reduced scale	1. Hart's mechanism
B. A straight line mechanism made up of turning pairs	2. Pantograph
C. Approximate straight line motion consisting of one sliding pair	3. Grasshopper mechanism
D. Exact straight line motion mechanism	4. Peaucellier's mechanism

Codes: A   B   C   D   A   B   C   D
(a)   3   1   2   4   (b) 2   1   3   4
(c)   3   4   2   1   (d) 2   4   3   1

Exact straight line motion mechanism

IAS-40. Which one of the following is an exact straight line mechanism using lower pairs? [IAS-2003]
(a) Watt's mechanism
(b) Grasshopper mechanism
(c) Robert's mechanism
(d) Paucellier's mechanism

Steering gear mechanism

Mechanism

Chapter 1

IAS-41. **Assertion (A):** Davis steering gear is preferred to Ackermann type in automobile applications. **[IAS-2001]**
Reason (R): Davis steering gear consists of sliding pairs as well as turning pairs.
(a) Both A and R are individually true and R is the correct explanation of A
(b) Both A and R are individually true but R is **not** the correct explanation of A
(c) A is true but R is false
(d) A is false but R is true

Answers with Explanation (Objective)

Previous 20-Years GATE Answers

GATE-1. Ans. (d)
GATE-2. Ans. (c)
GATE-3. Ans. (c)
    No. of links   I = 8
    No. of revolute joints,   J = 9
    No. of higher pair.   h = 0
    ∴ Number of degree of freedom
         n = 3 (I-1) -2J-h
         = 3 (8-1)-2 × 9-0
    ∴   n = 3

GATE-4. Ans. (b)
GATE-5. Ans. (c)
    Degrees of freedom
    m = 3(n-1) -2J_1-J_2
    where   n = number of links
    J_1 = number of single degree of freedom, and
    J_2 = number of two degree of freedom
    Given,   n = 5, J_1 = 5, J_2 = 0
    Hence   m = 3 (5-1) -2 × 5 - 0 = 2

GATR-6. Ans. (d)
GATE-7. Ans. (b)
Whatever may be the number of links and joints Grubler's criterion applies to mechanism with only single degree freedom. Subject to the condition 3l-2j-4=0 and it satisfy this condition.
Degree of freedom is given by
= 3 (1-1) - 2j
= 3 (8-1) – (2 × 10) = 1

GATE-8. Ans. (b)
1. D' Alembert's principal → Dynamic-static analysis
2. Grubler's criterion → Mobility (for plane mechanism)
3. Grashoff's law → Continuous relative rotation
4. Kennedy's theorem → Velocity and acceleration

GATE-9. Ans. (a)
According to Grashof's rule for complete relative rotation r/w links L + S ≤ p + q.

GATE-10. Ans. (a) According to Grashof's law for a four bar mechanism. The sum of shortest and longest link lengths should not be greater than the sum of the remaining two link length.

Mechanism

Chapter 1

i.e. $S + L \leq P + Q$

GATE-11. Ans. (c) No. of links of a slider crank mechanism = 4
So there are four inversion of slider crank mechanism.

GATE-12. Ans. (d) Quick return mechanism.

GATE-13. Ans. (d) Here $2 = \dfrac{360 - \alpha}{\alpha}$ and $AC = \dfrac{BC}{\cos(\alpha/2)}$

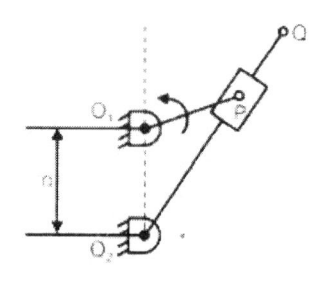

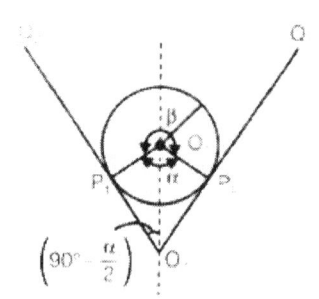

$O_1P = 125$ mm

Quick Return Mechanism

$\dfrac{\text{Time of working (Forward) Stroke}}{\text{Time of return stroke}} = \dfrac{\beta}{\alpha} = \dfrac{360 - \alpha}{\alpha}$

$\dfrac{2}{1} = \dfrac{360° - \alpha}{\alpha}$

$\Rightarrow 2\alpha = 360° - \alpha$

$\Rightarrow 3\alpha = 360°$

$\Rightarrow \alpha = 120°$

$\Rightarrow \dfrac{\alpha}{2} = 60°$

The extreme position of the crank (O_1P) are shown in figure.

From right triangle $O_2O_1P_1$, we find that $\sin(90° - \alpha/2) = \dfrac{O_1P_1}{O_1O_2}$

$\Rightarrow \sin(90° - 60°) = \dfrac{125}{O_1O_2} = \dfrac{125}{d}$

$\Rightarrow \sin 30° = \dfrac{125}{d}$

$\Rightarrow d = \dfrac{125}{\sin 30°} = 250$ mm

GATE-14 Ans. (c)

GATE-15. Ans. (b)

Mechanism

Chapter 1

$\dfrac{\text{Forward stroke}}{\text{Re turn stroke}}$

$= \dfrac{240}{120}$

$= 2.$

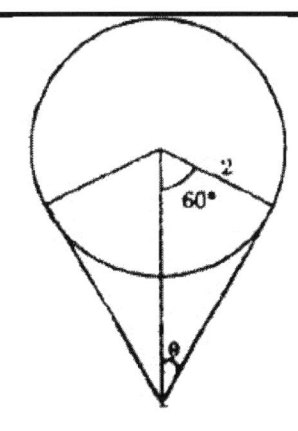

GATE-16. Ans. (d)

GATE-17. Ans. (c)

GATE-18. Ans. (c)

When $\angle O_4 O_2 P = 180°$

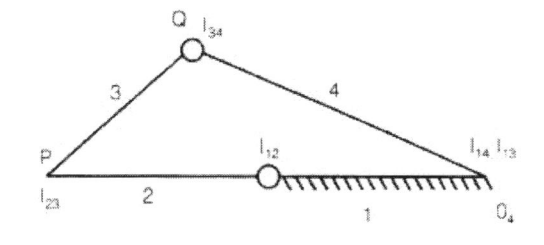

Now, $\dfrac{\omega_3}{\omega_2} = \dfrac{|_{12} \; |_{23}}{|_{13} \; |_{23}} = \dfrac{a}{2a}$

$I_{13} \langle \begin{smallmatrix} 12 & 23 \\ 14 & 34 \end{smallmatrix}$

$\dfrac{\omega_3}{2} = \dfrac{1}{2}$

$\therefore \; \omega_3 = 1 \, \text{rad/s}$

GATE-19. Ans. (b)

GATE-20. Ans. (b)
GATE-21. Ans. (d)

GATE-22 Ans. (b)

GATE-23. Ans. (c)

GATE-24. Ans. (d)

GATE-25. Ans. (b, d)
GATE-26. Ans. (a)
Velocity of link 1 with respect to 2

Mechanism

Chapter 1

$$V_{1,2} = 12 \text{ m/s}$$
$$\omega = \frac{2\pi N}{60} = \frac{2\pi \times 120}{60}$$
$$= 12.566 \text{ rad/s}$$

∴ Corioli's component of acceleration
$$= 2V_{1,2}\omega$$
$$= 2 \times 12 \times 12.566$$
$$= 302 \text{ m/s}^2$$

GATE-27. Ans. (b)
GATE-28. Ans. (d)

Previous 20-Years IES Answers

IES-1. Ans. (d)
Sliding pair → piston and cylinder
Revolute pair → Crank shaft in a journal bearing in an engine
Rolling → A road roller rolling over the ground
Spherical pair → Ball and socket joint

IES-2. Ans. (c) When two elements or links are connected in such a way that their relative motion is constrained they form a kinematic pair. The relative motion of a kinematic pair may be completely, incompletely or successfully constrained

IES-3. Ans. (c)

IES-4. Ans. (d) 4 links and 4 turning pairs satisfy the equation $L = \frac{3}{2}(j+2)$; It is case of complete constraint. 3 links and 3 turning pairs form rigid frame. Foot step bearing results in successful constraint and 5 links and 5 turning pairs provide incomplete constraint.

IES-5. Ans. (a)
IES-7. Ans. (a)
IES-8. Ans. (a)
IES-9. Ans. (b)
IES-10. Ans. (c)
IES-11. Ans. (d)
IES-12. Ans. (a) Elements of higher pairs must be force closed to provide completely constrained motion.
IEA-13. Ans. (c) When l number of links are joined at the same connection, the joint is equivalent to $(l-1)$ binary joints.
IES-14. Ans. (b)
IES-15. Ans. (d)
IES-16. Ans. (a)
IES-18. Ans. (d)
IES-19. Ans. (a) Two link system shown in the above figure has 2 degrees of freedom.
IES-20. Ans. (d) When supported on three points, following six degrees of freedom are arrested (two line movements along y-axis, two rotational movements each along x-axis and z-axis.)

Mechanism

Chapter 1

IES-21. Ans. (c)
IES-22. Ans. (a)
IES-24. Ans. (b)
IES-26. Ans. (c)
IES-27. Ans. (c)
IES-28. Ans. (c)
IES-29. Ans. (a)
IES-30. Ans. (a)
IES-31. Ans. (d)
IES-32. Ans. (b) $5l = 2(l + 30)$, $3l = 60$ and $l = 20$cm
IES-33. Ans. (a)
IES-34. Ans. (a)
IES-35. Ans. (c)
IES-36. Ans. (d)
IES-37. Ans. (b)
IES-38. Ans. (d)
IES-39. Ans. (d)
IES-40. Ans. (b)
IES-41. Ans. (c)
IES-43. Ans. (a)
IES-44. Ans. (c)
IES-45. Ans. (a)

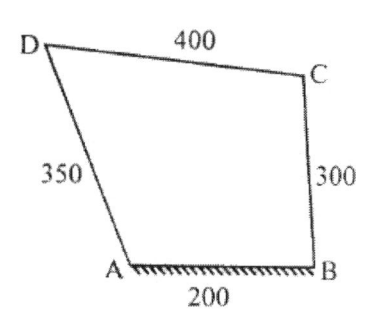

IES-46. Ans. (d)
IES-47. Ans. (d)
IES-49. Ans. (a)
Scotch Yoke mechanism: Here the constant rotation of the crank produces harmonic translation of the yoke. Its four binary links are:
1- Fixed Link
2- Crank
3- Sliding Block
4- Yoke
IES-51. Ans. (c) Double Slider Crank mechanism
It has four binary links, two revolute pairs, two sliding pairs. Its various types are:
1. Scotch Yoke mechanism
2. Oldhams Coupling
3. Elliptical Trammel
IES-52. Ans. (d)
IES-53. Ans. (c)
IES-54. Ans. (a)

Mechanism

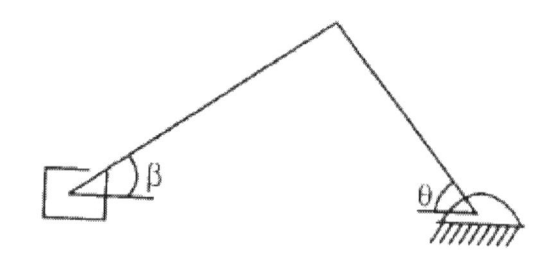

Chapter 1

IES-55. Ans. (c)
IES-56. Ans. (d)
IES-57. Ans. (a)
IES-59. Ans. (c) It is used for transmitting angular velocity between two parallel but eccentric shafts
IES-60. Ans. (c)

$$\frac{\omega_1}{\omega} = \frac{\cos\alpha}{1 - \cos^2\theta \sin^2\alpha}$$

For maximum speed ratio $\cos^2\theta = 1$

$$\therefore \quad \frac{\omega_1}{\omega} = \frac{1}{\cos\alpha}$$

IES-61. Ans. (a)
IES-62. Ans. (b)
IES-63. Ans. (d)

Angular velocity of both points A and B are same.
$V_A = 800$ m/s; $V_B = 800$ m/s; AB = 300 mm; OA + AB = OB

or $\dfrac{V_A}{OA} = \dfrac{V_B}{OB}$

or 80 × OB = 140 × OA = 140 × (OB−AB)

or OB = $\dfrac{140}{60}$ = 700 mm

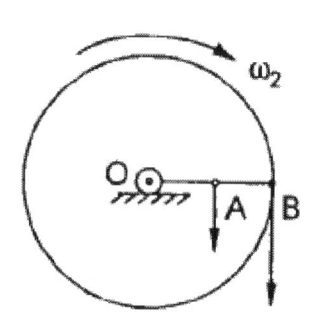

IES-64. Ans. (b)
IES-65. Ans. (b) When the piston will be in the middle of the spoke length
IES-65(i). Ans. (a)

$T = I\alpha$ Where, $I = \dfrac{1}{2}mr^2 = \dfrac{1}{2} \times 1 \times (0.2)^2 = 0.2$ kgm^2

$\therefore \alpha = \dfrac{T}{I} = \dfrac{(10-5) \times 0.2}{0.02} = \dfrac{5 \times 0.2}{0.02} = 50$ rad/sec^2

IES-66. Ans. (d)
IES-67. Ans. (b) Velocity of A = $\omega r = \dfrac{2\pi \times 100}{60} \times 30 = 314$ cm/s

IES-68. Ans. (a) Velocity of C = $\dfrac{45}{30}V = \dfrac{3}{2}V$

IES-69. Ans. (a)
IES-70. Ans. (b)

Mechanism

Chapter 1

$$\sin\beta = \frac{\sin\theta}{n}$$

$$\cos\beta \frac{d\beta}{dt} = \frac{\cos\theta}{n} \cdot \frac{d\theta}{dt}$$

$$\frac{d\beta}{dt} = \left(\frac{\cos\theta}{\cos\beta}\right)\left(\frac{1}{n}\right)\left(\frac{d\theta}{dt}\right)$$

$$\omega_{Cr} = \frac{\omega\cos\theta}{\sqrt{n^2 - \sin^2\theta}}$$

Since $\sin^2\theta$ is small as compared to n^2

∴ it may be neglected. $\omega_{Cr} = \dfrac{\omega\cos\theta}{n}$

$\omega(\text{crank}) = 3000 \text{ rev/min} = 50 \text{ rev/sec}$

$= 314 \text{ rad/sec}$

∴ $\omega_{Cr_{max}} = \dfrac{314}{5} = 62.8 \text{ rad/sec}$

IES-71. Ans. (c) Angular acceleration of Rigid body

$$= \frac{150 \text{ mm/s}^2 + 300 \text{ mm/s}^2}{90 \text{ mm}}$$

$$= \frac{450 \text{ mm/s}^2}{90 \text{ mm}} = 5.00 \text{ rad/sec}^2$$

IES-72 Ans. (d)

IES-73. Ans. (c)

IES-74. Ans. (b) The relative acceleration of two variable points on a moving link can be determined by using the instantaneous centre of rotation method.

IES-75. Ans. (a)

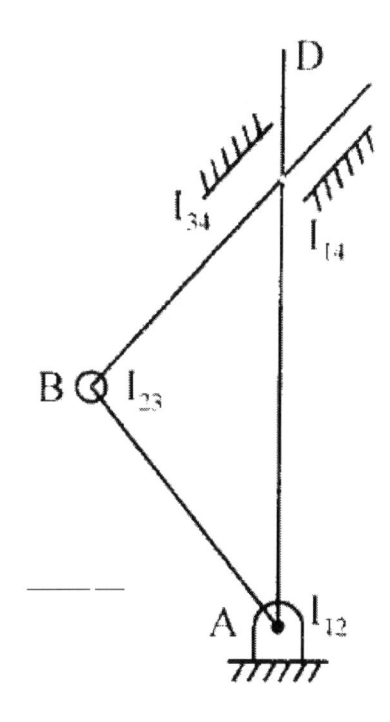

IES-76. Ans. (a)

IES-77. Ans. (d) $N = \dfrac{n(n-1)}{2} = \dfrac{6\times(6-1)}{2} = 15$

IES-78. Ans. (d)

IES-79. Ans. (c)

Mechanism

Chapter 1

Apply two equal and opposite forces F at CG. Thus inertia force F (→) acts at CG and inertia torque Fa (clockwise)

IES-80. Ans. (c) $\omega_{32} = \omega_3 - \omega_2 = 6 - 10 = -4 \text{ rad/s}$

IES-81. Ans. (d)

IES-82 Ans. (a)

IES-83. Ans. (a)

IES-84. Ans. (a)

IES-85. Ans. (a)

IES-86. Ans. (c)

IES-87. Ans. (d)

IES-88. Ans. (d) Link B experiences Coriolis acceleration relative to the fixed link.

IES-89. Ans. (c)

IES-90. Ans. (a)

IES-91. Ans. (c)

IES-92. Ans. (a)

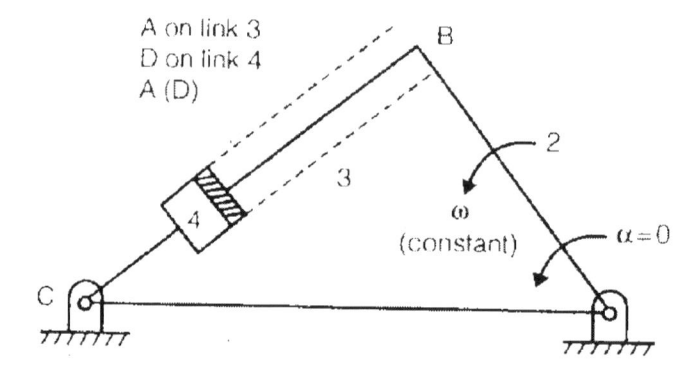

Radial acceleration = $\dfrac{V_B^2}{BO}$

Tangential acceleration = $(OB)\alpha_{OB} = 0$

Coriolis acceleration = $2\omega_{CD} \cdot V_{D/A}$

IES-93. Ans. (c) Simplex indicator is closely resembles to the pantograph copying mechanism.

IES-94. Ans. (c)

IES-95. Ans. (c)

IES-96. Ans. (c)

IES-97. Ans. (b)

IES-98. Ans. (b)

The main application of the universal or Hooke's coupling is found in the transmission from the gear box to the differential or back axle of the automobiles. In such a case, we use two Hooke's coupling, one at each end of the propeller shaft, connecting the gear box at one end and the differential on the other end.

IES-99. Ans. (a)

IES-100. Ans. (b) A Hooke's joint is used to *connect* two shafts, which are intersecting at a small angle.

IES-101. Ans. (d)

IES-102. Ans. (c)

IES-103. Ans. (a)

Mechanism

Chapter 1

Previous 20-Years IAS Questions

IAS-1. Ans. (c)

IAS-2. Ans. (c)

IAS-3. Ans. (d) (a) has only one DOF i.e. rotational

(b has only one DOF i.e. translational about z-axis

(c has only two DOF i.e. rotation and translation

IAS-4. Ans. (c)

IAS-5. Ans. (c)

IAS-6. Ans. (c)

Here $l = 5$, and $j = 5$

condition-1, $l = 2p - 4$ or $5 = 2 \times 5 - 4 = 6$ i.e. L.H.S < R.H.S

condition-2, $j = \frac{3}{2}l - 2$ or $5 = \frac{3}{2} \times 5 - 4 = 5.5$ i.e. LHS S < R.H.S

It is not a kinematic chain. L.H.S < R.H.S, such a type of chain is called unconstrained chain i.e. relative motion is not completely constrained.

IAS-6a. Ans. (b) Here notation of number of pairs (j) [our notation is p]

IAS-7. Ans. (d) *A is false.* Kinematic inversion is obtained different mechanisms by fixing different links *in a kinematic chain*. Here they change kinematic chain also.

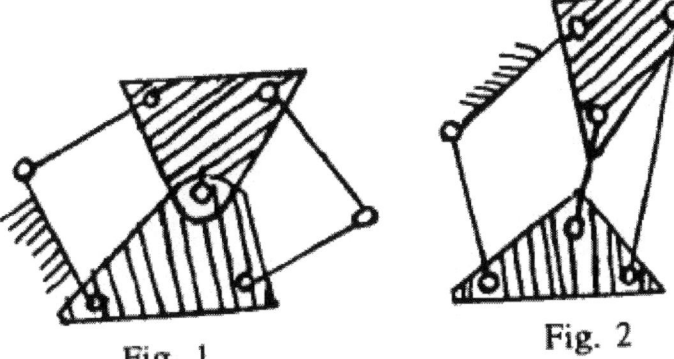

Fig. 1   Fig. 2

IAS-8. Ans. (d)

IAS-9. Ans. (d) From Grubler's criteria $1 = 3(l-1) - 2j$ or $j = \frac{3}{2}l - 2$ *for six link*

$j = \frac{3}{2} \times 6 - 2 = 7$ 1 ternay link ≡ 2 binary link

(a) j = 4 + 2 × 2 ≠ 7   (b) j = 4 × 2 + 2 ≠ 7

c) j = 3 × 2 + 2 ≠ 7   (d) j = 1 × 2 + 5 = 7 ans. is d

IAS-10. Ans. (b)

IAS-11. Ans. (a) In a kinematic inversion relative motion does not change but absolute motion change drastically.

IAS-12. Ans. (c)

IAS-13. Ans. (c)

IAS-14. Ans. (c)

IAS-15. Ans. (a) Point P being rigidly connected to point 3, will trace same path as point 3, *i.e.* ellipse.

IAS-16. Ans. (d)

$5l = 2(l + 30)$, $3l = 60$ and $l = 20$ cm

IAS-17. Ans. (b)

Mechanism

Chapter 1

IAS-18. Ans. (d)

IAS-19. Ans. (c) Time for forward stroke = T_f, Time for return stroke = T_r, $R = \dfrac{T_r}{T_f}$

$$\therefore \text{Time for only one cutting stroke}(T) = \dfrac{1}{N} \times \dfrac{T_f}{(T_f + T_r)}$$

$$\therefore \text{Average cutting speed} = \dfrac{S}{T} = SN \dfrac{(T_f + T_r)}{T_f} = SN(1+R)$$

IAS-20. Ans. (d)

IAS-21. Ans. (a) $T = \dfrac{96}{3} = 32$

IAS-23. Ans. (b) The inversions of double slider crank mechanism are
(i) First inversion-Elliptic Trammel,
(ii) Second inversion-Scotch Yoke
(iii) Third inversion-Oldham's coupling
Thus out of choices given, only 1 and 4 are correct.

IAS-24. Ans. (d)

IAS-25. Ans. (b)

IAS-26. Ans. (d) Through the process of inversion the relative motions between the various links is not changed in any manner but their absolute motions may be changed drastically.
Elliptical trammels have two sliding pairs and two turning pairs. It is an instruments used for drawing ellipse.

IAS-27. Ans. (d)

IAS-28. Ans. (c)

IAS-29. Ans. (d)

IAS-30. Ans. (d) Instantaneous centre method gives
$\dfrac{V_B}{EB} = \dfrac{V_C}{EC}$ or $V_C = \dfrac{V_B}{EB} \times EC = \dfrac{0.5}{0.25} \times 0.1 = 0.2 \text{ m/s}$

IAS-31. Ans. (c) Kennedy theorem says number of instantaneous centre (N) $= \dfrac{n(n-1)}{2}$

or $\dfrac{6 \times (6-1)}{2} = 15$

IAS-32. Ans. (b) $\dfrac{n(n-1)}{2} = \dfrac{8 \times 7}{2} = 28$

IAS-33. Ans. (c) $N = \dfrac{4(4-1)}{2} = 6$

IAS-34. Ans. (a) Its unit is m/s². Therefore translatory acceleration ($a^t = 2\omega V$). It does exist at infinity distance. Kennedy theorem says number of instantaneous centre (N) $= \dfrac{n(n-1)}{2}$. Count it.

IAS-35. Ans. (d)

IAS-36. Ans. (c) Radial component of acceleration $(\alpha^r) = \dfrac{V^2}{CD}$ or $5 = \dfrac{10^2}{CD}$ or $CD = 20$ cm

Mechanism

Chapter 1

IAS-37. Ans. (c) Coriolis acceleration $= 2\omega V = 2 \times \dfrac{2\pi N}{60} \times V = 2 \times \dfrac{2\pi \times 60}{60} \times 10 = 40\pi \text{ cm/s}^2$

IAS-38. Ans. (d)

IAS-39. Ans. (b & d) Exact straight line motion mechanisms made up of turning pairs are Peaucellier's mechanism and Hart's mechanism. Hart's mechanism consists of six links and Peaucellier's mechanism consists of eight links.

IAS-40. Ans. (d)

IAS-41. Ans. (d) Ackermann steerig gear is preferred to Devis as it consists of turning pairs.

Flywheel

Chapter 2

2. Flywheel

Previous 20-Years GATE Questions

GATE-1. Which of the following statement is correct? **[GATE-2001]**
(a) Flywheel reduces speed fluctuations during a cycle for a constant load, but flywheel does not control the mean speed of the engine if the load changes
(b) Flywheel does not educe speed fluctuations during a cycle for a constant load, but flywheel does control the mean speed of the engine if the load changes
(c) Governor control a speed fluctuations during a cycle for a constant load, but governor does not control the mean speed of the engine if the load change
(d) Governor controls speed fluctuations during a cycle for a constant load, and governor also controls the mean speed of the engine if the load changes

GATE-2. The speed of an engine varies from 210 rad/s to 190 rad/s. During a cycle the change in kinetic energy is found to be 400 Nm. The inertia of the flywheel in kgm² is **[GATE-2007]**
(a) 0.10 (b) 0.20 (c) 0.30 (d) 0.40

Coefficient of Fluctuation of speed

GATE-3. If C_f is the coefficient of speed fluctuation of a flywheel then the ratio of $\omega_{max}/\omega_{min}$ will be **[GATE-2006]**
(a) $\dfrac{1-2c_f}{1+2c_f}$ (b) $\dfrac{2-c_f}{1+2c_f}$ (c) $\dfrac{1+2c_f}{1-2c_f}$ (d) $\dfrac{2+c_f}{2-c_f}$

Energy stored in a flywheel

GATE-4. A fly wheel of moment of inertia 9.8 kgm² fluctuates by 30 rpm for a fluctuation in energy of 1936 Joules. The mean speed of the flywheel is (in rpm) **[GATE-1998]**
(a) 600 (b) 900 (c) 968 (d) 2940

Flywheel rim (Dimension)

GATE-5. For a certain engine having an average speed of 1200 rpm, a flywheel approximated as a solid disc, is required for keeping the fluctuation of speed within 2% about the average speed. The fluctuation of kinetic energy per cycle is found to be 2 kJ. What is the least possible mass of the flywheel if its diameter is not to exceed 1m? **[GATE-2003]**
(a) 40 kg (b) 51 kg (c) 62 kg (d) 73 kg

Flywheel

Chapter 2

Previous 20-Years IES Questions

IES-1. Consider the following statements: **[IES-2003]**
 1. Flywheel and governor of an engine are the examples of an open loop control system
 2. Governor is the example of closed loop control system
 3. The thermostat of a refrigerator and relief valve of a boiler are the examples of closed loop control system
 Which of these statements is/are correct?
 (a) 1 only (b) 2 and 3 (c) 3 only (d) 2 only

IES-2. Which of the following pairs of devices and their functions are correctly matched? **[IES-2001]**
 1. Flywheel..........................For storing kinetic energy
 2. Governors........................For controlling speeds
 3. Lead screw in lathe ....For providing feed to the slides
 4. Fixtures...........................For locating workpiece and guiding tools
 Select the correct answer using the codes given below:
 Codes: (a) 1, 3 and 4 (b) 2 and 3 (c) 1 and 2 (d) 2 and 4

IES-3. Assertion (A): In designing the size of the flywheel, the weight of the arms and the boss are neglected.
 Reason (R): The flywheel absorbs energy during those periods when the turning moment is greater than the resisting moment. **[IES-2000]**
 (a) Both A and R are individually true and R is the correct explanation of A
 (b) Both A and R are individually true but R is **not** the correct explanation of A
 (c) A is true but R is false
 (d) A is false but R is true

IES-4. A rotating shaft carries a flywheel which overhangs on the bearing as a cantilever. If this flywheel weight is reduced to half of its original weight, the whirling speed will
 (a) be double (b) increase by $\sqrt{2}$ times **[IES-1999]**
 (c) decrease by $\sqrt{2}$ times (d) be half

IES-5. Which one of the following engines will have heavier flywheel than the remaining ones?
 (a) 40 H.P. four-stroke petrol engine running at 1500 rpm.
 (b) 40 H.P. two-stroke petrol engine running at 1500 rpm. **[IES-1996]**
 (c) 40 H.P. two-stroke diesel engine running at 750 rpm.
 (d) 40 H.P. four-stroke diesel engine running at 750 rpm.

Coefficient of Fluctuation of speed

IES-6. The maximum fluctuation of energy E_f, during a cycle for a flywheel is

Flywheel

Chapter 2

(a) $I\left(\omega_{max}^2 - \omega_{min}^2\right)$

(b) $\frac{1}{2} . I . \omega_{av} . \left(\omega_{max} - \omega_{min}\right)$ [IES-2003]

(c) $\frac{1}{2} . I . K_{es} . \omega_{av}^2$

(d) $I . K_{es} . \omega_{av}^2$

(Where, I = Mass moment of inertia of the flywheel
ω_{av} = Average rotational speed
K_{es} = Coefficient of fluctuation of speed)

IES-7. Consider the following parameters: [IES-1999]
1. Limit of peripheral speed   2. Limit of centrifugal stress
3. Coefficient of fluctuation of speed   4. Weight of the rim
Which of these parameters are used in the calculation of the diameter of fly wheel rim?
(a) 1, 3 and 4   (b) 2, 3 and 4   (c) 1, 2 and 3   (d) 1, 2 and 4

IES-8. For minimizing speed fluctuations of an engine as a prime mover, it must have
(a) Only a flywheel fitted to the crankshaft [IES-2003]
(b) A governor provided in the system
(c) Both a flywheel and a governor provided in the system
(d) Neither a flywheel nor a governor

IES-9. In the case of a flywheel, the maximum fluctuation of energy is the
(a) sum of maximum and minimum energies [IES-1998]
(b) difference between the maximum and minimum energies
(c) ratio of the maximum and minimum energy
(d) ratio of the minimum and maximum energy

IES-10. Match List-I with List-II and select the correct answer using the codes given below the Lists:
List-I   List-II
A. Flywheel   1. Dunkerley Method   [IES-1997]
B. Governor   2. Turning Moment
C. Critical speed   3. D' Alembert's Principle
D. Inertia Force   4. Speed control on par with load

Code: A B C D   A B C D
(a)   4 2 3 1   (b)   4 2 1 3
(c)   2 4 3 1   (d)   2 4 1 3

Energy stored in a flywheel

IES-11. What is the value of the radius of gyration of disc type flywheel as compared to rim type flywheel for the same diameter? [IES-2004]
(a) $\sqrt{2}$ times   (b) $1/\sqrt{2}$ times   (c) 2 times   (d) 1/2 times

Flywheel

Chapter 2

IES-12. If the rotating mass of a rim type fly wheel is distributed on another rim type flywheel whose mean radius is half mean radius of the former, then energy stored in the latter at the same speed will be **[IES-1993]**
(a) four times the first one
(b) same as the first one
(c) one-fourth of the first one
(d) one and a half times the first one

IES-13. A flywheel is fitted to the crankshaft of an engine having 'E' amount of indicated work per revolution and permissible limits of co-efficient of fluctuation of energy and speed as K_e and K_s respectively. **[IES-1993]**
The kinetic energy of the flywheel is then given by
(a) $\dfrac{2K_e E}{K_s}$
(b) $\dfrac{K_e E}{2K_s}$
(c) $\dfrac{K_e E}{K_s}$
(d) $\dfrac{K_s E}{2K_e}$

IES-14. For the same indicated work per cycle, mean speed and permissible fluctuation of speed, what is the size of flywheel required for a multi-cylinder engine in comparison to a single cylinder engine? **[IES-2006]**
(a) Bigger
(b) Smaller
(c) Same
(d) depends on thermal efficiency of the engine

Flywheel rim (Dimension)

IES-15. Consider the following methods: **[IES-2004]**
1. Trifiler suspension
2. Torsional oscillation
3. Fluctuation of energy of engine
4. Weight measurement & measurement of radius of flywheel

Which of the above methods are used to determine the polar mass moment of inertia of an engine flywheel with arms?
(a) 1 and 4
(b) 2 and 3
(c) 1, 2 and 3
(d) 1, 2 and 4

IES-16. If the rotating mass of a rim type fly wheel is distributed on another rim type fly wheel whose mean radius is half the mean radius of the former, then energy stored in the latter at the same speed will be **[IES-2002]**
(a) four times the first one
(b) same as the first one
(c) one-fourth of the first one
(d) two times the first one

Turning moment diagram

IES-17. The turning moment diagram for a single cylinder double acting steam engine consists of +ve and −ve loops above and bellow the average torque line. For the +ve loop, the ratio of the speeds of the flywheel at the beginning and the end is which one of the following?
(a) less than unity
(b) Equal to unity **[IES 2007]**
(c) Greater than unity
(d) Zero

Flywheel

Chapter 2

IES-18. Consider the following statements regarding the turning moment diagram of a reciprocating engine shown in the above figure: (Scale 1 cm² = 100 N·m)
1. It is four stroke IC engine
2. The compression stroke is 0° to 180°
3. Mean turning moment $T_m = \dfrac{580}{\pi}$ N.m

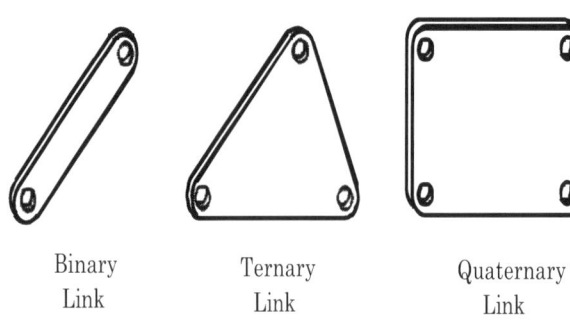

Binary Link   Ternary Link   Quaternary Link

[IES-2000]

4. It is a multi-cylinder engine.
Which of these statements are correct?
(a) 1, 2 and 3 (b) 1, 2 and 4 (c) 2, 3 and 4 (d) 1, 3 and 4

IES-19.

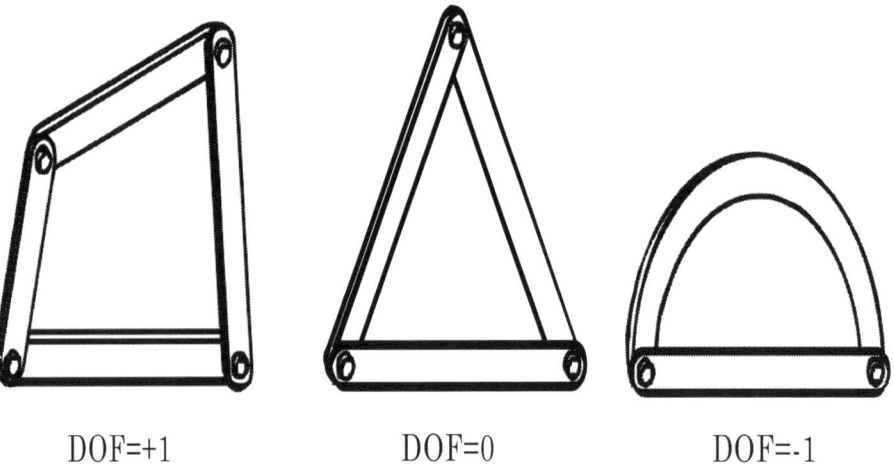

DOF=+1   DOF=0   DOF=-1

The crank of a slider-crank punching press has a mass moment of inertia of 1 kgm². The above figure shows the torque demand per revolution for a punching operation. If the speed of the crank is found to drop from 30 rad/s to 20 rad/s during punching, what is the maximum torque demand during the punching operation? **[IES-2005]**
(a) 95.4 Nm (b) 104.7 Nm (c) 477.2 Nm (d) 523.8 Nm

IES-20. A certain machine requires a torque of $(500 + 50\sin\theta)$ KNm to derive it, θ where is the angle of rotation of shaft measured from certain datum. The machine is directly coupled to an engine which produces a toques $(500 + 50\sin\theta)$ KNm in a cycle how many times the value of torque of machine and engine will be identical **[IES-1992]**
(a) 1 (b) 2 (c) 4 (d) 8

Flywheel

Chapter 2

IES-21. The given figure shows the output torque plotted against crank positions for a single cylinder four-stroke-cycle engine. The areas lying above the zero-torque line represent positive work and the areas below represent negative work. The engine drives a machine which offers a resisting torque equal to the average torque. The relative magnitudes of the hatched areas are given by the numbers (in the areas) as shown:

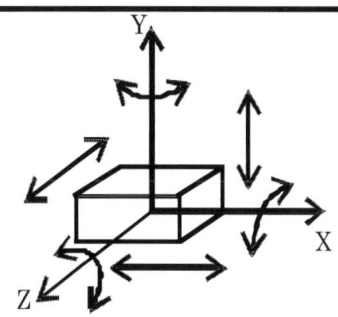

Figure 4-2 Degrees of freedom of a rigid body in space

During the cycle, the minimum speed occurs in the engine at [IES-1995]
(a) B (b) D (c) H (d) F

IES-22. In which of the following case, the turning moment diagram will have least variations:
(a) Double acting steam engine
(b) Four stroke single cylinder patrol engine
(c) 8 cylinder, 4 stroke diesel engine
(d) Pelton wheel [IES-1992]

IES-23. In a 4-stroke I.C. engine, the turning moment during the compression stroke is
(a) positive throughout
(b) negative throughout
(c) positive during major portion of the stroke [IES-1996]
(d) negative during major portion of the stroke.

Previous 20-Years IAS Questions

IAS-1. Consider the following statements: [IAS-2001]
If the fluctuation of speed during a cycle is ± 5% of mean speed of a flywheel, the coefficient of fluctuation of speed will
1. increase with increase of mean speed of prime mover
2. decrease with increase of mean speed of prime mover
3. remain same with increase of mean speed of prime mover
Which of these statement(s) is/are correct?
(a) 1 and 3 (b) 1 and 2 (c) 3 alone (d) 2 alone

Flywheel

Chapter 2

Energy stored in a flywheel

IAS-2. With usual notations for different parameters involved, the maximum fluctuation of energy for a flywheel is given by
[IAS-2002]

(a) $2EC_s$ (b) $\dfrac{EC_s}{2}$ (c) $2EC_s^2$ (d) $2E^2C_s$

IAS-3. The amount of energy absorbed by a flywheel is determined from the [IAS-2000]
(a) torque-crank angle diagram   (b) acceleration-crank angle diagram
(c) speed-space diagram   (d) speed-energy diagram

IAS-4. In the case of a flywheel of mass moment of inertia 'I' rotating at an angular velocity 'ω', the expression $\dfrac{1}{2}I\omega^2$ represents the

[IAS-1999]
(a) centrifugal force   (b) angular momentum   (c) torque   (d) kinetic energy

IAS-5. The moment of inertia of a flywheel is 2000 kg m². Starting from rest, it is moving with a uniform acceleration of 0.5 rad/s². After 10 seconds from the start, its kinetic energy will be
(a) 250 Nm   (b) 500 Nm   (c) 5,000 Nm   (d) 25,000 Nm   [IAS-1997]

IAS-6. Consider the following statements: [IAS-1997]
The flywheel in an IC engine
1. acts as a reservoir of energy   2. minimizes cyclic fluctuations in the engine speed.
3. takes care of load fluctuations in the engine and controls speed variation.
Of these statements:
(a) 1 and 2 are correct   (b) 1 and 3 are correct
(c) 2 and 3 are correct   (d) 1, 2 and 3 are correct

IAS-7. The radius of gyration of a solid disc type flywheel of diameter 'D' is [IAS-1996]
(a) D   (b) D/2   (c) $D/2\sqrt{2}$ (d) $\left[\dfrac{\sqrt{3}}{2}\right]D$

IAS-8. The safe rim velocity of a flywheel is influenced by the
(a) centrifugal stresses   (b) fluctuation of energy
[IAS-1998]
(c) fluctuation of speed   (d) mass of the flywheel

Flywheel

Chapter 2

IAS-9. Consider the following statements relating to the curve for the inertia torque v/s crank angle for a horizontal, single cylinder petrol engine shown in the given figure:
1. $\theta_1 + \theta_2 = 180°$ 2. $T_1 = T_2$
3. $\theta_1 \neq \theta_2$ 4. $A_1 = A_2$

Of these statements:
(a) 1 and 3 are correct   (b) 2 and 3 are correc
(c) 1, 2 and 4 are correct   (d) 1, 3 and 4 are correct

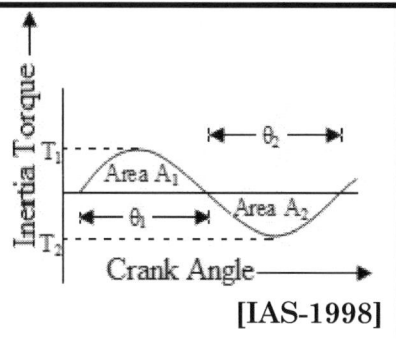

[IAS-1998]

IAS-10. A simplified turning moment diagram of a four-stroke engine is shown in the given figure. If the mean torque 'T_m' is 10 Nm, the estimated peak torque 'T_p' will be (assuming negative torque demand is negligible)
(a) 80 Nm (b) 120 Nm
(c) 60 Nm (d) 40 Nm

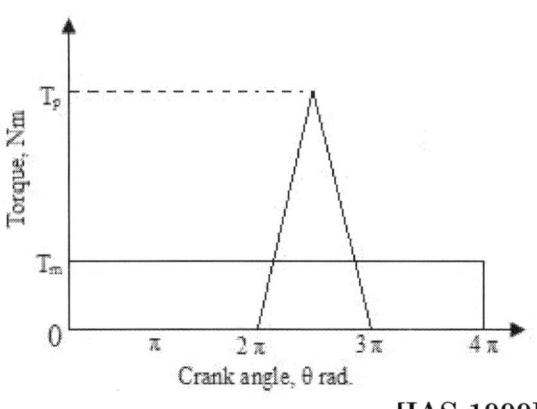

[IAS-1999]

Answers with Explanation (Objective)

Previous 20-Years GATE Answers

GATE-1. Ans. (a)
GATE-2. Ans. (a)
GATE-3. Ans. (d) $c_f = \dfrac{\omega_{max} - \omega_{min}}{\left(\dfrac{\omega_{max} + \omega_{min}}{2}\right)}$ or $\dfrac{\omega_{max}}{\omega_{min}} = \dfrac{2 + c_f}{2 - c_f}$

GATE-4. Ans. (a)

GATE-5. Ans. (b)

Flywheel
Chapter 2

Average speed, N = 1200 rpm

Co-efficient of fluctuation of speed = $c_s = \dfrac{\omega_1 - \omega_2}{\omega} = 2\% = 0.02$

Fluctuation of kinetic energy = $\Delta E = 2 \times 10^3$ J

Now
$$\Delta E = \frac{1}{2} I \omega_1^2 - \frac{1}{2} I \omega_2^2$$
$$= \frac{1}{2} I (\omega_1^2 - \omega_2^2)$$

Since $\dfrac{\omega_1 + \omega_2}{2} = \omega$

$$= I \left(\frac{\omega_1 + \omega_2}{\omega} \right)(\omega_1 - \omega_2)$$
$$= I\omega \cdot \frac{(\omega_1 - \omega_2)}{\omega} \cdot \omega$$
$$= I\omega^2 c_s$$

$\Rightarrow \quad 2 \times 10^2 = \dfrac{1}{2} MR^2 \cdot \omega^2 \cdot c_s^2$, where R = Radius of disc

$$= \frac{1}{2} M \times \left(\frac{1}{2}\right)^2 \times \left(\frac{2176200}{60}\right) \times 0.02$$

$\therefore \quad M = \dfrac{2 \times 10^3 \times 60 \times 60 \times 8}{0.02 \times (2 \times \pi \times 1200)^2}$

$\quad = 50.65 \approx 51$ kg.

Previous 20-Years IES Answers

IES-1. Ans. (c)
IES-2. Ans. (c)
IES-3. Ans. (b)

IES-4. Ans. (b) Whirling speed $\propto \sqrt{\dfrac{1}{I}}$

IES-5. Ans. (d) The four stroke engine running at lower speed needs heavier fly wheel.

IES-6. Ans. (d)

Maximum fluctuation of energy = $\dfrac{1}{2} I \left(\omega_{max}^2 - \omega_{min}^2 \right)$
$$= \frac{1}{2} I \left(\omega_{max} + \omega_{min} \right)\left(\omega_{max} - \omega_{min} \right)$$
$$= I \omega_{avg} \left(\omega_{max} - \omega_{min} \right) = I \left(\omega_{avg} \right)^2 K_{es}$$

IES-7. Ans. (a) Limit of centrifugal stress is not considered.
IES-8. Ans. (c)
IES-9. Ans. (b)
IES-10. Ans. (d)

Flywheel

Chapter 2

IES-11. Ans. (b) Moment of gyration of a disc = $\dfrac{d}{\sqrt{8}}$

Moment of gyration of a rim = $\dfrac{d}{2}$

IES-12. Ans. (c) Energy stored $\propto I\omega^2$, also $I \propto k^2$ (k = radius of gyration which is function of radius of wheel)
∴ If radius is reduced to half, then energy stored will be reduced to one-fourth.

IES-13. Ans. (b)

IES-14. Ans. (b)
IES-15. Ans. (c)

IES-16. Ans. (c)
IES-17. Ans. (a)

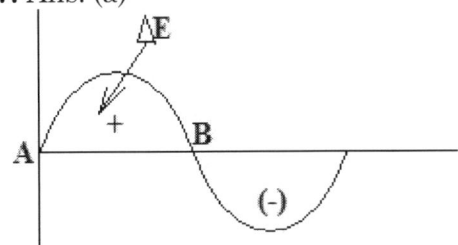

Energy at B = Energy at A + Δ E

or $\dfrac{1}{2}I\omega_B^2 = \dfrac{1}{2}I\omega_A^2 + \Delta E$

∴ $\omega_A > \omega_A$ or $\dfrac{\omega_A}{\omega_B} < 1$

Flywheel

Chapter 2

IES-19. Ans. (c) Energy needed for punching = $\frac{1}{2}I(\omega_1^2 - \omega_2^2) = \frac{1}{2} \times 1 \times (30^2 - 20^2) J = 250 J$

From graph $\frac{1}{2} \times T_{max} \times \Delta\theta = 250$ or $T_{max} = \frac{250 \times 2}{\Delta Q} = \frac{500 \times 2}{\Delta\theta} = \frac{500}{\left(\frac{2\pi}{3} - \pi/3\right)} = \frac{1500}{\pi} = 477.2 Nm$

IES-20. Ans. (c)

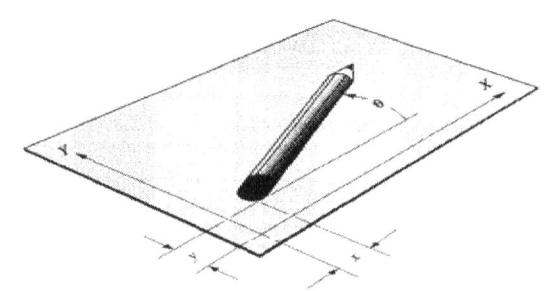

IES-21. Ans. (d) Minimum speed occurs at point where cumulative torque is least, i.e. -23 at F.

IES-22. Ans. (d)

IES-23. Ans. (a)

Previous 20-Years IAS Answers

IAS-1. Ans. (c)

IAS-2. Ans. (a)

IAS-3. Ans. (a)

IAS-4. Ans. (d)

IAS-5. Ans. (c) $\omega = \omega_o + \alpha t = 0 + 0.5 \times 10 = 5 rad/s$

$K.E = \frac{1}{2}I\omega^2 = \frac{1}{2} \times 2000 \times 5 = 5000 Nm$

IAS-6. Ans. (a) Flywheel has no effect on load fluctuations

IAS-7. Ans. (c) Moment of inertia = $\frac{mr^2}{2} = mk^2$ or $k = \frac{r}{\sqrt{2}} = \frac{D}{2\sqrt{2}}$

IAS-8. Ans. (a) centrifugal stresses $(\sigma) = \rho v^2$

IAS-9. Ans. (d)

IAS-10. Ans. (a) Area $T_m \times (4\pi - 0) =$ Area $T_p \times \left(\frac{3\pi - 2\pi}{2}\right)$ or $T_p = 8T_m = 8 \times 10 = 80 Nm$

Governor

Chapter 3

3. Governor

Objective Questions (IES, IAS, GATE)

Previous 20-Years GATE Questions

Not a single question in 20 years

Previous 20-Years IES Questions

IES-1. For a governor running at constant speed, what is the value of the force acting on the sleeve?
(a) Zero
(b) Variable depending upon the load
(c) Maximum
(d) Minimum
[IES 2007]

Watt Governor

IES-2. Consider the following speed governors: **[IES-1999]**
1. Porter governor 2. Hartnell governor 3. Watt governor 4. Proell Governor
The correct sequence of development of these governors is
(a) 1, 3, 2, 4
(b) 3, 1, 4, 2
(c) 3, 1, 2, 4
(d) 1, 3, 4, 2

IES-3. Given that m = mass of the ball of the governor, **[IES-1998]**
ω = angular velocity of the governor and
g = acceleration due to gravity,
The height of Watt's governor is given by
(a) $\dfrac{g}{2\omega^2}$
(b) $\dfrac{g}{\omega^2}$
(c) $\dfrac{\sqrt{2}g}{\omega^2}$
(d) $\dfrac{2g}{\omega^2}$

Porter Governor

IES-4. Consider the given figure:
Assertion (A): In order to have the same equilibrium speed for the given values of w, W and h, the masses of balls used in the Proell governor are less than those of balls used in the Porter governor.
Reason (R): The ball is fixed to an extension link in Proell governor.
(a) Both A and R are individually true and R is the correct explanation of A
(b) Both A and R are individually true but R is **not** the correct explanation of A
(c) A is true but R is false
(d) A is false but R is true

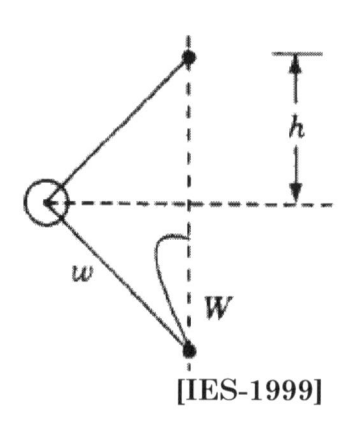

[IES-1999]

Governor

Chapter 3

IES-5 Which one of the following equation is valid with reference to the given figure?

(a) $\omega^2 = \left(\dfrac{W}{w}\right)\left(\dfrac{g}{h}\right)$
(b) $\omega^2 = \left(\dfrac{W+w}{w}\right)\left(\dfrac{g}{h}\right)^{1/2}$
(c) $\omega^2 = \left(\dfrac{w}{W+w}\right)\left(\dfrac{h}{g}\right)^{1/2}$
(d) $\omega^2 = \left(\dfrac{W+w}{w}\right)\left(\dfrac{g}{h}\right)$

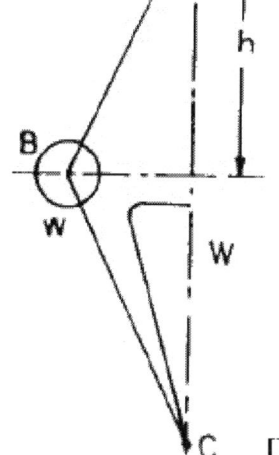

[IES-1996]

Hartnell Governor

IES-6. In a Hartnell governor, the mass of each ball is 2.5 kg. Maximum and minimum speeds of rotation are 10 rad/s and 8 rad/s respectively. Maximum and minimum radii of rotation are 20 cm and 14 cm respectively. The lengths of horizontal and vertical arms of bell crank levers are 10 cm and 20 cm respectively. Neglecting obliquity and gravitational effects, the lift of the sleeve is
 (a) 1.5 cm (b) 3.0 cm (c) 6.0 cm (d) 12.0 cm **[IES-2002]**

Pickering Governor

IES-7. Which one of the following governors is used to drive a gramophone?
 (a) Watt governor (b) Porter governor **[IES-2005]**
 (c) Pickering governor (d) Hartnell governor

Sensitiveness of Governor

IES-8. Sensitiveness of a governor is defined as **[IES-2000]**

(a) $\dfrac{\text{Range of speed}}{2\times \text{Mean speed}}$
(b) $\dfrac{2\times \text{Mean speed}}{\text{Range of speed}}$
(c) Mean speed × Range of speed
(d) $\dfrac{\text{Range of speed}}{\text{Mean speed}}$

IES-9. Which one of the following expresses the sensitiveness of a governor? **[IES-2005]**

(a) $\dfrac{N_1+N_2}{2N_1 N_2}$
(b) $\dfrac{N_1-N_2}{2N_1 N_2}$
(c) $\dfrac{2(N_1+N_2)}{N_1-N_2}$
(d) $\dfrac{2(N_1-N_2)}{N_1+N_2}$

(Where N_1 = Maximum equilibrium speed, N_2 = Minimum equilibrium speed)

IES-10. For a given fractional change of speed, if the displacement of the sleeve is high, then the governor is said to be **[IES-1998]**
 (a) hunting (b) isochronous (c) sensitive (d) stable

IES-11. Effect of friction, at the sleeve of a centrifugal governor is to make it **[IES-2003]**

Governor

Chapter 3

(a) More sensitive (b) More stable
(c) Insensitive over a small range of speed (d) Unstable

IES-12. Which one of the following statement is correct? [IES-2004]
A governor will be stable if the radius of rotation of the balls
(a) increases as the equilibrium speed decreases
(b) decreases as the equilibrium speed increases
(c) increases as the equilibrium speed increases
(d) remains unaltered with the change in equilibrium speed

IES-13. The sensitivity of an isochronous governor is [IES-1997]
(a) zero (b) one (c) two (d) infinity

IES-14. A spring controlled governor is found unstable. It can be made stable by [IES-1994]
(a) increasing the spring stiffness (b) decreasing the spring stiffness
(c) increasing the ball weight (d) decreasing the ball weight.

Isochronous Governor

IES-15. The nature of the governors is shown by the graph between radius (r) of rotation and controlling force (F). Which of the following is an isochronous governor? [IES-2002]

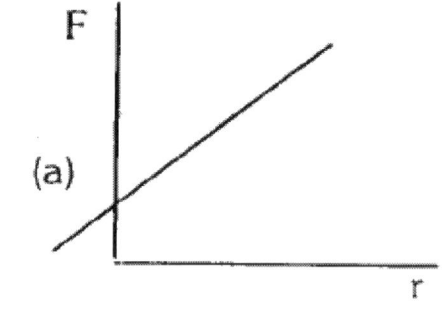

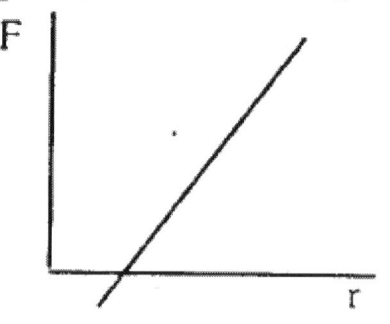

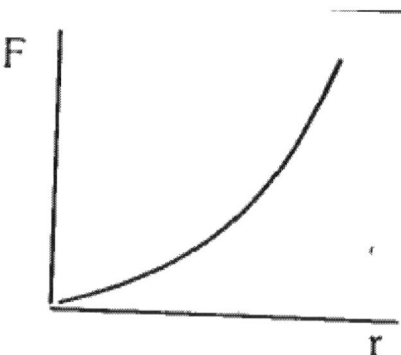

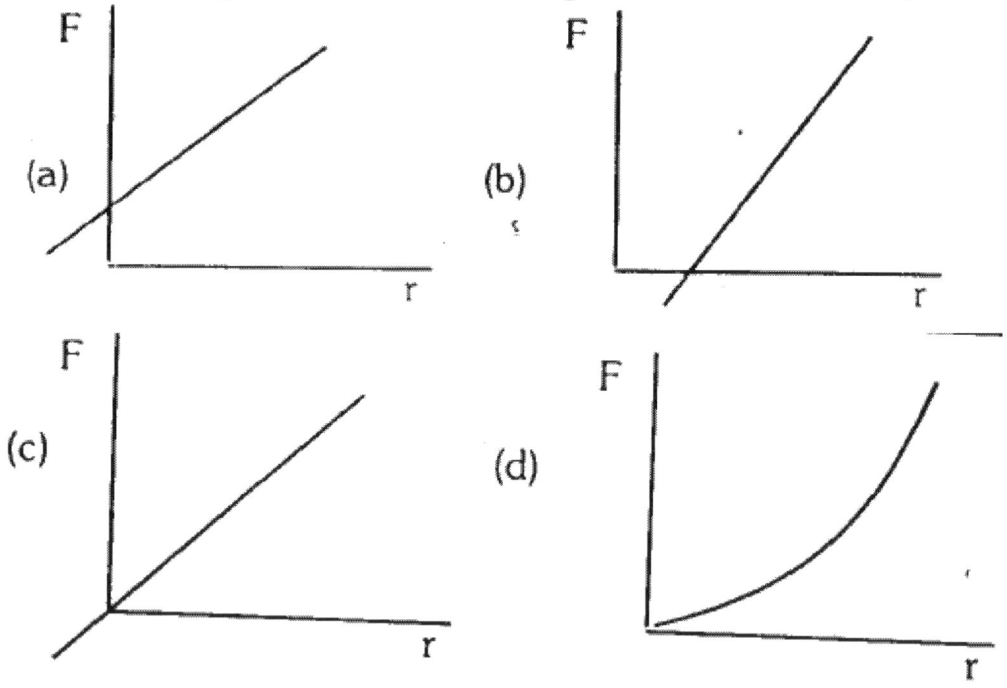

IES-16. Assertion (A): The degree of hunting with an unstable governor will be less than with an isochronous governor. [IES-1997]
Reason (R): With an unstable governor, once the sleeve has moved from one extreme position to the other, a finite change of speed is required to cause it to move back again.
(a) Both A and R are individually true and R is the correct explanation of A
(b) Both A and R are individually true but R is **not** the correct explanation of A
(c) A is true but R is false
(d) A is false but R is true

IES-17. A Hartnell governor has its controlling force F given by [IES-1993]

Governor

Chapter 3

$F = p + qr$

Where, is the radius of the balls and p and q are constants.
The governor becomes isochronous when
(a) P = 0 and q is positive (b) p is positive and q = 0
(c) p is negative and q is positive (d) P is positive and q is also positive

Hunting

IES-18. Match List I with List II and select the correct answer

List I	List II	[IES-1996]
A. Hunting	1. One radius rotation for each speed	
B. Isochronism	2. Too sensitive	
C. Stability	3. Mean force exerted at the sleeve during change of speed.	
D. Effort	4. Constant equilibrium speed for all radii of rotation	

Codes: A B C D   A B C D
(a)   2 4 1 3   (b) 3 1 4 2
(c)   2 1 4 3   (d) 1 2 3 4

Controlling force

IES-19.

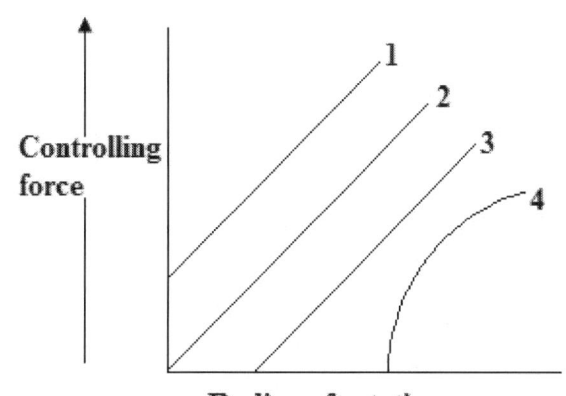

The controlling force curves for a spring-controlled governor are shown in the above figure. Which curve represents a stable governor? **[IES 2007]**
(a) 1   (b) 2   (c) 3   (d) 4

IES-20. Consider the following statements: **[IES-2006]**
1. The condition of stability of a governor requires that the slope of the controlling force curve should be less than that of the line representing the centripetal force at the equilibrium speed under consideration.
2. For a centrifugal governor when the load on the prime mover drops suddenly, the sleeve should at once reach the lower-most position.
Which of the statements given above is/are correct?
(a) Only 1   (b) Only 2   (c) Both 1 and 2   (d) Neither 1 nor 2

IES-21. Consider the following statements concerning centrifugal governors: **[IES-2005]**
1. The slope of the controlling force curve should be less than that of the straight line representing the centripetal force at the speed considered for the stability of a centrifugal governor.
2. Isochronism for a centrifugal governor can be achieved only at the expense of stability.

Governor

Chapter 3

3. When sleeve of a centrifugal governor reaches its topmost position, the engine should develop maximum power.
Which of the statements given above is/are correct?
(a) 1 and 2   (b) 2 and 3   (c) 2 only   (d) 3 only

IES-22. For a spring controlled governor to be stable, the controlling force (F) is related to the radius (r) by the equation. **[IES-1995]**
(a) F = ar - b   (b) F = ar + b   (c) F = ar   (d) F = a/r + b

IES-23. The plots of controlling force versus radii of rotation of the balls of spring controlled governors are shown in the given diagram. A stable governor is characterised by the curve labelled
(a) I
(b) II
(c) III
(d) IV

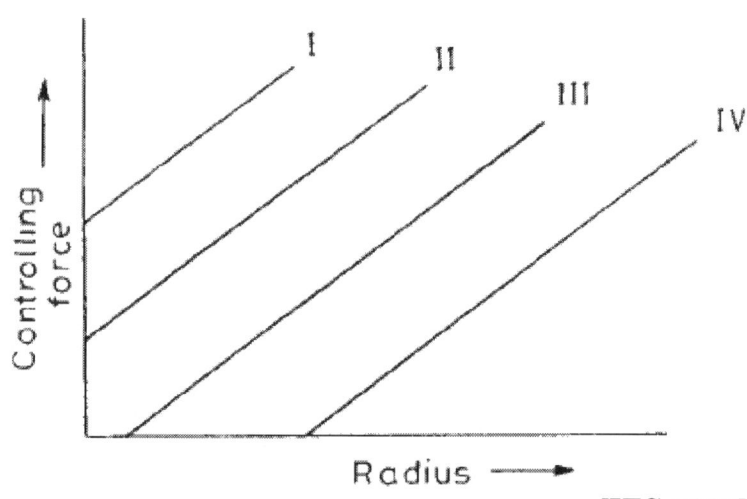

[IES-1993]

Previous 20-Years IAS Questions

Watt Governor

IAS-1. The height of a simple Watt governor running at a speed 'N' is proportional to
(a) N   (b) 1/N   (c) N^2 (d) $1/N^2$ **[IAS-1999]**

IAS-2. Match List I (Feature or application) with List-II (Governor) and select the correct answer using the codes given below the lists: **[IAS-1999]**

List 1
A. Gas engines
B. Rate of change of engine speed
C. Low speeds
D. Gramophone mechanism

List II
1. Quantity governing
2. Isochronous governor
3. Pickering governor
4. Watt governor
5. Inertia governor

Codes: A   B   C   D   A   B   C   D
(a)   1   5   4   3   (b)   2   5   4   3
(c)   3   4   5   2   (d)   1   2   5   3

IAS-3. The height of Watt's governor is **[IAS-2003]**
(a) directly proportional to the speed
(b) directly proportional to the (speed)^2
(c) inversely proportional to the speed
(d) inversely proportional to the (speed)^2

Governor

Chapter 3

Porter Governor

IAS-4. The height h of Porter governor with equal arms pivoted at equal distance from axis of rotation is expressed as (where m = mass of balls of the governor, M = mass of sleeve of the governor and N = rpm) [IAS-1998]

(a) $h = 91.2 \left[\dfrac{m+M}{m}\right] \dfrac{g}{N^2}$
(b) $h = 91.2 \left[\dfrac{mg - Mg}{mg}\right] \dfrac{g}{N^2}$
(c) $h = 91.2 \left[\dfrac{m}{mM}\right] \dfrac{g}{N^2}$
(d) $h = 91.2 \left[\dfrac{M}{m}\right] \dfrac{g}{N^2}$

IAS-5. The sensitivity dh/dN of a given Porter Governor. Where 'h' is the height of the pin point A from the sleeve and N is the r.p.m., is proportional to

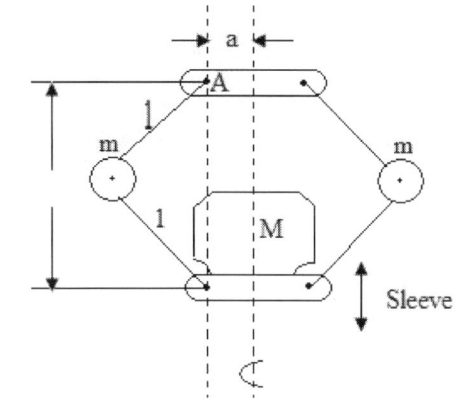

(a) N^2 (b) N^3
(c) $\dfrac{1}{N^2}$ (d) $\dfrac{1}{N^3}$

[IAS-1995]

Hartnell Governor

IAS-6. A Hartnell governor is a governor of the [IAS-1996]
(a) inertia type (b) pedulum type (c) centrifugal type (d) dead weight type

IAS-7. The stiffness of spring k used in the Hartnell governor as shown in the given figure (F_1 and F_2 are centrifugal forces at maximum and minimum radii of rotation r_1 and r_2 respectively) is

(a) $2\left(\dfrac{b}{a}\right)^2 \left(\dfrac{F_1 - F_2}{r_1 - r_2}\right)$
(b) $2\left(\dfrac{a}{b}\right)^2 \left(\dfrac{F_1 - F_2}{r_1 - r_2}\right)$
(c) $\left(\dfrac{b}{a}\right)^2 \left(\dfrac{F_1 - F_2}{r_1 - r_2}\right)$
(d) $\left(\dfrac{a}{b}\right)^2 \left(\dfrac{F_1 - F_2}{r_1 - r_2}\right)$

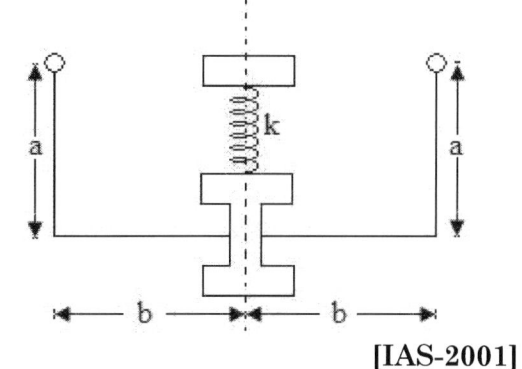

[IAS-2001]

Sensitiveness of Governor

IAS-8. If a centrifugal governor operates between speed limits ω_1 and ω_2 then what is its sensitivity equal to? [IAS-2007]

(a) $\dfrac{\omega_1 + \omega_2}{\omega_2 - \omega_1}$
(b) $\dfrac{\omega_1 + \omega_2}{2(\omega_2 - \omega_1)}$
(c) $\dfrac{\omega_2 - \omega_1}{2(\omega_2 + \omega_1)}$
(d) $\dfrac{\omega_2 - \omega_1}{\omega_2 + \omega_1}$

Governor

Chapter 3

IAS-9. Sensitiveness of a governor is defined as the ratio of the
(a) maximum equilibrium speed to the minimum equilibrium speed   [IAS-2000]
(b) difference between maximum and minimum equilibrium speeds to the mean equilibrium speed
(c) difference between maximum and minimum equilibrium speeds to the maximum equilibrium speed
(d) minimum difference in speeds to the minimum equilibrium speed

Isochronous Governor

IAS-10. A governor is said to be isochronous when the equilibrium speed for all radii or rotation the balls within the working range   [IAS-1996]
(a) is not constant   (b) is constant   (c) varies uniformly   (d) has uniform acceleration

IAS-11. Match List - I (Type of Governor) with List-II (Characteristics) and select the correct answer using the codes given below the lists:

List-I	List-II
A. Isochronous governor	1. Continuously fluctuates above and below mean speed
B. Sensitive governor	2. For each given speed there is only one radius of rotation
C. Hunting governor	3. Higher displacement of sleeve for fractional change of speed
D. Stable governor	4. Equilibrium speed is constant for all radii of rotation [IAS-1998]

Codes: A   B   C   D   A   B   C   D
(a)   4   3   2   1   (b)   2   4   1   3
(c)   2   4   3   1   (d)   4   3   1   2

IAS-12. In a spring-controlled governor, the controlling force curve is straight line. The balls are 400mm apart when the controlling force is 1600 N, and they are 240mm apart when the force is 800 N. To make the governor isochronous, the initial tension must be increased by   [IAS-1997]
(a) 100 N   (b) 200 N   (c) 400 N   (d) 800 N

Controlling force

IAS-13. The controlling force curve of spring-loaded governor is given by the equation
F = ar - c, (where r is the radius of rotation of the governor balls and a, c are constants). The governor is   [IAS-1999]
(a) stable   (b) unstable   (c) isochronous   (d) insensitive

Governor

Chapter 3

Answers with Explanation (Objective)

Previous 20-Years IES Answers

IES-1. Ans. (a)
IES-2. Ans. (b) Watt, Porter, Proell, Hartnell.
IES-3. Ans. (b)
IES-4. Ans. (a)
IES-5. Ans. (d)
IES-6. Ans. (b)
IES-7. Ans. (c)
IES-8. Ans. (d)
IES-9. Ans. (d)
IES-10. Ans. (c)
IES-11. Ans. (c)
IES-12. Ans. (c)
IES-13. Ans. (d) Sensitivity = $\dfrac{N_1 + N_2}{2(N_1 - N_2)}$, since $N_1 \simeq N_2$ for isochronous governor,

sensitivity = α.

IES-14. Ans. (b) A spring controlled governor can be made stable by decreasing the spring stiffness.
IES-15. Ans. (c)
IES-16. Ans. (a)
IES-17. Ans. (a) For isochronous governor F = qr
So P should be zero and q be + ve.
IES-18. Ans. (a)
IES-19. Ans. (c)
IES-20. Ans. (b)
IES-21. Ans. (a)
IES-22. Ans. (a)
IES-23. Ans. (d) For stable governor, F = qr - p which is possible with curve IV.

Previous 20-Years IAS Answers

IAS-1. Ans. (d) For Watt governor, height (h) = $\dfrac{895}{N^2}$ metre

IAS-2. Ans. (a)
IAS-3. Ans. (d)
IAS-4. Ans. (a)
IAS-5. Ans. (d)
IAS-6. Ans. (c) It is a spring loaded centrifugal governor.
IAS-7. Ans. (b)

Governor

Chapter 3

IAS-8. Ans. (c) [no one is correct] because correct expression is $\dfrac{2(\omega_2 - \omega_1)}{(\omega_2 + \omega_1)}$

IAS-9. Ans. (b) Sensitiveness of a governor = $\dfrac{N_2 - N_1}{N} = \dfrac{2(N_2 - N_1)}{(N_2 + N_1)}$

IAS-10. Ans. (b)

IAS-11. Ans. (d)

IAS-12. Ans. (c) It is a stable governor so F = ar – b
Or 1600 = a x 0.4 – b and 800 = a x 0.24 – b
Solving we get a = 5000 and 400
For isochronous governor, F = a.r i.e. b must be zero. i.e. initial tension must increase by 400 N.

IAS-13. Ans. (a)

CAM

Chapter 4

4. CAM

Objective Questions (IES, IAS, GATE)

Previous 20-Years GATE Questions

Classification of follower

GATE-1. In a plate cam mechanism with reciprocating roller follower, the follower has a constant acceleration in the case of **[GATE-1993]**
(a) cycloidal motion (b) simple harmonic motion
(c) parabolic motion (d) 3-4-5 polynomial motion

Pressure angle

GATE-2. For a spring-loaded roller-follower driven with a disc cam, [GATE-2001]
(a) the pressure angle should be larger during rise than that during return for ease of transmitting motion.
(b) the pressure angle should be smaller during rise than that during return for ease of transmitting motion.
(c) the pressure angle should be large during rise as well as during return for ease of transmitting motion.
(d) the pressure angle does not affect the ease of transmitting motion.

Pitch point

GATE-3. The profile of a cam in a particular zone is given by $x = \sqrt{3}\cos\theta$ and $y = \sin\theta$. The normal to the cam profile at $\theta = \pi/4$ is at an angle (with respect to x axis)
(a) $\dfrac{\pi}{4}$ (b) $\dfrac{\pi}{2}$ (c) $\dfrac{\pi}{3}$ (d) 0   **[GATE-1998]**

Displacement, Velocity, Acceleration and Jerk (Follower moves in uniform velocity)

GATE-4. In a cam-follower mechanism, the follower needs to rise through 20 mm during 60° of cam rotation, the first 30° with a constant acceleration and then with a deceleration of the same magnitude. The initial and final speeds of the follower are zero. The cam rotates at a uniform speed of 300 rpm. The maximum speed of the follower is
(a) 0.60m/s   (b) 1.20m/s   (c) 1.68m/s   (d) 2.40m/s **[GATE-2005]**

CAM

Chapter 4

Displacement, Velocity, Acceleration and Jerk (Follower moves in SHM)

GATE-6. In a cam design, the rise motion is given by a simple harmonic motion (SHM) $s = \dfrac{h}{2}\left(1 - \cos\dfrac{\pi\theta}{\beta}\right)$ where h is total rise, θ is camshaft angle, β is the total angle of the rise interval. The jerk is given by **[GATE-2008]**

(a) $\dfrac{h}{2}\left(1 - \cos\dfrac{\pi\theta}{\beta}\right)$
(b) $\dfrac{\pi}{\beta}\dfrac{h}{2}\sin\left(\dfrac{\pi\theta}{\beta}\right)$
(c) $\dfrac{\pi^2 h}{\beta^2}\dfrac{}{2}\cos\left(\dfrac{\pi\theta}{\beta}\right)$
(d) $-\dfrac{\pi^3 h}{\beta^3}\dfrac{}{2}\sin\left(\dfrac{\pi\theta}{\beta}\right)$

Displacement, Velocity, Acceleration and jerk (Follower moves in cycloidal motion)

GATE-7. In an experiment to find the velocity and acceleration of a particular cam rotating at 10 rad/s, the values of displacements and velocities are recorded. The slope of displacement curve at an angle of 'θ' is 1.5 m/s and the slope of velocity curve at the same angle is -0.5 m/s². The velocity and acceleration of the cam at the instant are respectively **[GATE-2000]**
(a) 15 m/s and – 5 m/s²
(b) 15 m/s and 5 m/s²
(c) 1.2 m/s and - 0.5 m/s²
(d) 1.2 m/s and 0.5 m/s²

Previous 20-Years IES Questions

Classification of follower

IES-1. In a circular arc cam with roller follower, the acceleration in any position of the lift would depend only upon **[IES-1994]**
(a) total lift, total angle of lift, minimum radius of earn and earn speed.
(b) radius of circular are, earn speed, location of centre of circular arc and roller diameter.
(c) weight of earn follower linkage, spring stiffness and earn speed.
(d) total lift, centre of gravity of the earn and earn speed.

IES-2. In a single spindle automatic lathe two tools are mounted on the turret, one form tool on the front slide and the other, a parting tool on the rear slide. The parting tool operation is much longer than form tool operation and they operate simultaneously (overlap). The number of cams required for this job is **[IES-1994]**
(a) one
(b) two
(c) three
(d) four

IES-3. Consider the following statements: **[IES-2006]**
Cam followers are generally classified according to
1. the nature of its motion
2. the nature of its surface in contact with the cam
3. the speed of the cam
Which of the statements given above are correct?
(a) 1, 2 and 3
(b) Only 1 and 2
(c) Only 2 and 3
(d) Only 1 and 3

CAM
S K Mondal's   **Chapter 4**

IES-4. 4. The above figure shows a cam with a circular profile, rotating with a uniform angular velocity of ω rad/s.
What is the nature of displacement of the follower?
(a) Uniform   (b) Parabolic
(c) Simple harmonic   (d) Cycloidal

[IES-2005]

IES-5. In a plate cam mechanism with reciprocating roller follower, in which one of the following cases the follower has constant acceleration? **[IES-2004]**
(a) Cycloidal motion   (b) Simple harmonic motion
(c) Parabolic motion   (d) 3 - 4 - 5 polynomial motion

IES-6. The choice of displacement diagram during rise or return of a follower of a cam-follower mechanism is based on dynamic considerations. For high speed cam follower mechanism, the most suitable displacement for the follower is **[IES-2002]**
(a) Cycloidal motion   (b) simple harmonic motion
(c) parabolic or uniform acceleration motion
(d) uniform motion or constant velocity motion

IES-7. Which one of the following sets of elements are quick acting clamping elements for fixtures? **[IES-2001]**
(a) Wedge and Cam   (b) Cam and Toggle
(c) Toggle and Wedge   (d) Wedge, Cam and Toggle

IES-8. Assertion (A): Cam of a specified contour is preferred to a cam with a specified follower motion.
Reason (R): Cam of a specified contour has superior performance. **[IES-2000]**
(a) Both A and R are individually true and R is the correct explanation of A
(b) Both A and R are individually true but R is **not** the correct explanation of A
(c) A is true but R is false
(d) A is false but R is true

IES-9. In a cam drive, it is essential to off-set the axis of a follower to   **[IES-1998]**
(a) decrease the side thrust between the follower and guide
(b) decrease the wear between follower and cam surface
(c) take care of space limitation   (d) reduce the cost

IES-10. Which one of the following is an Open Pair? **[IES-1996]**
(a) Ball and socket joint   (b) Journal bearing
(c) Lead screw and nut   (d) Cam and follower.

CAM

Chapter 4

Displacement, Velocity, Acceleration and Jerk (Follower moves in uniform velocity)

IES-11. In a cam drive with uniform velocity follower, the slope of the displacement must be as shown in Fig. I. But in actual practice it is as shown in Fig. II (i.e. rounded at the corners). This is because of

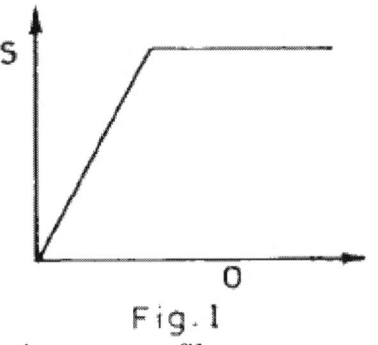

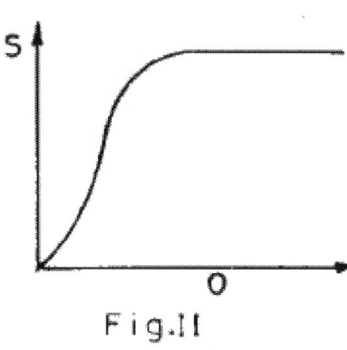

(a) the difficulty in manufacturing cam profile [IES-1996]
(b) loose contact of follower with cam surface
(c) the acceleration in the beginning and retardation at the end of stroke would require to be infinitely high
(d) uniform velocity motion is a partial parabolic motion.

IES-12. For a given lift of the follower in a given angular motion of the cam, the acceleration/retardation of the follower will be the least when the profile of the cam during the rise portion is
(a) such that the follower motion is simple harmonic [IES-1999]
(b) such that the follower motion has a constant velocity from start to end
(c) a straight line, it being a tangent cam
(d) such that the follower velocity increases linearly for half the rise portion and then decreases linearly for the remaining half of the rise portion.

Displacement, Velocity, Acceleration and Jerk (Follower moves in SHM)

IES-13. What is the maximum acceleration of a cam follower undergoing simple harmonic motion? [IES-2006]

(a) $\dfrac{h}{2}\left(\dfrac{\pi\omega}{\phi}\right)^2$
(b) $4h\left(\dfrac{\omega}{\phi}\right)^2$
(c) $4h\left(\dfrac{\omega^2}{\phi}\right)$
(d) $\dfrac{2h\pi\omega^2}{\phi^2}$

Where, h = Stroke of the follower; (ω) = Angular velocity of the cam; ϕ = Cam rotation angle for the maximum follower displacement.

Displacement, Velocity, Acceleration and jerk (Follower moves in cycloidal motion)

IES-14. Consider the following follower motions in respect of a given lift, speed of rotation and angle of stroke of a cam:
1. Cycloidal motion.
2. Simple harmonic motion.
3. Uniform velocity motion.
Which one of the following is the correct sequence of the above in the descending order of maximum velocity? [IES 2007]
(a) 3-2-1
(b) 1-2-3
(c) 2-3-1
(d) 3-1-2

CAM

Chapter 4

Previous 20-Years IAS Questions

Pitch point

IAS-1. Consider the following statements:
1. For a radial-translating roller follower, parabolic motion of the follower is very suitable for high speed cams.
2. Pitch point on pitch circle of a cam corresponds to the point of maximum pressure angle. [IAS-2007]

Which of the statements given above is/are correct?
(a) 1 only (b) 2 only   (c) Both 1 and 2   (d) Neither 1 nor 2

Answers with Explanation (Objective)

Previous 20-Years GATE Answers

GATE-1. Ans. (c) For uniform acceleration and retardation, the velocity of the follower must change at a constant rate and hence the velocity diagram of the follower consists of sloping straight lines. The velocity diagram represents everywhere the slope of the displacement diagram, the later must be a curve whose slope changes at a constant rate. Hence the displacement diagram consists of double parabola.

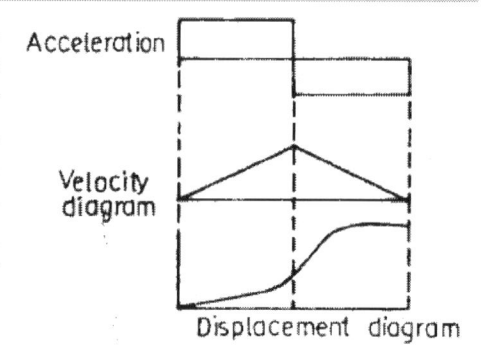

GATE-2. Ans. (c)

GATE-3. Ans. (c)

Explanation. $\sin^2 \theta + \cos^2 \theta = 1$

Equation of curve is

$$y^2 + \frac{x^2}{3} = 1$$

or $\quad x^2 + 3y^2 = 3$

To find Slope at required point $\left(\frac{\sqrt{3}}{2}, \frac{1}{\sqrt{2}}\right)$, differentiating we get

$$2x + 6y \frac{dy}{dx} = 0$$

or $\quad \dfrac{dy}{dx} = -\dfrac{x}{3y} = -\dfrac{1}{\sqrt{3}}$

–ve sign indicates that slope is – ve.

∴ $\quad \dfrac{dy}{dx} = 2x + 6y \tan \theta = \dfrac{1}{\sqrt{3}}$

or $\quad \theta = 30°$

∴ Angle made by normal = 60° = π/3 **radians.**

CAM

Chapter 4

GATE-4. Ans. (d)

$$\omega = \frac{2\pi N}{60}$$

Time taken to move 30° = $\dfrac{\frac{\pi}{180} \times 30}{\omega} = \dfrac{\pi}{6} \times \dfrac{60}{2\pi N} = \dfrac{5}{N}$ sec

During this period follower move 20 mm = .02 m and u = 0

From $s = ut + \dfrac{1}{2}at^2$

$$0.02 = 0 + \dfrac{1}{2} a \times \left(\dfrac{5}{N}\right)^2$$

$\Rightarrow \quad a = 144 \text{ m/sec}^2$

$\therefore \quad v = u + at$

$v = 0 + \left(144 \times \dfrac{5}{300}\right) = 2.4$ m/sec

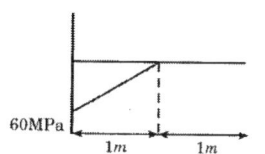

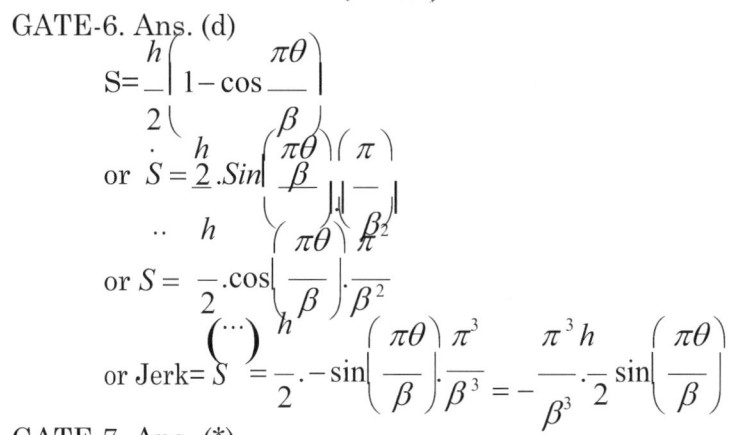

GATE-6. Ans. (d)

$$S = \dfrac{h}{2}\left(1 - \cos\dfrac{\pi\theta}{\beta}\right)$$

or $\dot{S} = \dfrac{h}{2}.\sin\left(\dfrac{\pi\theta}{\beta}\right)\left(\dfrac{\pi}{\beta}\right)$

or $\ddot{S} = \dfrac{h}{2}.\cos\left(\dfrac{\pi\theta}{\beta}\right).\dfrac{\pi^2}{\beta^2}$

or Jerk = $\dddot{S} = \dfrac{h}{2}.-\sin\left(\dfrac{\pi\theta}{\beta}\right).\dfrac{\pi^3}{\beta^3} = -\dfrac{\pi^3 h}{\beta^3}.\dfrac{1}{2}\sin\left(\dfrac{\pi\theta}{\beta}\right)$

GATE-7. Ans. (*)

Previous 20-Years IES Answers

IES-1. Ans. (b)
IES-2. Ans. (a) One cam is required.
IES-3. Ans. (b)
IES-4. Ans. (c)
IES-5. Ans. (*)
IES-6. Ans. (a)
IES-7. Ans. (a)
IES-8. Ans. (d)
IES-9. Ans. (b)
IES-10. Ans. (d) Cam and follower is open pair.
IES-11. Ans. (c)
IES-12. Ans. (b)
IES-13. Ans. (a)
IES-14. Ans. (d) $1 \to 2h\left(\dfrac{\omega}{\phi}\right), \quad 2 \to \dfrac{h}{2}\pi\left(\dfrac{\omega}{\phi}\right), \quad 3 \to 2h\left(\dfrac{\omega}{\phi}\right)$

Previous 20-Years IAS Answers

IAS-1. Ans. (d) For high speed use cycloidal motion.
Pitch point on pitch circle of a cam corresponds to the point of maximum pressure angle.

CAM

Chapter 4

5. Balancing of Rigid Rotors and Field Balancing

Theory at a glance (IES, GATE & PSU)

Balancing
A machine consists of a number of moving part. The motion of moving parts may be of rotary or reciprocating type. These m/c parts are subjected to different forces.

1. **Static Force:** The static forces acting on the machine components are due to the weight of components.

2. **Inertia or Dynamic force:** The inertia force is due to acceleration of various components or members of the m/c.

   In comparison with the static forces, the dynamic forces are very large in magnitude.

   Balancing is the process of correcting or eliminating, either partially or completely. The effects due to resultant inertia forces and couples acting on the machine parts.

Static balancing
The system is said to be statically balanced if the centre of mass (C. Cx. Of the system of masses lies on the axis of rotation.

For the system to be statically balanced, the resultant of all the dynamic forces (centrifugal forces) acting on the system during rotation must be zero.

Dynamic Balancing
The system is said to be dynamically (complete) balanced, if it satisfies following two condition.
(i) The resultant of all the dynamic forces acting on the system during rotation must be zero.
(ii) The resultant couples due to all the dynamic forces acting on the system during rotation, about any plane must be zero.

Balancing of Rotating Masses
In any rotating system, having one or more rotating masses, if the centre of mass of the system does not lie on the axis of rotation, then the system is unbalanced.
The **unbalanced** in rotating system is mainly due to the following factors.

(i) Errors and tolerances in manufacturing and assembly.
(ii) Non-homogeneity of material
(iii) Unsymmetrical shape of the rotors due to functional requirement.

Balancing of Rigid Rotors and field Balancing

Chapter 5

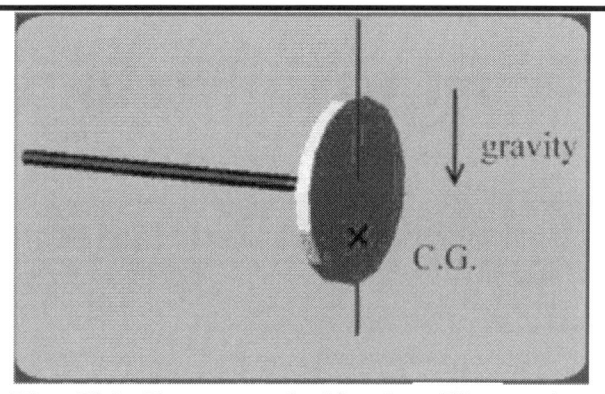

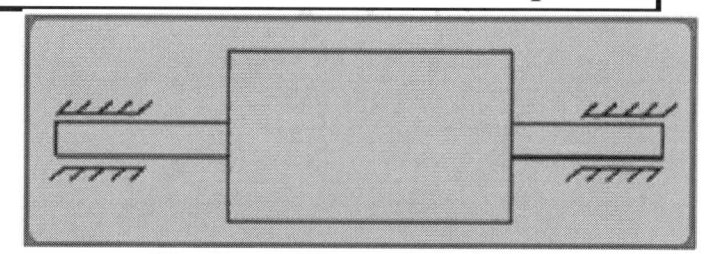

Fig. Thin Rotor on a knife edge. Illustration of Static Unbalance

Fig. Case of Dynamic Unbalance

Balancing of mass rotating in single plane

1. Balancing of single rotating mass

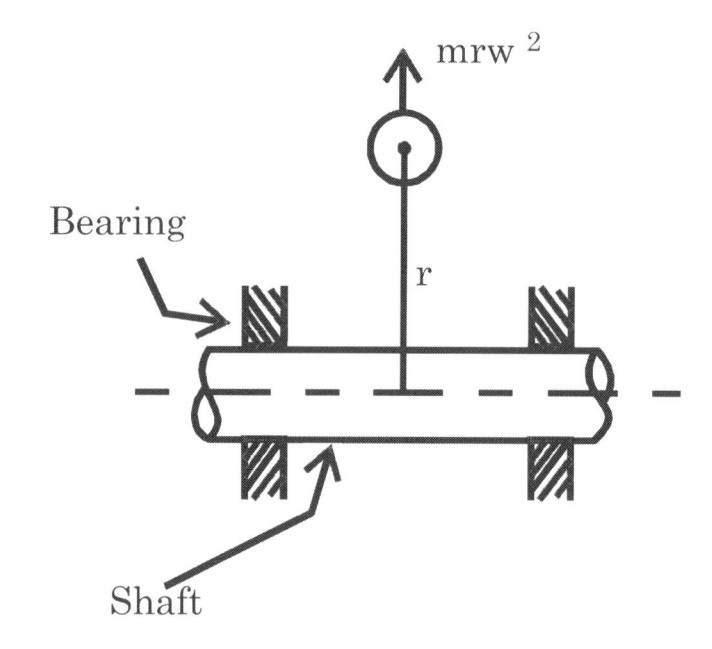

Consider a single mass m attached to a shaft which is rotating with an angular velocity ω'. Let r be the distance of C.G of mass m from the axis of rotation.

During the rotation of shaft a dynamic force equal to $mr\omega^2$ acts in radial outward direction. This dynamic force can be balanced by either of the following two methods.

(i) Balancing by single mass rotating in same plane (Internal balancing):
To balance the single rotating mass, a counter mass m_b is placed in the plane of rotation at a radius r_b and exactly opposite to it, such that centrifugal force due to the two masses is equal and opposite.

Balancing of Rigid Rotors and field Balancing
Chapter 5

$$\therefore \quad mr\omega^2 = m_b r_b \omega^2$$

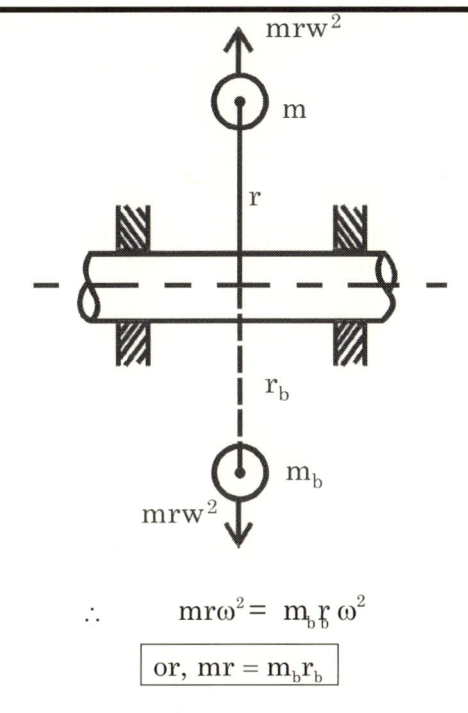

$$\boxed{\text{or, } mr = m_b r_b}$$

(ii) Balancing by two masses rotating in different planes (External Balancing):
If the balancing mass cannot be placed in the plane of rotation of the disturbing mass then it is not possible to balance the disturbing mass by single balancing mass.

If a single balancing mass is placed in a plane parallel to the plane of rotation of disturbing mass, the dynamic force can be balanced, while it will produce an unbalanced couple.
In order to achieve the complete balancing of the system, at least two balancing masses are required to be placed in two planes parallel to the plane of disturbing mass.
The balancing masses may be arranged in one of the two ways.

Case I: Balancing masses are placed in two different planes on opposite side of the plane of rotation of disturbing mass.

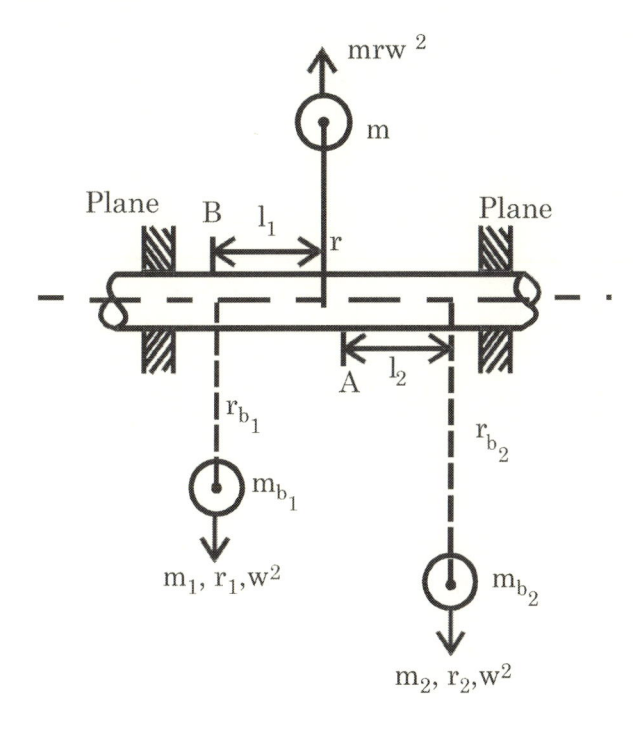

Balancing of Rigid Rotors and field Balancing

Chapter 5

Balancing mass m_{b_1} are at radius r_{b_1} from shaft axis at plane B from distance l_1 to disturbing mass, and balancing mass m_{b_2} at radius r_{b_2} from shaft axis at plane C from distance to disturbing mass. To achieve complete balancing following two conditions must satisfy

(i) Centrifugal force of disturbing mass 'm' must be equal to the sum of the centrifugal forces of the balancing masses m_{b_1} and m_{b_2}.

$$\therefore mr\omega^2 = m_{b_1} r_{b_1} \omega^2 + m_{b_2} r_{b_2} \omega^2$$

Or $\boxed{mr = m_{b_1} r_{b_1} + m_{b_2} r_{b_2}}$

(ii) The sum of moments due to centrifugal forces about any point must be zero. Thus taking moments about plane A,

$$m_{b_1} r_{b_1} \omega^2 l_1 = m_{b_2} r_{b_2} \omega^2 l_2$$

$$\therefore \boxed{m_{b_1} r_{b_1} l_1 = m_{b_2} r_{b_2} l_2}$$

Case II: Balancing mass is placed opposite to each other in two difference planes.

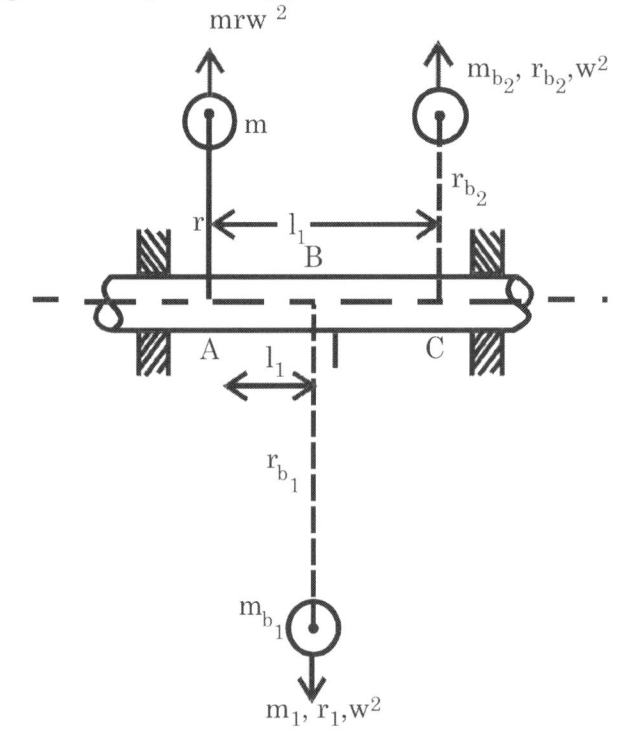

The balancing mass are m_{b_1} and m_{b_2} at radius r_{b_1} and r_{b_2} from shaft axis and distance. l_1 And l_2 from plane A.

In order to achieve the complete balancing of the system the following two conditions must satisfied.

(i) $\quad mr\omega^2 + m_{b_2} r_{b_2} \omega^2 = m_{b_1} r_{b_1} \omega^2$

Or, $\boxed{mr + m_{b_2} r_{b_2} = m_{b_1} r_{b_1}}$

(ii) $\quad m_{b_1} r_{b_1} \omega^2 l_1 = m_{b_2} r_{b_2} \omega^2 l_2$

Or, $\boxed{m_{b_1} r_{b_1} l_1 = m_{b_2} r_{b_2} l_2}$

Balancing of Rigid Rotors and field Balancing

Chapter 5

2. Balancing of several masses rotating in same plane

Consider the number of masses (say four) m_1, m_2, m_3 and m_4 attached to a shaft at a distance of r_1, r_2, r_3 and r_4 from the axis of rotation of shaft.

The masses are at angular position θ_1, θ_2, θ_3, and θ_4 with horizontal line OX. The angular position is measured in anticlockwise direction.

*Since ω^2 is same for all masses, the magnitude of centrifugal force for each mass is proportion to the product of mass and radius of rotation. Hence in subsequent section, the centrifugal force is taken as term 'mr', kg-m and the couple due to centrifugal force is taken as term 'mrl', kg-m².

Such type of balancing can be done by the following two methods.

1. Graphical Method

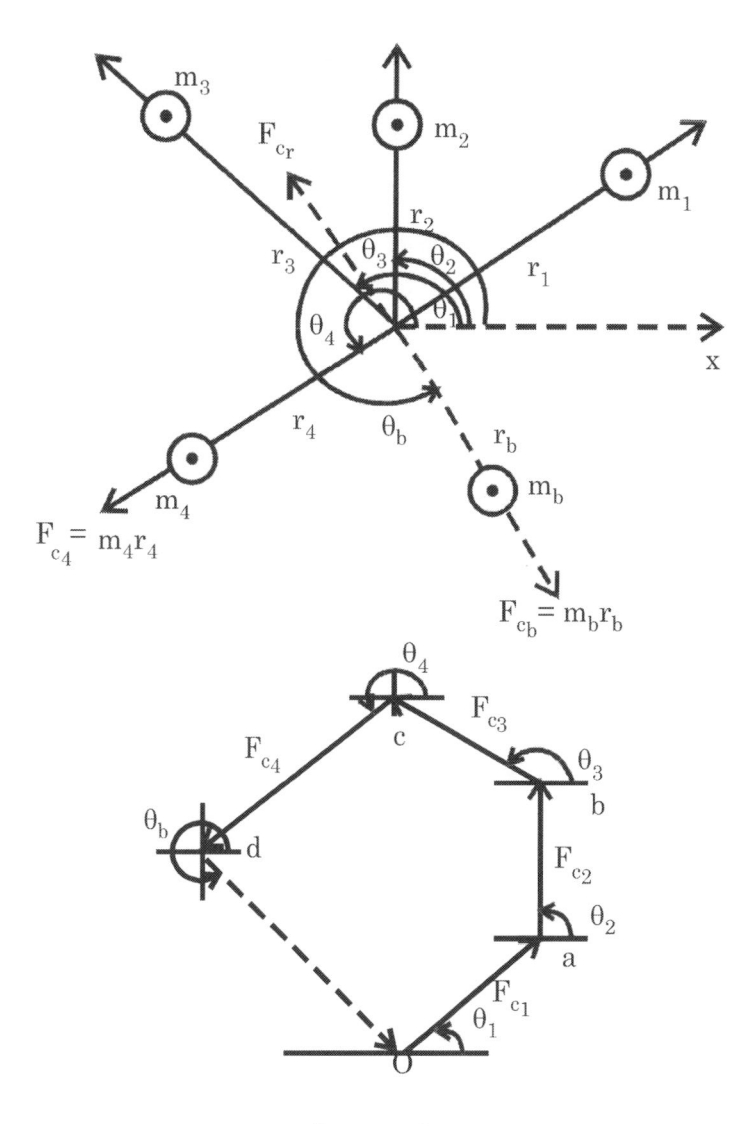

Force polygon

(i) First draw configuration diagram with given position.
(ii) Calculate centrifugal force by individual masses on the rotating shaft.

Balancing of Rigid Rotors and field Balancing

Chapter 5

$$F_{C_1} = m_1 r_1, \quad F_{C_2} = m_2 r_2$$
$$F_{C_3} = m_3 r_3, \quad F_{C_4} = m_4 r_4$$

(iii) Draw the force polygon such that $\vec{oa}$ represent the centrifugal force exerted by the mass m_1 m given direction with some suitable scale. Similar draw $\vec{ab}$, $\vec{bc}$, and $\vec{cd}$.

(iv) The closing side of the force polygon $\vec{do}$ represents the balancing centrifugal force.

(v) Determine the magnitude of balancing mass m_b at a given radius of rotation 'r_b' such that balancing centrifugal force $= m_b r_b = \vec{do} \times$ scale

2. Analytical Method

Refer configuration diagram.

(i) Resolve the centrifugal force horizontally and vertically and find their sums.
$$\Sigma F_H = m_1 r_1 \cos\theta_1 + m_2 r_2 \cos\theta_2 + m_2 r_3 \cos\theta_3 + m_4 r_4 \cos\theta_4$$
and
$$\Sigma F_V = m_1 r_1 \sin\theta_1 + m_2 r_2 \sin\theta_2 + m_3 r_3 \sin\theta_3 + m_4 r_4 \sin\theta_4$$

(ii) Calculate the magnitude of resultant centrifugal force.
$$F_{C_r} = \sqrt{(\Sigma F_H)^2 + (\Sigma F_V)^2}$$

(iii) Calculate the angle made by the resultant centrifugal force with horizontal OX.
$$\tan\theta_r = \frac{\Sigma F_V}{\Sigma F_H}$$

(iv) The balancing centrifugal force F_{C_b} should be equal to magnitude of the F_{C_r} but opposite in direction (θ_b).

(v) Determine the magnitude and radius of rotation of balance mass by using relation
$$m_b r_b = F_{C_r}$$

Balancing of several masses rotating in different plane

Consider mass m_1, m_2, m_3 and m_4 revolving in planes A, B, C and D respectively. The relative angular position shown in fig.

Balancing of Rigid Rotors and field Balancing

Chapter 5

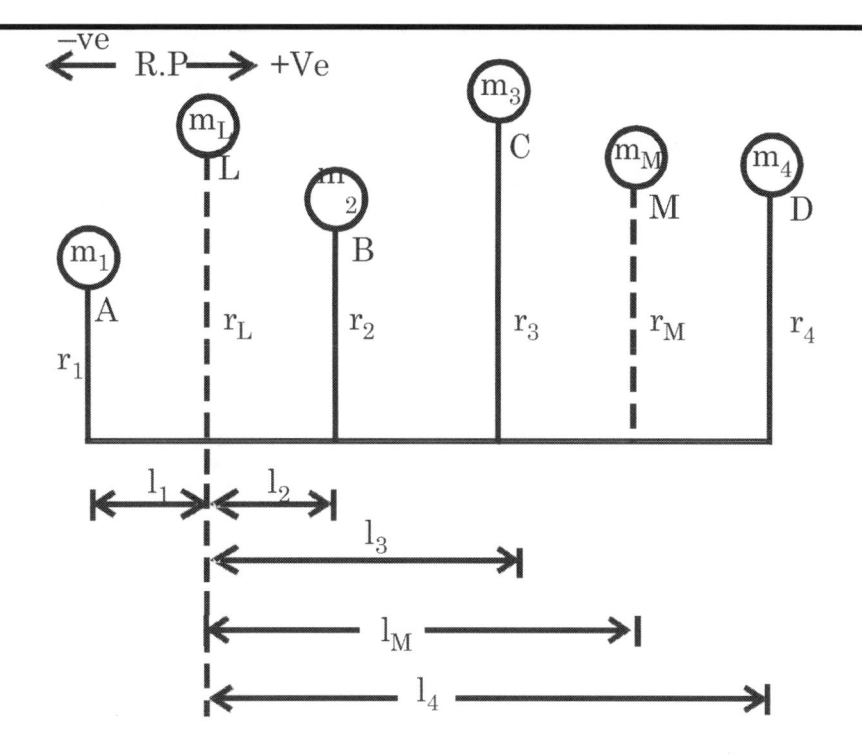

Such system is balanced by two masses m_L and m_M which are put in planes L and M respectively.

Position of Planes of Masses

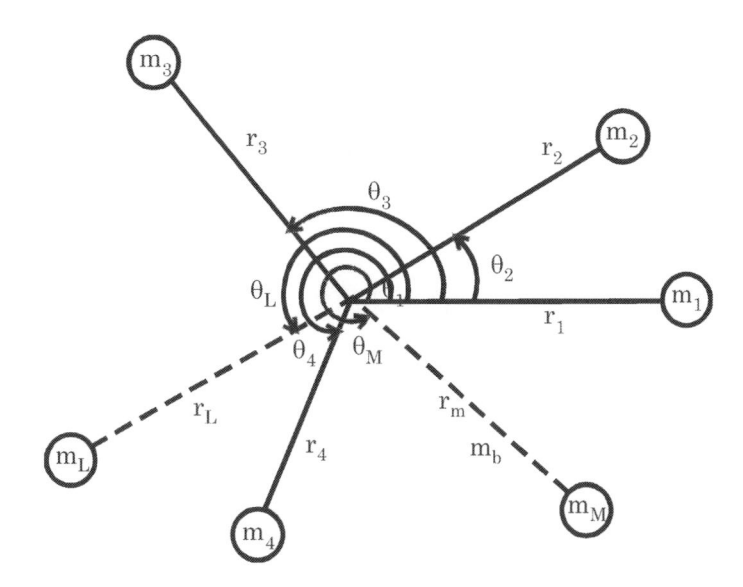

The complete balancing of such system can be done by following two methods:

1. Graphical Method

(i) Take one of the planes from balancing plane. Say 'L' as reference plane (R.P). The distance of other planes to the right of R.P are taken as positive and the left of R.P are taken as negative.

(ii) Tabulate the centrifugal forces and couples due to centrifugal forces.

Balancing of Rigid Rotors and field Balancing

Chapter 5

Plane	Mass (kg)	Radius (m)	C.F ÷ ω^2 (kg-m)	Dist. from R.P.1 (m)	Couple ÷ ω^2 (kg-m^2)
A	m_1	r_1	$m_1 r_1$	$-l_1$	$-m_1 r_1 l_1$
L(R.P)	M_L	r_L	$m_L r_L$	O	O
B	m_2	r_2	$m_2 r_2$	l_2	$m_2 r_2 l_2$
C	m_3	r_3	$m_3 r_3$	l_3	$m_3 r_3 l_3$
M	m_M	r_M	$m_M r_M$	l_M	$m_M r_M l_M$
D	m_4	r_4	$m_4 r_4$	l_4	$m_4 r_4 l_4$

Since ω^2 is same for all masses, the centrifugal forces and centrifugal couple are taken as terms (mr) and (mr l) respectively.

(iii) Draw the couple polygon, taking some suitable scale. Closing vector d'o' represent the couple $m_M r_M l_M$ which is called balancing couple.

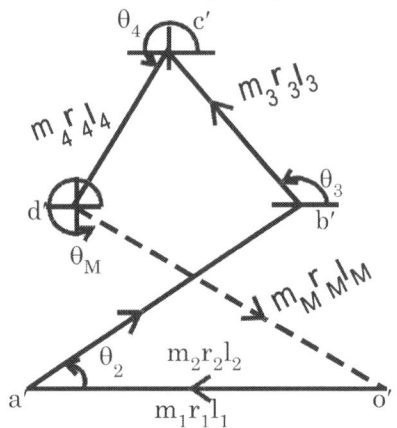

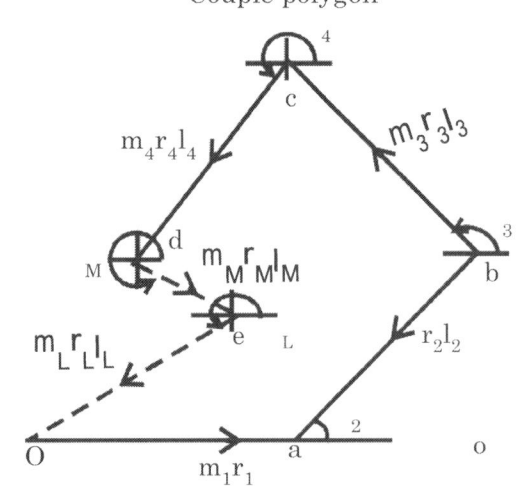

Couple polygon

Force polygon

(iv) Now draw force polygon, $\vec{eo}$ is balancing force.

2. Analytical Method

Refer previous table

Balancing of Rigid Rotors and field Balancing
Chapter 5

(i) $\Sigma(\text{couples}) = 0$

i.e. $\Sigma(m\ r\ l) = 0$

Now resolve the couple horizontally and vertically and find their summation.

$\Sigma(m\ r\ l)_H = 0$

And find $m_M r_M l_M \cos\theta_M$ ... (i)

$\Sigma(m\ r\ l)_V = 0$

and find $m_M r_M l_M \sin\theta_M$ ... (ii)

by the use of (i) and (ii)

$m_M r_M$ Can be finding and also θ_M can be find.

Similarly above process can be done for centrifugal forces, and $m_L r_L$, θ_L can be find.

Objective Questions (IES, IAS, GATE)

Previous 20-Years GATE Questions

GATE-1. A cantilever type gate hinged at Q is shown in the figure. P and R are the centers of gravity of the cantilever part and the counterweight respectively. The mass of the cantilever part is 75 kg. The mass of the counterweight, for static balance, is [GATE-2008]

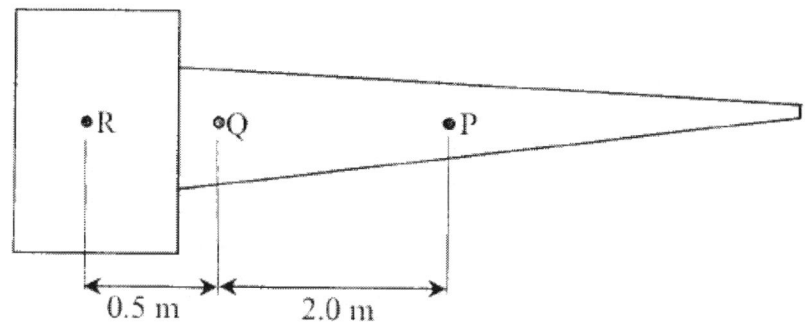

Balancing of a single rotating mass by a single mass rotating in a same plane

GATE-2. A rotating disc of 1 m diameter has two eccentric masses of 0.5 kg each at radii of 50 mm and 60 mm at angular positions of 0° and 150°, respectively. a balancing mass of 0.1 kg is to be used to balance the rotor. What is the radial position of the balancing mass? [GATE-2005]

(a) 50 mm (b) 120 mm (c) 150 mm (d) 280 mm

Balancing of Rigid Rotors and field Balancing

Chapter 5

GATE-3. A rigid body shown in the Fig. (a) has a mass of 10 kg. It rotates with a uniform angular velocity 'ω'. A balancing mass of 20 kg is attached as shown in Fig. (b). The percentage increase in mass moment of inertia as a result of this addition is
(a) 25%   (b) 50%
(c) 100% (d) 200%

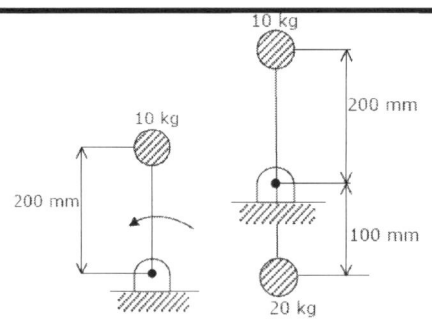

[GATE-2004]

Previous 20-Years IES Questions

IES-1. What is the condition for dynamic balancing of a shaft-rotor system?
(a) $\sum M = 0$ and $\sum F = 0$ (b) $\sum M = 0$ [IES 2007]
(c) $\sum F = 0$ (d) $\sum M + \sum F = 0$

IES-2. **Assertion (A):** A dynamically balanced system of multiple rotors on a shaft can rotate smoothly at the critical speeds of the system. [IES-2002]
Reason (R): Dynamic balancing eliminates all the unbalanced forces and couples from the system.
(a) Both A and R are individually true and R is the correct explanation of A
(b) Both A and R are individually true but R is **not** the correct explanation of A
(c) A is true but R is false

IES-3. A system in dynamic balance implies that [IES-1993]
(a) the system is critically damped
(b) there is no critical speed in the system
(c) the system is also statically balanced
(d) there will be absolutely no wear of bearings.

Balancing of a single rotating mass by a single mass rotating in a same plane

IES-4. Consider the following statements for completely balancing a single rotating mass: [IES-2002]
1. Another rotating mass placed diametrically opposite in the same plane balances the unbalanced mass.
2. Another rotating mass placed diametrically opposite in a parallel plane balances the unbalanced mass.
3. Two masses placed in two different parallel planes balance the unbalanced mass.
Which of the above statements is/are correct?
(a) 1 only   (b) 1 and 2   (c) 2 and 3   (d) 1 and 3

Balancing of a single rotating mass by two masses rotating in different planes

IES-4. If a two-mass system is dynamically equivalent to a rigid body, then the system will NOT satisfy the condition that the [IES-1999]

Balancing of Rigid Rotors and field Balancing

Chapter 5

(a) Sum of the two masses must be equal to that of the rigid body
(b) Polar moment of inertia of the system should be equal to that of the rigid body
(c) Centre of gravity (c.g.) of the system should coincide with that of the rigid body
(d) Total moment of inertia about the axis through c.g. must be equal to that of the rigid body

IES-5. A system of masses rotating in different parallel planes is in dynamic balance if the resultant. [IES-1996]
(a) Force is equal to zero
(b) Couple is equal to zero
(c) Force and the resultant couple are both equal to zero
(d) Force is numerically equal to the resultant couple, but neither of them need necessarily be zero.

IES-6. A rotor supported at A and B, carries two masses as shown in the given figure. The rotor is
(a) dynamically balanced
(b) statically balanced
(c) statically and dynamically balanced
(d) not balanced.

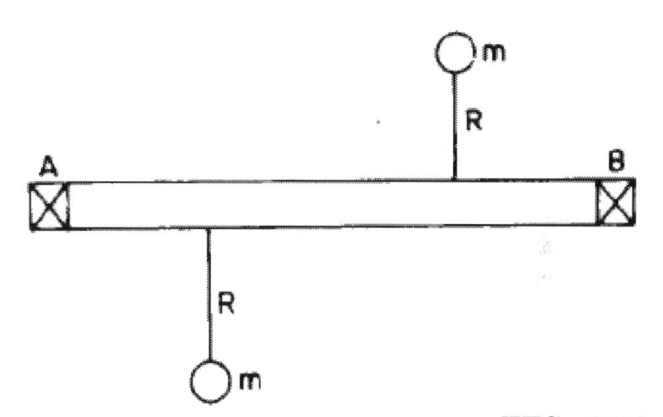

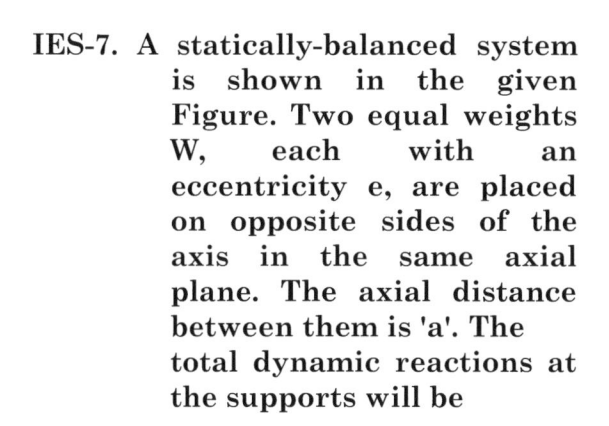

[IES-1995]

IES-7. A statically-balanced system is shown in the given Figure. Two equal weights W, each with an eccentricity e, are placed on opposite sides of the axis in the same axial plane. The axial distance between them is 'a'. The total dynamic reactions at the supports will be

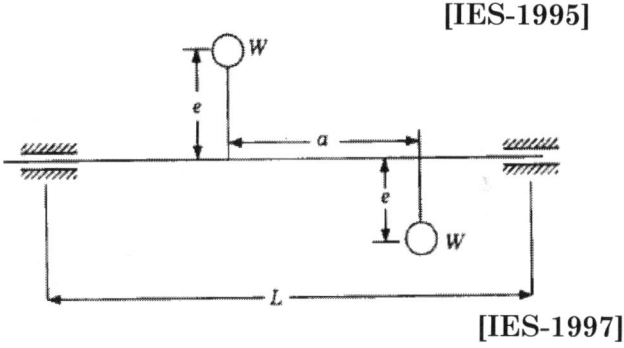

[IES-1997]

(a) zero (b) $\dfrac{W}{g}\omega^2 e \dfrac{a}{L}$ (c) $\dfrac{2W}{g}\omega^2 e \dfrac{a}{L}$ (d) $\dfrac{W}{g}\omega^2 e \dfrac{L}{a}$

Balancing of several masses rotating in a same plane

IES-8.

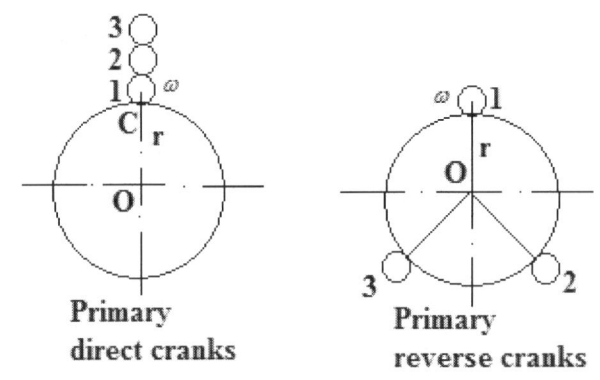

Balancing of Rigid Rotors and field Balancing

Chapter 5

(W = Weight of reciprocating parts per cylinder) [IES 2007]
For a three-cylinder radial engine, the primary and direct reverse cranks are as shown in the above figures.
Which one of the following pairs is **not** correctly matched in this regard?

(a) Primary direct force... $\dfrac{3W}{2g}\omega^2.r$ (b) Primary reverse force... Zero

(c) Primary direct crank speed... ω (d) Primary reverse crank speed... 2ω

Balancing of several masses rotating in different planes

IES-9. What is the number of nodes in a shaft carrying three rotors? [IES-2006]
(a) Zero (b) 2 (c) 3 (d) 4

IES-10. Which one of the following can completely balance several masses revolving in different planes on a shaft? [IES-2005]
(a) A single mass in one of the planes of the revolving masses
(b) A single mass in any one plane
(c) Two masses in any two planes
(d) Two equal masses in any two planes.

IES-11. Ans. (c)28. Masses B₁, B₂ and 9 kg are attached to a shaft in parallel planes as shown in the figure. If the shaft is rotating at 100 rpm, the mass B₂ is
(a) 3 kg
(b) 6 kg
(c) 9 kg
(d) 27 kg

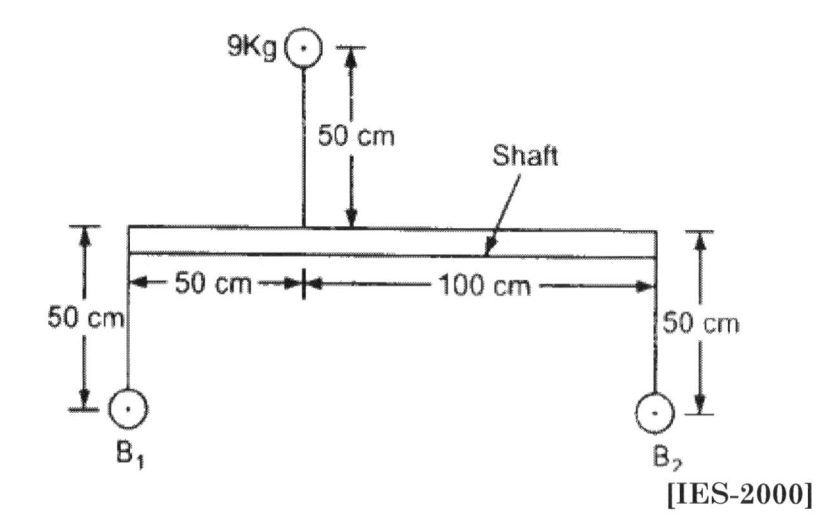

[IES-2000]

IES-12. Which one of the following can completely balance several masses revolving in different planes on a shaft? [IES-1993]
(a) A single mass in one of the planes of the revolving masses
(b) A single mass in a different plane
(c) Two masses in any two planes
(d) Two equal masses in any two planes

Balancing of Rigid Rotors and field Balancing

Chapter 5

Previous 20-Years IAS Questions

IAS-1. The figures given on right show different schemes suggested to transmit continuous rotary motion from axis A to axis B. Which of these schemes are not dynamically balanced?
(a) 1 and 3   (b) 2 and 3
(c) 1 and 2   (d) 1, 2 and 3

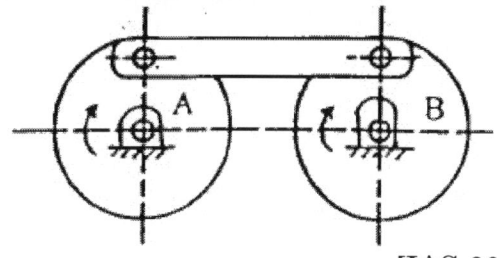

[IAS-2004]

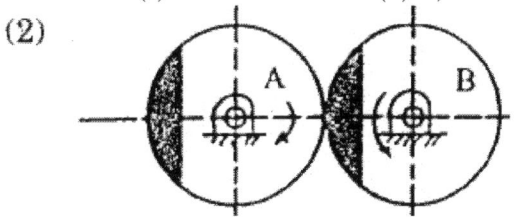

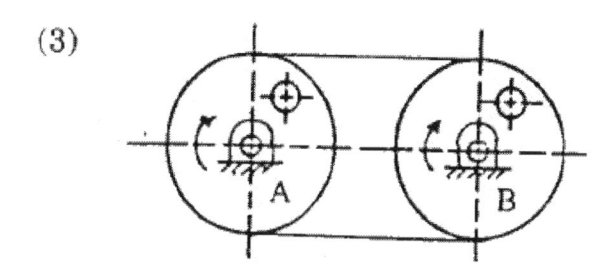

IAS-2. Static balancing is satisfactory for low speed rotors but with increasing speeds, dynamic balancing becomes necessary. This is because, the
(a) Unbalanced couples are caused only at higher speeds   [IAS 1994]
(b) Unbalanced forces are not dangerous at higher speeds
(c) Effects of unbalances are proportional to the square of the speed
(d) Effects of unbalances are directly proportional to the speed

Balancing of a single rotating mass by two masses rotating in different planes

IAS-3. Which of the following conditions are to be satisfied by a two-mass system which is dynamically equivalent to a rigid body?   [IAS-1997]
1. The total mass should be equal to that of the rigid body.
2. The centre of gravity should coincide with that of the rigid body.
3. The total moment of inertia about an axis through the centre of gravity must be equal to that of the rigid body.
Select the correct answer using the codes given below:
Codes:
(a) 1 and 2   (b) 2 and 3   (c) 1 and 3   (d) 1 and 3

IAS-4. Consider the following necessary and sufficient conditions for replacing a rigid body by a dynamical equivalent system of two masses: [IAS-2002]
1. Total mass must be equal to that of the rigid body.
2. Sum of the squares of radii of gyration of two masses about the c.g. of the rigid body must be equal to square of its radius of gyration about the same point.
3. The c.g. of two masses must coincide with that of the rigid body.
4. The total moment of inertia of two masses about an axis through the c.g. must be equal to that of the rigid body.
Which of the above conditions are correct?
(a) 1, 2 and 3   (b) 1, 3 and 4   (c) 2, 3 and 4   (d) 1, 2 and 4

Balancing of Rigid Rotors and field Balancing

Chapter 5

IAS-5. A rigid rotor consists of a system of two masses located as shown in the given figure. The system is
(a) statically balanced
(b) dynamically balanced
(c) statically unbalanced
(d) both statically and dynamically unbalanced

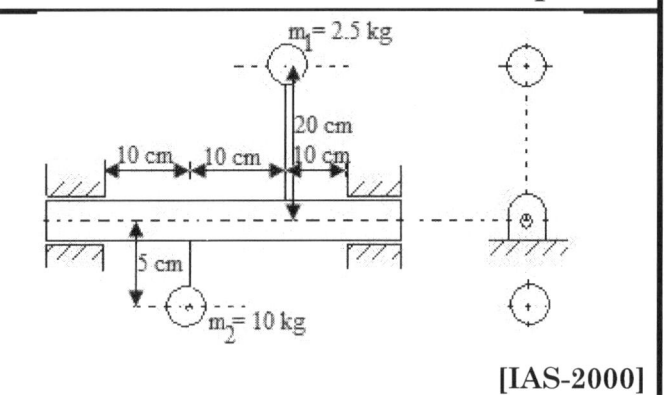

[IAS-2000]

IAS-6. For the rotor system shown in figure, the mass required for its complete balancing is
(a) 1.5 kg at 2 m radius and at 225⁰ from reference
(b) 3 kg at 1m radius and at 45⁰ from reference
(c) 8 kg at 1 m radius and at 225⁰ from reference
(d) 4 kg at 2 m radius and at 45⁰ from reference

[IAS-2004]

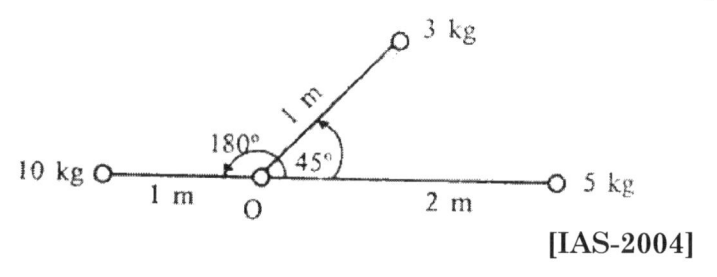

IAS-7. Balancing of a rigid rotor can be achieved by appropriately placing balancing weights in [IAS-1995]
(a) a single plane (b) two planes (c) three planes (d) four planes

IAS-8. The shaft-rotor system given above is
(a) Statically balanced only
(b) Dynamically balanced only
(c) Both statically and dynamically balanced
(d) Neither statically nor dynamically balanced

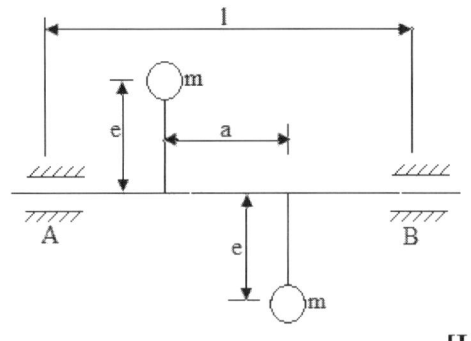

[IAS-2007]

IAS-9. Consider the following statements: [IAS-2003]
Two rotors mounted on a single shaft can be considered to be equivalent to a geared shaft system having two rotors provided.
1. The kinetic energy of the equivalent system is equal to that of the original system.
2. The strain energy of the equivalent system is equal to that of the original system.
3. The shaft diameters of the two systems are equal
Which of these statements are correct?
(a) 1, 2 and 3   (b) 1and 2   (c) 2 and 3   (d) 1 and 3

Balancing of Rigid Rotors and field Balancing

Chapter 5

IAS-10. Two rotors are mounted on a shaft. If the unbalanced force due to one rotor is equal in magnitude to the unbalanced force due to the other rotor, but positioned exactly 180⁰ apart, then the system will be balanced
(a) statically (b) dynamically [IAS-1999]
(c) statically as well as dynamically (d) neither statically nor dynamically

Balancing of several masses rotating in a same plane

Balancing of several masses rotating in different planes

IAS-12. The balancing weights are introduced in planes parallel to the plane of rotation of the disturbing mass. To obtain complete dynamic balance, the minimum number of balancing weights to be introduced in different planes is [IAS-2001]
(a) 1 (b) 2 (c) 3 (d) 4

Answers with Explanation (Objective)

Previous 20-Years GATE Answers

GATE-1. Ans. (d) Taking Moment about 'Q' $75 \times 2.0 = R \times 0.5$ or $R = 300$ kg

GATE-2. Ans. (c)
Along x-axis, $0.5(-60 \times 10^{-3} \cos 30° + 50 \times 10^{-3})\omega^2 = 0.1\omega^2 \times x \times 10^{-3}$
∴ $x = -9.8076$ mm
Along y-axis, $0.5(60 \times 10^{-3} \sin 30°)\omega^2 = 0.1\omega^2 y$
∴ $y = 150$ mm
∴ $r = \sqrt{x^2 + y^2} = 150.32$ mm

GATE-3. Ans. (b)
$I_1 = 10 \times (0.2)^2 = 0.4 \text{ kgm}^2$
$I_2 = 10 \times (0.2)^2 + 20 \times 0.1^2 = 0.6 \text{ kg-m}^2$
%Increase $= \dfrac{I_2 - I_1}{I_1} \times 100 = 50\%$

Previous 20-Years IES Answers

IES-1. Ans. (a)
IES-2. Ans. (b)
IES-3. Ans. (c) A system in dynamic balance implies that the system is also statically balanced.
IES-3(i). Ans. (d)
IES-4. Ans. (d)
IES-5. Ans. (c)
IES-6. Ans. (b)
IES-7. Ans. (c)
Moment about A

Balancing of Rigid Rotors and field Balancing

Chapter 5

$$\Rightarrow \frac{W}{g}e\omega^2 \times \frac{(L-2)}{2} - \frac{W}{g}e\omega^2\frac{(L+2)}{2}$$

$$+ R_B \times L = 0$$

$$L \times R_B = \frac{W}{g}\frac{e\omega^2}{2}\left[L+a-(L-a)\right]$$

$$R_B = \frac{W}{g}e\omega^2\frac{a}{L}$$

$$R_A = \frac{W}{g}e\omega^2\frac{a}{L}$$

Total dynamic force (magnitude) = $R_A + R_B$

$$= \frac{2W}{g}e\omega^2\frac{a}{L}$$

IES-8. Ans. (d)
IES-9. Ans. (b)
IES-10. Ans. (c)
IES-11. Ans. (a)
IES-12. Ans. (c)

Previous 20-Years IAS Answers

IAS-1. Ans. (a)
IAS-2. Ans. (c)
IAS-3. Ans. (d) A is false. The centre of gravity of the two masses should coincide with that of the rigid body.
IAS-4. Ans. (b)
IAS-5. Ans. (a) As centre of masses lie on the axis of rotation.
IAS-6. Ans. (a) 10×1 and 2×5 are balanced each other
Unbalance mass is 3 kg at 45^0
∴ Balanced system given in figure

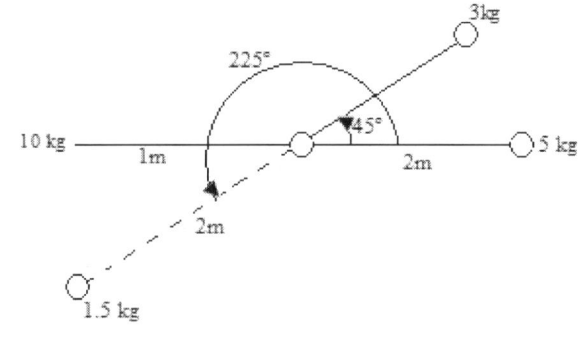

IAS-7. Ans. (b) An unbalance rigid rotor behaves as if several masses are there in different planes. Such a situation can be handled by fixing balancing weights in two planes.
IAS-8. Ans. (a)
IAS-9. Ans. (b)
IAS-10. Ans. (a)
IAS. 12. Ans. (b)

Balancing of Rigid Rotors and field Balancing
Chapter 5

6. Balancing of single and multi-cylinder engines

Theory at a glance (IES, GATE & PSU)

Balancing of Reciprocating Masses

In application like IC engine, reciprocating compressor and reciprocating pumps the reciprocating parts are subjected to continuous acceleration and retardation.

Due to this accelerate and retardation, the inertia force acts on the reciprocating parts which are in a direction opposite to the direction of acceleration. This inertia force is unbalanced dynamic force acting on the reciprocating parts.

Primary and Secondary Unbalanced Forces of Reciprocating masses

Consider a reciprocating engine mechanism as shown in Figure.

Let m = Mass of the reciprocating parts.
  l = length of the connecting rod PC
  r = radius of the crank OC.
  θ = Angle of inclination of the crank with the line of stroke PO.
  ω = Angular speed of the crank.
  n = Ratio of length of the connecting rod to the crank radius = l/r.

We have already discussed in Art. 15.8 that the acceleration, of the reciprocating parts is approximately given by the expression.

$$a_R = \omega^2 \cdot r \left(\cos\theta + \frac{\cos 2\theta}{n} \right)$$

∴ Inertia force due to reciprocating parts or force required to accelerate the reciprocating parts.

$$Fl = FR = Mass \times Acceleration = m \cdot \omega^2 \cdot r \left(\cos\theta + \frac{\cos 2\theta}{n} \right)$$

We have discussed in the previous article that the horizontal component of the force exerted on the crank shaft bearing (i.e. F_{BH}) is equal opposite to inertia force exerted and opposite to inertia force (F_I). This force is an unbalanced one and is denoted by F_V.

∴ Unbalanced force,
$$F_V = m \cdot \omega^2 \cdot r \left(\cos\theta + \frac{\cos 2\theta}{n} \right) = m \cdot \omega^2 \cdot r \cos\theta + m \cdot \omega^2 \cdot r \times \frac{\cos 2\theta}{n}.$$

Balancing of single and multi-cylinder engines

Chapter 6

The expression (m.ω².r cos θ) is known as primary unbalanced force and $\left(m.\omega^2.r \times \dfrac{\cos 2\theta}{n}\right)$ is called secondary unbalanced force.

∴ Primary unbalanced force, $F_P = m.\omega^2.r \cos\theta$

And secondary unbalanced force, $F_S = m.\omega^2.r \times \dfrac{\cos 2\theta}{n}$

Note: 1. The primary unbalanced force is maximum, when θ = 0° or 180°. Thus, the primary force is maximum twice in one revolution of the crank. The maximum Primary unbalanced force is given by

$$F_P = m.\omega^2.r$$

2. The secondary unbalanced force is maximum, when θ = 0°, 90° or 180° and 360°. Thus, the secondary force is maximum four times in one revolution of the crank. The maximum secondary unbalanced force is given by

$$F_S = m.\omega^2 \times \dfrac{r}{n}$$

Balancing of Reciprocating Mass in single cylinder engine

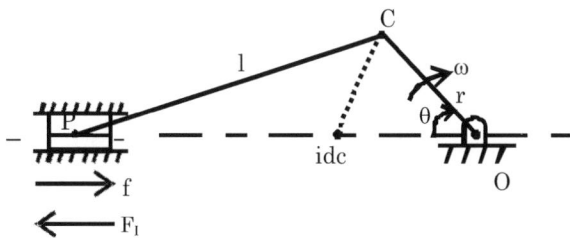

Let $\overline{m}$ = mass of reciprocating parts
 ω = angular speed of the crank
 r = radius of crank
 l = length of connecting rod
 n = obliquity ratio l/r
 θ = angle made by crank with i. d. c.
 f = acceleration of the reciprocating mass.
 F_I = inertia force due to reciprocating mass

Now
$$f = r\omega^2\left(\cos\theta + \dfrac{\cos 2\theta}{n}\right)$$

∴ $F_I = mr\omega^2\left(\cos\theta + \dfrac{\cos 2\theta}{n}\right)$

Hence unbalanced force
$$F_U = F_I = mr\omega^2\left(\cos\theta + \dfrac{\cos 2\theta}{n}\right)$$

$$= mr\omega^2\cos\theta + mr\omega^2\dfrac{\cos 2\theta}{n}$$

$$= F_P + F_S$$

Where F_P = primary unbalanced force
  = $mr\omega^2\cos\theta$
F_S = secondary unbalanced force

Balancing of single and multi-cylinder engines

Chapter 6

$$= mr\omega^2 \frac{\cos 2\theta}{n}$$

The primary unbalanced force is maximum when $\theta = 0°$ or $180°$ i.e. twice in one rotation of the crank.
The secondary unbalanced force is maximum. When $\theta = 0°, 90°, 180°$ and 270 i.e. four times in one rotating of the crank.

Thus, the frequency of secondary unbalanced force is twice as that of the primary unbalanced force. However, the magnitude of the secondary unbalanced force is $\frac{1}{n}$ times the of the primary unbalanced force. Therefore in case of low or moderate speed engines, the secondary unbalanced force is small and is generally neglected. But in case high speed engine the secondary unbalanced force is taken into account.

It is important to note that, in single cylinder reciprocating engines. All reciprocating and rotating masses are in the single plane. Hence, for complete balance only forces have to be balanced and the is no unbalanced couple.

The unbalanced force due to reciprocating mass varies in magnitude but constant in direction, while the unbalanced force due to rotating mass is constant in magnitude but varies in direction. Therefore a single rotating mass cannot be used to balance a reciprocating mass completely; however a single rotating mass can be used to partially balance the reciprocating mass.

The primary unbalanced force acting on the reciprocating engine is given by
$$F_P = mr\omega^2 \cos\theta$$

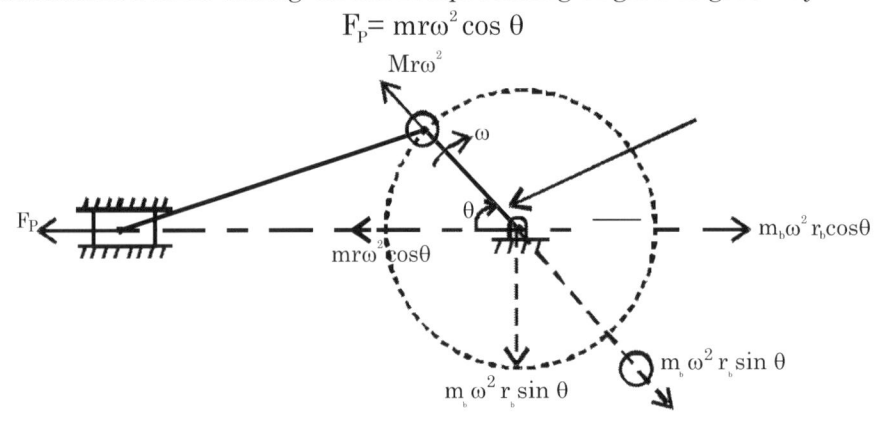

The primary unbalanced force may be treated as the component of centrifugal force along the line of stroke produced by an imaginary rotating mass m placed at crank pin _C'.
Now balancing can be done by attaching a balancing mass m_1 at radius r_b diametrically opposite to the cram.
Due to balancing mass m_b two component of force

(i) $m_b r_b \omega^2 \cos\theta$ Along the line of stroke.
(ii) $m_b r_b \omega^2 \sin\theta$ Perpendicular to line of stroke $m_b r_b \omega^2 \cos\theta$ is responsible for balancing the primary unbalanced force.

$$\therefore mr\omega^2 \cos\theta = m_b r_b \omega^2 \cos\theta$$

$$\boxed{mr = m_b r_b}$$

$m_b \omega^2 r_b \sin\theta$ is perpendicular to the line of stroke. This is the unbalanced force due to balancing of primary force (F_P).

Balancing of single and multi-cylinder engines
Chapter 6

In such case the maximum unbalanced force perpendicular to the line of stroke is $m_b r_b \omega^2$ which is same as the maximum primary unbalanced force.

Thus the effect of the above method of balancing is to change the direction of maximum unbalanced force.

In practice, a compromise is achieved by balancing only a portion of the primary unbalanced force.

The balancing mass m_b is placed diametrically opposite to the crank pin such that.

$$m_b r_b \omega^2 \cos\theta = cmr\omega^2 \cos\theta$$

$$\boxed{m_b r_b = cmr}$$

This is known as partial primary balancing. Now, unbalanced primary force due to partial balancing.

The unbalancing force along the line of stroke

$$F_H = m\omega^2 r \cos\theta - cmr\omega^2 \cos\theta$$
$$= (1-c) mr\omega^2 \cos\theta$$

The unbalanced force perpendicular to line of stroke

$$F_V = cmr\omega^2 \sin\theta$$

Resultant unbalanced force

$$F_R = \sqrt{F_H^2 + F_V^2}$$

$$\boxed{F_R = mr\omega^2 \sqrt{(1-c)^2 \cos^2\theta + c^2 \sin^2\theta}}$$

If the balancing mass m_b has to balance the rotating mass m_r rotating at radius r_r as well as give partial balancing of reciprocating part,

$$\therefore \boxed{m_b r_b = m_r r_r + cmr}$$

Balancing of reciprocating masses in multi cylinder inline engine

The multi cylinder engines having the axis of the entire cylinder in the same plane and on the same side of the axis of crank shaft, is known as incline engine.

Consider the multi cylinder inline engine having two inner cranks and two outer cranks.

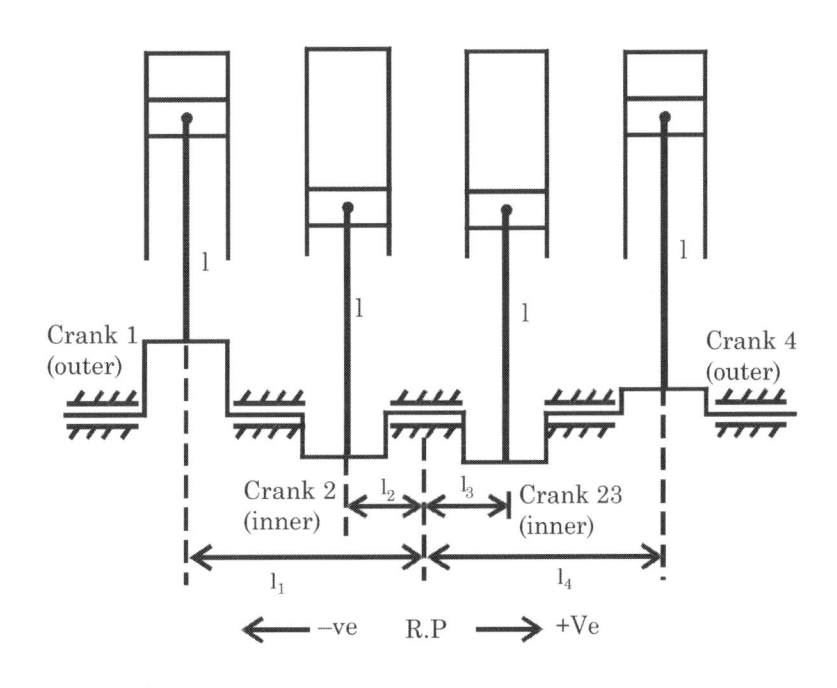

Balancing of single and multi-cylinder engines
Chapter 6

Thus unbalanced force due to reciprocating mass of each cylinder
$$F_p = mr\omega^2 \cos\theta$$
$$F_s = mr\omega^2 \frac{\cos\theta}{n}$$

Unbalanced couple
$$C_p = mr\omega^2 l \cos\theta$$
$$C_s = mr\omega^2 l \frac{\cos 2\theta}{n}$$

Analytical Method:
For complete primary balancing

(i) $\sum mr \cos\theta = 0$

(ii) $\sum mr \sin\theta = 0$

(iii) $\sum mrl \cos\theta = 0$

(iv) $\sum mrl \sin\theta = 0$

Up to three cylinder inline engines, complete primary balancing is not possible. However, for multi cylinder inline with four or more cylinders, complete primary balancing can be achieved by suitably arranging the cranks.

Secondary Balancing

(i) $\sum mr\omega^2 \frac{\cos 2\theta}{n} = 0$

(ii) $\sum mrw^2 l \frac{\cos 2\theta}{n} = 0$

The above condition can be written as

(i) $\sum m(2\omega)^2 \left(\frac{r}{4n}\right) \cos 2\theta = 0$

(ii) $\sum m(2\omega)^2 \left(\frac{r}{4n}\right) l \cos 2\theta = 0$

This condition are equivalent to the condition of primary balancing for an imaginary crank of length $\left(\frac{r}{4n}\right)$, rotating at speed 2ω and inclined at an angle 2θ to idc. This imaginary crank is known as secondary crank.

To ensure that the above two conditions are satisfied for all angular positions of the crank shaft. The above condition are modified as

(i) $\sum \frac{mr}{n} = 0$

(ii) $\sum \frac{mrl}{n}$

Analytical Method

For complete secondary balancing:

(i) $\sum \frac{mr}{n} \cos 2\theta = 0$

Balancing of single and multi-cylinder engines
Chapter 6

(ii) $\sum \dfrac{mr}{n} \sin 2\theta = 0$

(iii) $\sum \dfrac{mrl}{n} \cos 2\theta = 0$

(iv) $\sum \dfrac{mrl}{n} \sin 2\theta = 0$

If n is same for all cylinder.

(i) $\sum mr \cos 2\theta = 0$

(ii) $\sum mr \sin 2\theta = 0$

(iii) $\sum mrl \cos 2\theta$

(iv) $\sum mrl \sin 2\theta = 0$

Balancing of V-engine

V-engine is two cylinder radial engine in which the connecting rod are fixed to common crank.

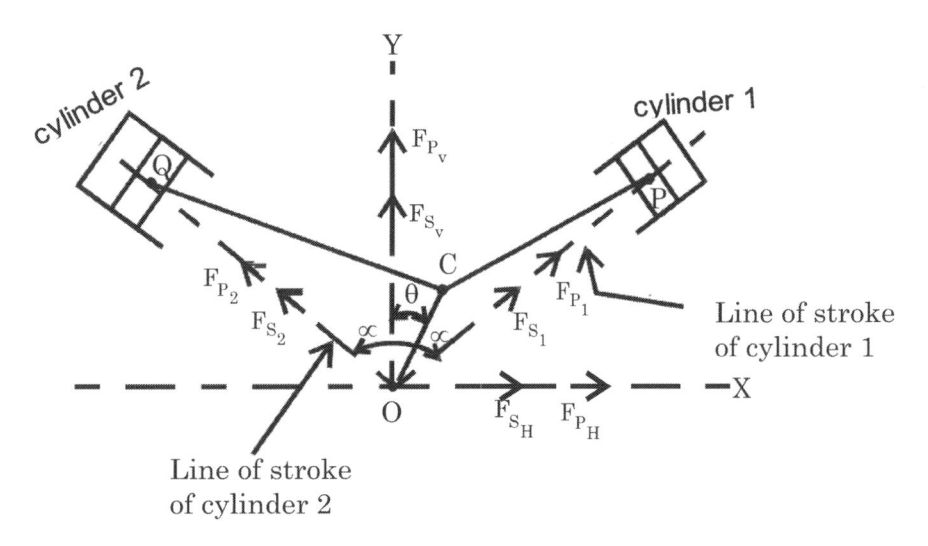

2α = V– angle

Primary Forces:

The primary unbalanced force acting along the line of stroke of cylinder 1 is
$$F_{P_1} = mr\omega^2 \cos(\alpha - \theta)$$

Similarly,
$$F_{P_2} = mr\omega^2 \cos(\alpha + \theta)$$

Total primary force along vertical axis OY is
$$F_{P_V} = F_{P_1} \cos\alpha + F_{P_2} \cos\alpha$$
$$= 2mr\omega^2 \cos^2\alpha \cos\theta$$

Total primary force along horizontal line OX is
$$F_{P_H} = F_{P_1} \sin\alpha - F_{P_2} \sin\alpha$$
$$F_{P_H} = 2mr\omega^2 \sin^2\alpha \sin\theta$$

∵ Resultant primary force
$$F_P = \sqrt{(F_{PV})^2 + (F_{PH})^2}$$
$$= 2mr\omega^2 \sqrt{(\cos^2\alpha \cos\theta)^2 + (\sin^2\alpha \sin\theta)^2}$$

Balancing of single and multi-cylinder engines

Chapter 6

The angle made by resultant force F_P with vertical axis OY (measured in clockwise direction)

$$\beta_P = \tan^{-1}\left(\frac{F_{PH}}{F_{PV}}\right)$$

Secondary Forces:

$$F_{S_1} = mr\omega^2 \frac{\cos 2(\alpha - \theta)}{n}$$

$$F_{S_2} = mr\omega^2 \frac{\cos 2(\alpha + \theta)}{n}$$

$$F_{S_V} = F_{S_1} \cos\alpha + F_{S_2} \cos\alpha$$

$$= \frac{2}{n} mr\omega^2 \cos\alpha \cos 2\alpha \cos 2\theta$$

$$F_{SH} = F_{S_1} \sin\alpha - F_{S_2} \sin\alpha$$

$$= \frac{2}{n} mr\omega^2 \sin\alpha \sin 2\alpha \sin 2\theta$$

Resultant secondary forces

$$F_S = \sqrt{(F_{SH})^2 + (F_{SV})^2}$$

and angle

$$\beta_S = \tan^{-1} \frac{F_{SH}}{F_{SV}}$$

Direct and Reverse Crank

In a radial engine and V-engine all the connecting rods are connected to a common crank and this crank revolves in one plane. Hence there is no couple.

The Primary force

$F_P = mr\omega^2 \cos\theta$ can be imagining as.

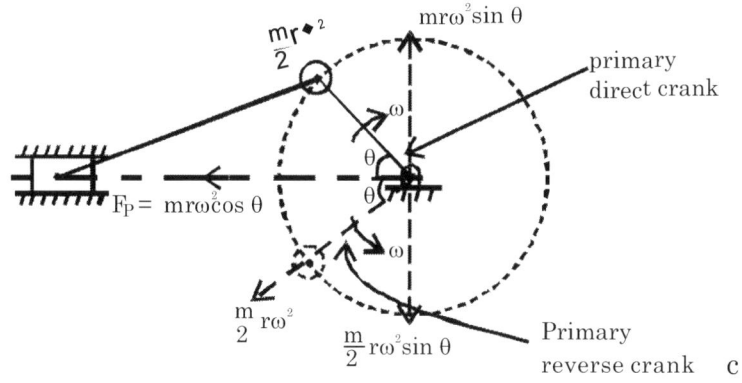

Secondary Force

$F_S = mr\omega^2 \dfrac{\cos 2\theta}{n}$ can be assumed as

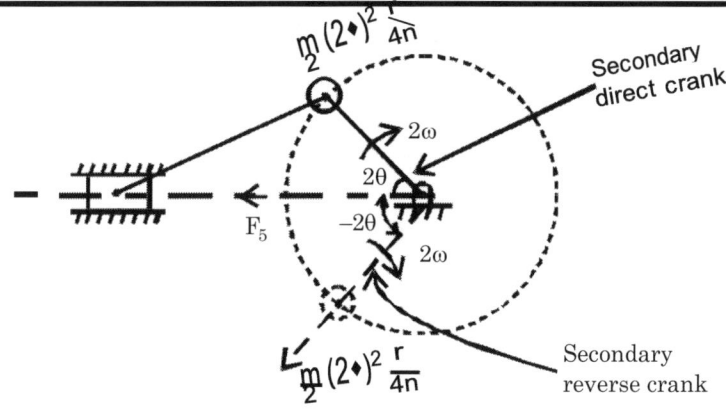

Balancing of single and multi-cylinder engines
Chapter 6

Partial balancing of locomotive:
The locomotive usually have two cylinder with crank placed at right angles to each other. Previous are see that the reciprocating parts are only partially balanced. Due to this partial balancing of reciprocating parts there is an unbalanced primary force along the line of stroke and also an unbalanced primary force perpendicular to the line of stroke. The effect of an unbalanced primary force along the line of stroke is to produce.
(i) Variation in tractive force.
(ii) Swaying couple

The maximum magnitude of the unbalanced force along the perpendicular to the line of stroke is known as hammer blow.

(i) Variation of tractive force
The resultant unbalanced force due to the two cylinders, along the line of stroke is known as **tractive force**. Let the crank of the for the first cylinder be inclined at an angle θ with the line of stroke. Since the crank for the second cylinder is at right angle to the first crank. Therefore the angle of inclination for the second crank will be (90 +θ).
   Let m = Mass of reciprocating parts per cylinder.
   C = Fraction of the reciprocating parts to be balanced.
∴ Unbalanced force along the line of stroke for cylinder 1
   $= (1 - C) \, mr\omega^2 \cos\theta$
Similarly for 2nd cylinder $= (1 - C) \, mr\omega^2 \cos(90 + \theta)$
∴ Resultant unbalanced force along the line of stroke (F_T) =
   $(1 - C) \, mr\omega^2 \cos\theta + (1 - C) \, mr\omega^2 \cos(90 + \theta)$

   $\boxed{F_T = (1 - C)mr\omega^2 (\cos\theta - \sin\theta)}$

F_T will be maximum or minimum at
$\frac{d}{d\theta} F_T = 0$ $F_{T_{max}} = \pm\sqrt{2}(1 - C)mrw^2$.

∴ Maximum or minimum F_T at θ = 135° or 315°

(ii) Swaying Couple
The unbalanced force along the line of stroke for two cylinder constitute a couple about the centre line y – y between the cylinder. This is known as swaying couple.

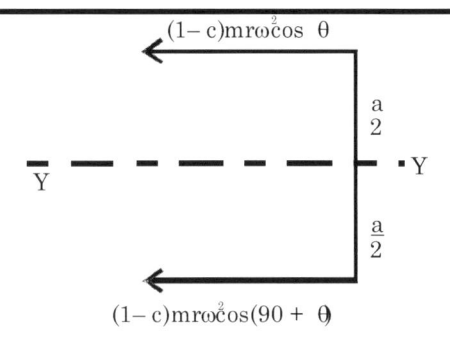

Balancing of single and multi-cylinder engines

Chapter 6

$$\text{Swaying couple} = (1-C)\,mr\omega^2 \cos\theta \times \frac{a}{2} - (1-C)\,mr\omega^2 \cos(\theta+90) \times \frac{a}{2}$$

$$= (1-C)\,mr\omega^2 \times \frac{a}{2}(\cos\theta + \sin\theta)$$

It will be maximum at $v = 45°$ or $225°$.

$$\text{Max value} = \pm \frac{1}{\sqrt{2}}(1-C)\,mr\omega^2 a$$

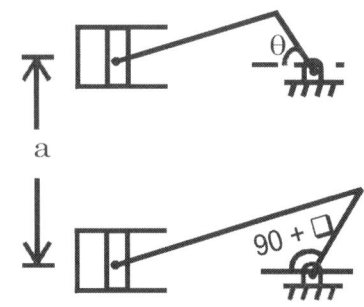

Hammer Blow

We have already discussed that the maximum magnitude of the unbalanced force along the perpendicular to the line of stroke is known as hammer blow.

D-Alembert's Principle

Thus D-Alembert's principle states that the resultant force acting on a body together with the reversed effective force (or inertia force), are in equilibrium.
This principle is used to reduce a dynamic problem into an equivalent static problem.

Velocity and Acceleration of the Reciprocating Parts in Engines

The velocity and acceleration of the reciprocating parts of the steam engine or internal combustion engine (briefly called as I.C. engine) may be determined by graphical method or analytical method. The velocity and acceleration, by graphical method, may be determined by one of the following constructions:
1. Klien's construction .2.Ritterhaus's construction, and 3. Bennett's construction.

Klien's Construction

Balancing of single and multi-cylinder engines

Chapter 6

Let *OC* be crank and *PC* the connecting rod of a reciprocating stream engine, as shown in Figure. Let the crank marks an angle θ with the line of stroke *PO* and rotates with uniform

Angular velocity ω rad /s in a clockwise direction. The Klien's velocity and acceleration diagrams are drawn discussed below:

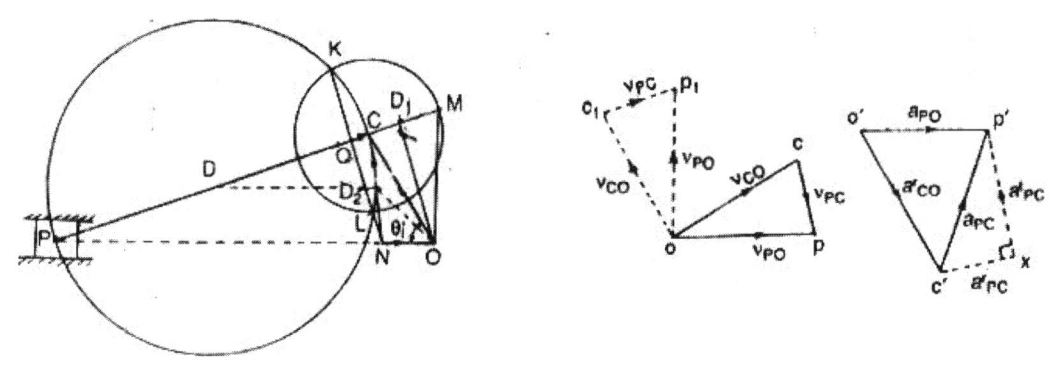

$a_{PC} = \omega^2 \times CN$

Velocity of D, $\quad V_D = \omega \times OD_1$

∴ Acceleration of D, $a_D = \omega^2 \times OD_2$

Q. The bearing of a shaft A and B are 5 m apart. The shaft carries three eccentric masses C, D and E. Which are 160 kg, 170 kg and 85 kg respectively. The respective eccentricity of each masses measured from axis of rotation is 0.5 cm, 0.3 cm and 0.6 cm and distance from A is 1.3 m, 3m and 4m respectively. Determine angular position of each mass with respect to C so that no dynamic force is exerted at B and also find dynamic force at A for this arrangement when the shaft runs at 100 rpm. [IES-2002]

Solution:

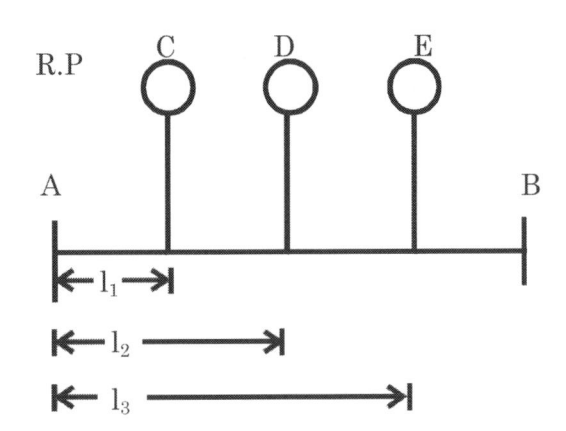

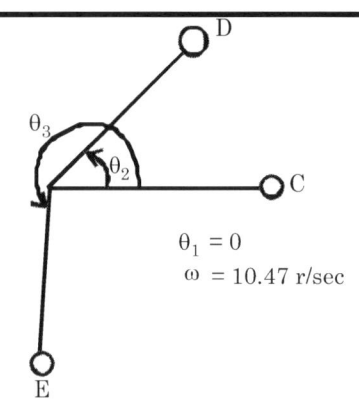

Balancing of single and multi-cylinder engines

Chapter 6

$m_1 = 160$ kg
$m_2 = 170$ kg
$m_3 = 85$ kg
$r_1 = 0.5$ cm
$r_2 = 0.3$ cm
$r_3 = 0.6$ cm
$\theta_1 = 0$
$\omega = 10.47$ r/sec

Couple balancing

X-comp

$m_1 r_1 l_1 \cos\theta_1 + m_2 r_2 l_2 \cos\theta_2 + m_3 r_3 l_3 \cos\theta_2 + 0 = 0$

$104 + 153 \cos\theta_2 + 204 \cos\theta_3 = 0$ ...(i)

Y-comp

$m_1 r_1 l_1 \sin\theta_1 + m_2 r_2 l_2 \sin\theta_2 + m_3 r_3 l_3 \sin\theta_3 = 0$

$153 \sin\theta_2 + 204 \sin\theta_3 = 0$ ... (ii)

Squaring and adding

$\cos(\theta_3 - \theta_2) = -0.868$

$\theta_3 - \theta_2 = 150.2°$ ... (iii)

Putting in (ii)

$\tan\theta_2 = \dfrac{101.18}{24.15} \Rightarrow \theta_2 = 76.6°$

$\theta_3 = 226.85°$

Let the dynamic force at A is F

∴ Force balance

$F\cos\theta + m_1 r_1 \cos\theta\, \omega^2 + m_2 r_2 \omega^2 \cos\theta_2 + m_3 r_3 \omega^2 \cos\theta_3 = 0$

$F \cos\theta = -62.41$

y-comp

$F\sin\theta + m_1 r_1 \sin\theta\, \omega^2 + m_2 r_2 \omega^2 \sin\theta_2 + m_3 r_3 \omega^2 \sin\theta_3 = 0$

$F \sin\theta = -13.6$

∴ $F = 63.88$ N

$\tan\theta = \dfrac{-13.6}{-62.41}$

$\theta = 192.3°$ (with C)

Q. The following data relate to a single cylinder reciprocating engine.
Mass of reciprocating part = 40 kg
Mass of revolving parts = 30 kg at crank pin

Balancing of single and multi-cylinder engines

Chapter 6

Speed = 150 rpm
Stroke = 350 mm
If 60% of the reciprocating parts and all the revolving parts are to be balanced, determine the
(i) Balanced mass required at a radius of 320 mm.
(ii) Unbalanced force when the crank has turned 45° from the top. Dead centre.

Solution: $\omega = 15.7$ rad/sec.
$r = 175$ mm

(i) Mass to be balanced at crank pin
$M = 0.6 \times 40 + 30$
$= 54$ kg

$\therefore \quad m_b \cdot r_b = m \cdot r$
$m_b \cdot 320 = 54 \times 175$
$m_b = 29.53$ kg.

(ii) Unbalanced force at $\theta = 45°$
$= \sqrt{[(1-C) mr\omega^2 \cos\theta]^2 + [cmr \omega^2 \sin\theta]^2}$

$= \sqrt{[(1-0.6) \times 40 \times 0.175 \times 15.7^2 \cos 45]^2 [0.6 \times 40 \times 0.175 \times 15.7)^2 \sin 45°]^2}$
$= 880.7$ N

Q. The following data refer to two cylinder locomotive with crank 90°.
Reciprocating mass per cylinder = 300 kg
Crank radius = 0.3 m
Driving wheel dia. = 1.8 m
Dist. between cylinder centre lines = 0.65 m
Dist. between the driving wheel = 1.55 m
Determine
(i) The fraction of the reciprocating masses to be balanced, if the hammer blow is not to exceed 46×10^3 N at 96.5 km/hr.
(ii) The variation in tractive effort
(iii) The maximum swaying couple.

Solution: m = 300 kg, r = 0.3 m, D = 1.8 m.
a = 0.65 m, Hammer blow = 46×10^3 N,
V = 96.5 km/hr = 26.8 m/sec. $\therefore$ w = 29.78 r/sec.
Let C = fraction of reciprocating masses to be balanced
$\therefore$ Mass to be balanced = 300C kg.
Let M_b = balancing mass placed at each driving wheel at a radius r_b.

Balancing of single and multi-cylinder engines

Chapter 6

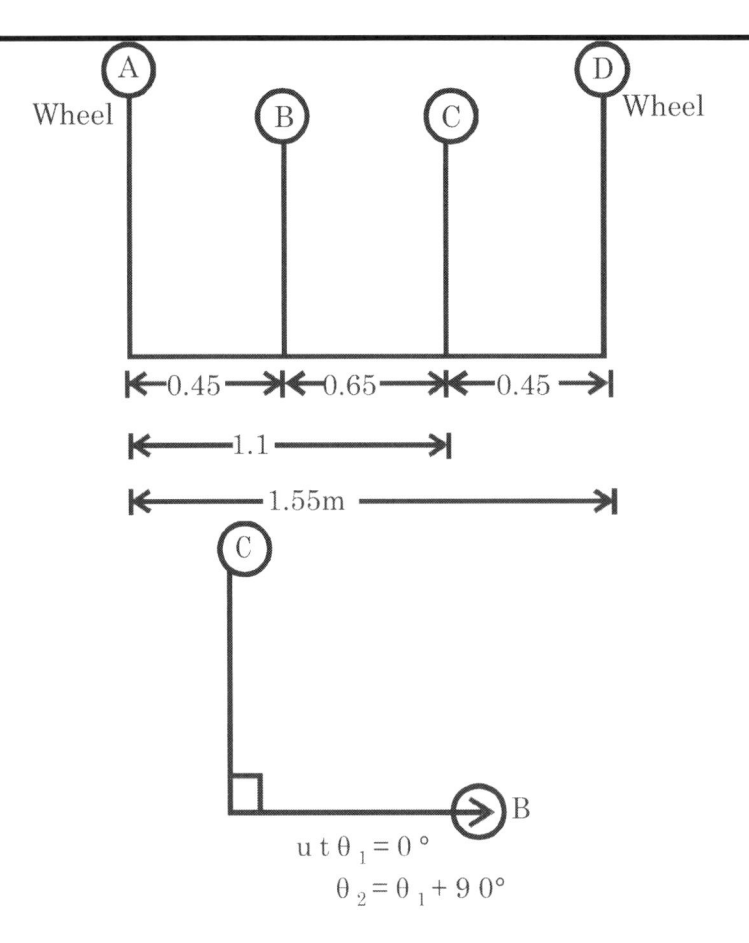

For couple balance

X-comp:

$m_1 r_1 l_1 \cos\theta_1 + m_2 r_2 l_2 \cos\theta_2 + m_b r_b l_b \cos\theta = 0$

$300 C \cdot 0.3 \cdot 0.45 + 0 + m_b r_b \cdot 1.55 \cos\theta = 0$

$\quad m_b r_b \cos\theta = -26.13 C$ ... (i)

Y-comp:

$m_1 r_1 l_1 \sin\theta_1 + m_2 r_2 l_2 \sin\theta_2 + m_b r_b l_b \sin\theta = 0$

$0 + 300 C \times 0.3 \times 1.1 + m_b r_b \times 1.55 \sin\theta = 0$

$\therefore m_b r_b \sin\theta = -63.87 C$ ... (ii)

From (i) and (ii)

VIBRATORY MOTIONS in machinery arise when variable forces act on elastic parts. Usually these motions are undesirable although in some cases (vibratory conveyors, for example) they are deliberately designed into the machine.

ANALYSING VIBRATIONS requires the following general procedure:
1. Evaluating masses and elasticity of parts involved.
2. Evaluating amount of friction involved.
3. Idealizing the actual mechanical device, replacing it by an approximately equivalent system of masses, springs, and dampers.
4. Writing differential equations of motion for the idealized system.
5. Solving the equations and interpreting the results.

$$m_b r_b = 69 \, C$$

Now

(i) Hammer blow = $m_b r_b \omega^2$

Balancing of single and multi-cylinder engines

Chapter 6

$$46 \times 10^3 = 69 \, C \times 29.78^2$$

∴ $\boxed{C = 0.75}$

(ii) Variation of tractive effort
$$= \pm \sqrt{2}(1 - C) \, mr\omega^2$$
$$= \pm \sqrt{2}(1 - 0.75) \times 300 \times 0.3 \times 29.78^2$$
$$= 28.14 \text{ kN}$$

(iii) Maximum swaying couple $= \dfrac{1}{\sqrt{2}}(1 - C) \, mr\omega^2$

Objective Questions (IES, IAS, GATE)

Previous 20-Years GATE Questions

GATE-1. Consider the triangle formed by the connecting rod and the crank of an IC engine as the two sides of the triangle. If the maximum area of this triangle occurs when the crank angle is 75°, the ratio of connecting rod length to crank radius is [GATE-1998]
(a) 5   (b) 4   (c) 3.73   (d) 3

Primary unbalanced forces

GATE-2. Match 4 correct pairs between list I and List II for the questions

List I	List II
(a) Collision of bodies	1. Kinetics
(b) Minimum potential energy	2. Reciprocating unbalance
(c) Degree of freedom	3. Dynamics
(d) Prony brake	4. Coefficient of restitution
(e) Hammer blow	5. Stability [GATE-1994]

Previous 20-Years IES Questions

D-Alembert's Principle

IES-1. Assertion (A): The supply of fuel is automatically regulated by governor according to the engine speed. [IES-2001]
Reason (R): The automatic function is the application of d' Alembert's principle.
(a) Both A and R are individually true and R is the correct explanation of A
(b) Both A and R are individually true but R is **not** the correct explanation of A
(c) A is true but R is false
(d) A is false but R is true

Balancing of single and multi-cylinder engines

Chapter 6

Klein's Construction

IES-2. The given figure shows the Klein's construction for acceleration of the slider-crank mechanism Which one of the following quadrilaterals represents the required acceleration diagram?
  (a) ORST   (b) OPST
  (c) ORWT   (d) ORPT

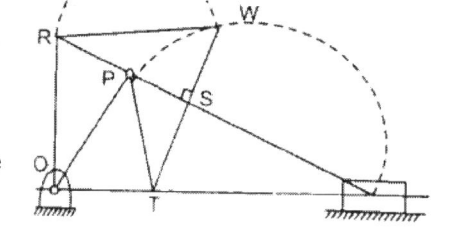

[IES-2001]

IES-3. The Klein's method of construction for reciprocating engine mechanism.
  (a) is a simplified version of instantaneous centre method   [IES-1994]
  (b) utilizes a quadrilateral similar to the diagram of mechanism for reciprocating engine
  (c) enables determination of Corioli' s component.
  (d) is based on the acceleration diagram.

IES-4. Figure shows Klein's construction for slider-crank mechanism OCP drawn to full scale. What velocity does CD represent?
  (a) Velocity of the crank pin
  (b) Velocity of the piston
  (c) Velocity of the piston with respect to crank pin
  (d) Angular velocity of the connecting rod

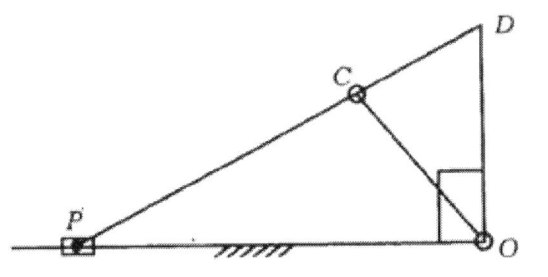

[IES-2003]

IES-5. Klein's construction for determining the acceleration of piston P is shown in the given figure. When N coincides with K
  (a) acceleration of piston is zero and its velocity is zero.
  (b) acceleration is maximum and velocity is maximum.
  (c) acceleration is maximum and velocity is zero
  (d) acceleration is zero and velocity is maximum.

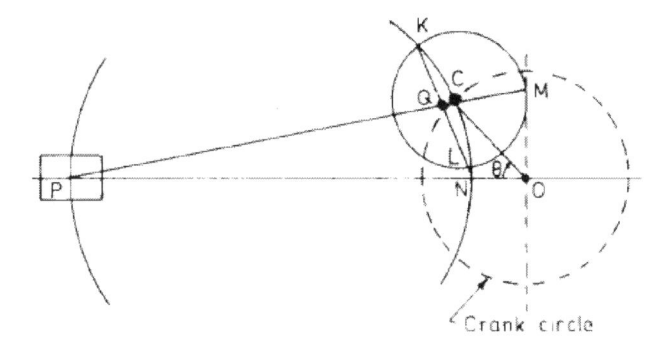

[IES-1995]

Velocity and Acceleration of the Piston

IES-6. For a slider-crank mechanism with radius of crank r, length of connecting rod I, obliquity ratio n, crank rotating at an angular velocity ω; for any angle θ of the crank, match List-I (Kinematic Variable) with List-II (Equation) and select the correct answer using the codes given below-the Lists: [IES-2003]

List-I
Kinematic Variable)

List II
(Equation)

Balancing of single and nulti-cylinder engines

Chapter 6

A. Velocity of piston	1.	$\dfrac{\omega}{n}.\cos\theta$
B. Acceleration of piston	2.	$\omega^2 r.\left(\cos\theta + \dfrac{\cos 2\theta}{n}\right)$
C. Angular velocity of connecting rod	3.	$-\dfrac{\omega^2}{n}.\sin\theta$
D. Angular acceleration of connecting rod	4.	$\omega r.\left(\sin\theta + \dfrac{\sin 2\theta}{2n}\right)$

Codes: A   B   C   D   A   B   C   D
(a)   4   2   3   1   (b)   2   4   3   1
(c)   4   2   1   3   (d)   2   4   1   3

IES-7. The above figure shows the schematic diagram of an IC engine producing a torque T = 40 N-m at the given instant. The Coulomb friction coefficient between the cylinder and the piston is 0.08. If the mass of the piston is 0.5 kg and the crank radius is 0.1 m, the Coulomb friction force occurring at the piston cylinder interface is
(a) 16 N   (b) 0.4 N
(c) 4 N   (d) 16.4 N

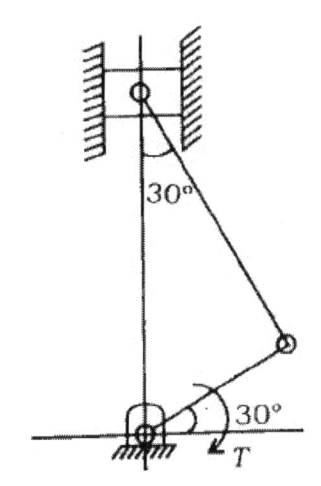

[IES-2003]

IES-8. In a slider-crank mechanism the maximum acceleration of slider is obtained when the crank is   [IES-2001]
(a) at the inner dead centre position
(b) at the outer dead centre position
(c) exactly midway position between the two dead centers
(d) slightly in advance of the midway position between the two dead centers

IES-9. Match List-I with List-II and select the correct answer using the code given below the lists:

List-I
(Corresponding Application)
A. Klein's construction
B. Kennedy's theorem
C. Alembert's principle
D. Grubler's rule

List-II
(Principle/Method)
1. Instantaneous centres in linkages
2. Relative acceleration of linkages
3. Mobility of linkages
4. Dynamic forces in linkages   IES-2008]

Code: A   B   C   D
(a)   4   1   2   3
(b)   2   3   4   1
(c)   4   3   2   1
(d)   2   1   4   3

Balancing of single and multi-cylinder engines

Chapter 6

Angular velocity and acceleration of connecting rod

IES-10. In a slider-bar mechanism, when does the connecting rod have zero angular velocity? [IES 2007]
(a) When crank angle = 0°
(b) When crank angle = 90°
(c) When crank angle = 45°
(d) Never

IES-11. In a reciprocating engine mechanism, the crank and connecting rod of same length r meters are at right angles to each other at a given instant, when the crank makes an angle of 45° with IDC. If the crank rotates with a uniform velocity of ω rad/s, the angular acceleration of the connecting rod will be [IAS-1999]

(a) $2\omega^2 r$
(b) $\omega^2 r$
(c) $\dfrac{\omega^2}{r}$
(d) zero

Forces on the reciprocating parts of an engine

IES-12. With reference to the engine mechanism shown in the given figure, match List I with List II and select the correct answer

List I	List II
A. F_Q	1. Inertia force of reciprocating mass
B. F_R	2. Inertia force of connecting rod
C. F_W	3. Crank effort
D. F_C	4. Piston side thrust

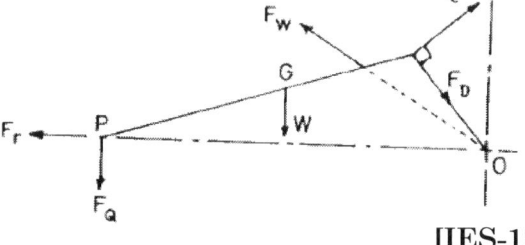

[IES-1996]

Code: A B C D   A B C D
(a)   1 2 4 3   (b)   1 2 3 4

IES-13. Consider the following statements for a 4-cylinder inline engine whose cranks are arranged at regular intervals of 90°: [IES-2005]
1. There are 8 possible firing orders for the engine.
2. Primary force will remain unbalanced for some firing orders.
Which of the statements given above is/are correct?
(a) 1 only
(b) 2 only
(c) Both 1 and 2
(d) Neither 1 nor 2

IES-14. Which one of the following statements in the context of balancing in engines is correct? [IES-2004]
(a) Magnitude of the primary unbalancing force is less than the secondary unbalancing force
(b) The primary unbalancing force attains its maximum value twice in one revolution of the crank
(c) The hammer blow in the locomotive engines occurs due to unbalanced force along the line of stroke of the piston
(d) The unbalanced force due to reciprocating masses varies in magnitude and direction

IES-15. In case of partial balancing of single-cylinder reciprocating engine, what is the primary disturbing force along the line of stroke? [IES-2006]
(a) $cmr\omega^2 \cos\theta$
(b) $(1-c^2)mr\omega^2 \cos\theta$

Balancing of single and multi-cylinder engines

Chapter 6

(c) $(1-c)mr\omega^2 \cos\theta$ (d) $(1-c)mr\omega^2 \cos 2\theta$

Where, c = Fraction of reciprocating mass to be balanced; ω = Angular velocity of crankshaft; θ = Crank angle.

IES-16. The primary disturbing force due to inertia of reciprocating parts of mass m at radius r moving with an angular velocity ω is given by [IES-1999]
(a) $m\omega^2 r \sin\theta$ (b) $m\omega^2 r \cos\theta$ (c) $m\omega^2 r \sin\left(\dfrac{2\theta}{n}\right)$ (d) $m\omega^2 r \left(\dfrac{2\theta}{n}\right)$ Ans. (b)

IES-17. A four-cylinder symmetrical in-line engine is shown in the given figure. Reciprocating weights per cylinder are R_1 and R_2, and the corresponding angular disposition of the crank are α and β.

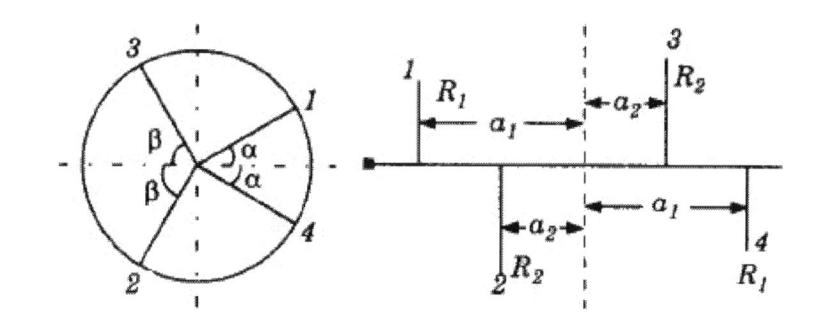

Which one of the following equations should be satisfied for its primary force balance?   [IES-1998]

(a) $a_1 \tan\alpha = a_2 \tan\beta$ (b) $\cos\alpha = \dfrac{1}{2}\sec\beta$

(c) $R_1 a_1 \sin 2\alpha = -R_2 a_2 \sin 2\beta$ (d) $a_1 \cos\alpha = R_2 \cos\beta$

Secondary unbalanced forces

IES-18. If the ratio of the length of connecting rod to the crank radius increases, then   [IES-1999]
(a) Primary unbalanced forces will increase
(b) Primary unbalanced forces will decrease
(c) Secondary unbalanced forces will increase
(d) Secondary unbalanced forces will decrease

IES-19. A single cylinder, four-stroke I.C. engine rotating at 900 rpm has a crank length of 50 mm and a connecting rod length of 200 mm. If the effective reciprocating mass of the engine is 1.2 kg, what is the approximate magnitude of the maximum 'secondary force' created by the engine?
(a) 533 N   (b) 666 N   [IES-2005]
(c) 133 N   (d) None of the above

IES-20. A four-cylinder in-line reciprocating engine is shown in the diagram given below. The cylinders are numbered 1 to 4 and the firing order is 1-4-2-3:   [IES-2004]

Balancing of single and multi-cylinder engines

Chapter 6

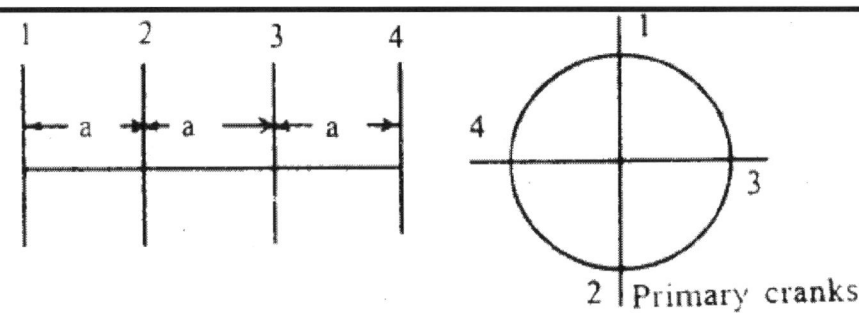

Which one of the following statements is correct?
(a) Both primary and secondary forces are balanced
(b) Only primary force is balanced
(c) Only secondary force is balanced
(d) Both primary and secondary forces are unbalanced

IES-23. Assertion (A): For a radial engine containing four or more cylinders, the secondary forces are in complete balance,
Reason (R): The secondary direct and reverse cranks form a balanced system in the radial engines. [IES-2000]
(a) Both A and R are individually true and R is the correct explanation of A
(b) Both A and R are individually true but R is **not** the correct explanation of A
(c) A is true but R is false
(d) A is false but R is true

IES-24. In a multi-cylinder in-line internal combustion engine, even number of cylinders is chosen so that [IES-1998]
(a) uniform firing order is obtained   (b) the couples are balanced
(c) primary forces are balanced   (d) secondary forces are balanced

IES-25. When the primary direct crank of a reciprocating engine is positioned at 30° clockwise, the secondary reverse crank for balancing will be at
(a) 30° anticlockwise   (b) 60° anticlockwise   [IES-1997]
(c) 30° clockwise   (d) 60° clockwise

Hammer Blow

IES-26. Which of the following pair(s) is/are correctly matched? [IES-1998]
I. Four bar chain… …………………….Oscillating - Oscillating converter
II. Inertia governor …………………….Rate of change of engine speed
III. Hammer blow…………………….Reciprocating unbalance.
Select the correct answer using the codes given below:
Codes:
(a) I alone   (b) I, II and III   (c) II and III   (d) I and III

IES-27. Assertion (A): In locomotive engines, the reciprocating masses are only partially balanced. [IES-1999]
Reason (R): Full balancing might lead to lifting the locomotive engine off the rails.
(a) Both A and R are individually true and R is the correct explanation of A
(b) Both A and R are individually true but R is **not** the correct explanation of A
(c) A is true but R is false
(d) A is false but R is true

Balancing of single and multi-cylinder engines

Chapter 6

Previous 20-Years IAS Questions

IAS-1. If s, v, t, .F, m and a represent displacement, velocity, time, force, mass and acceleration respectively, match List I (Expression) with List II (Feature / Principle) and select the correct answer using the codes given below the lists: [IAS-2003]

List-I
(Expression)
(a) v = 6t² - 9t
(b) v = 9t + 12
(c) s = 4t
(d) F- ma = 0

List-II
(Feature/Principle)
1. Constant acceleration
2. Variable acceleration
3. D' Alembert's principle
4. Uniform motion

Codes: A   B   C   D   A   B   C   D
    (a)   2   1   4   3   (b)   4   3   2   1
    (c)   2   3   4   1   (d)   4   1   2   3

IAS-2. Assertion (A): d' Alembert's principle is known as the principle of dynamic equilibrium.
Reason(R): d' Alembert's principle converts a dynamic problem into a static Problem. [IAS-2000]
(a) Both A and R are individually true and R is the correct explanation of A
(b) Both A and R are individually true but R is **not** the correct explanation of A
(c) A is true but R is false
(d) A is false but R is true

IAS-3. Consider the following statements:
The Klein's construction for slider crank mechanism with crank rotating at constant angular velocity provides values of [IAS-1998]
1. Piston velocity.
2. Piston acceleration.
3. Normal acceleration of crank pin Of these statements:
4. Angular acceleration of the connecting rod.
(a) 1 and 2 are correct
(b) 1, 2, 3 and 4 are correct
(c) 1, 2 and 4 are correct
(d) 3 and 3 are correct

Velocity and Acceleration of the Piston

IAS-4. In the figure given above, when is the absolute velocity of end B of the coupler equal to the absolute velocity of the end A of the coupler?
(a) θ₂ = 90°   (b) θ₂ = 45°
(c) θ₂ = 0°   (d) Never

IAS-2007]

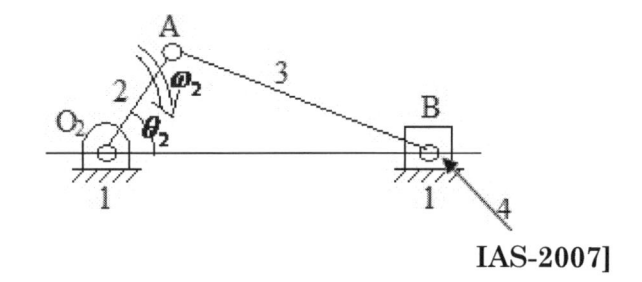

IAS-5. In a reciprocating engine mechanism, the crank and connecting rod of same length r meters are at right angles to each other at a given instant, when the crank makes an angle of 45° with IDC. If the crank rotates with a uniform velocity of ω rad/s, the angular acceleration of the connecting rod will be [IAS-1999]

Balancing of single and multi-cylinder engines
Chapter 6

(a) $2\omega^2 r$ (b) $\omega^2 r$ (c)(c) $\dfrac{\omega^2}{r}$ (d) zero

Forces on the reciprocating parts of an engine

IAS-6. A slider crank mechanism is shown in the given figure.

1. $F_Q.\sin(\theta+\phi)$ 2. $F_S.\sin\theta + \left(\dfrac{\sin 2\theta}{n}\right)$

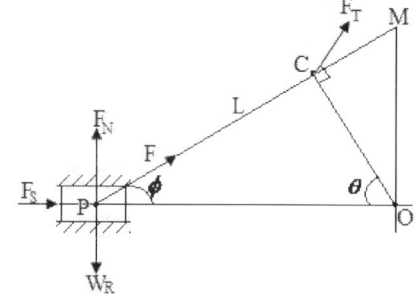

3. $F_S.OM$ 4. $F_T.r$

[IAS-1996]

Which of the following expressions stand for crank effort?
Select the correct answer using the codes given below:
Codes:
(a) 1 and 3   (b) 1, 2 and 4   (c) 1, 2 and 3   (d) 2, 3 and 4

Primary unbalanced forces

IAS-7. In reciprocating engines primary forces   [IAS-1996]
(a) are completely balanced   (b) are partially balanced
(c) are balanced by secondary forces   (d) cannot be balanced

IAS-8. The primary direct crank of a reciprocating engine is located at an angle θ clockwise. The secondary direct crank will be located at an angle
(a) 2θ clockwise   (b) 2θ anticlockwise   (c) θ clockwise   (d) θ anticlockwise
[IAS-1999]

Secondary unbalanced forces

IAS-9. Consider the following statements: [IAS-1998]
An in-line four-cylinder four-stroke engine is completely balanced for
1. primary forces   2. secondary forces
3. primary couples   4. secondary couples
Of these statements:
(a) 1, 3 and 4 are correct   (b) 1, 2 and 4 are correct
(c) 1 and 3 are correct   (d) 2 and 4 are correct

IAS-10. An in-line four-cylinder four-stroke engine is balanced in which of the following?   [IAS-1997]
1. Primary forces.   2. Primary couples
3. Secondary forces.   4. Secondary couples
Select the correct answer using the codes given below:
Codes:
(a) 1 and 4   (b) 2, 3 and 4   (c) 1 and 2   (d) 1, 2 and 4

Partial balancing Primary unbalanced forces

IAS-11. The method of direct and reverse cranks is used in engines for   [IAS-2003]
(a) the control of speed fluctuations   (b) balancing of forces and couples
(c) kinematic analysis   (d) vibration analysis

Balancing of single and multi-cylinder engines

Chapter 6

IAS-12. Consider the following statements: [IAS-2001]
The unbalanced force in a single-cylinder reciprocating engine is
1. Equal to inertia force of the reciprocating masses
2. Equal to gas force
3. Always fully balanced
Which of the statement(s) is/are correct?
(a) 1 alone   (b) 2 alone   (c) 1 and 3   (d) 2 and 3

Tractive force

IAS-13. What causes a variation in the tractive effort of an engine?
(a) Unbalanced portion of the primary force, along the line of stroke
(b) Unbalanced portion of the primary force, perpendicular to the line of stroke
(c) The secondary force
(d) Both primary and secondary unbalanced forces [IAS-2007]

Swaying couple

IAS-14. Hammer blow [IAS-2002]
(a) is the maximum horizontal unbalanced force caused by the mass provided to balance the reciprocating masses
(b) is the maximum vertical unbalanced force caused by the mass added to balance the reciprocating masses
(c) varies as the square root of the speed
(d) varies inversely with the square of the speed

Answers with Explanation (Objective)

Previous 20-Years GATE Answers

GATE-1. Ans. (c)
Explanation. $\Delta = \frac{1}{2}(PQ)(PR)\sin A$
Area will be maximum when $A = 90°$
i.e. PQR is a right angled triangle.
∴ Ratio of connecting rot length to crank radius,
$\frac{l}{r} = \tan 75° = 3.732;$
$l = 3.732\ r$

GATE-2. Ans. (a) – 4, (b) – 5, (e) – 2, (f) – 1

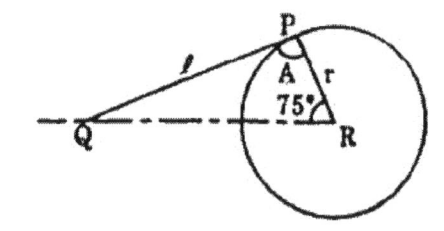

Previous 20-Years IES Answers

IES-1. Ans. (c)
IES-2. Ans. (b)
IES-3. Ans. (d) Klein's method of construction for reciprocating engine mechanism is based on the acceleration diagram.

Balancing of single and multi-cylinder engines
Chapter 6

IES-4. Ans. (c)

Velocity of crank pin $(V_c) = OC$

Velocity of piston $(V_p) = OD$

Velocity of piston with respect to crank pin $(V_{pc}) = CD$

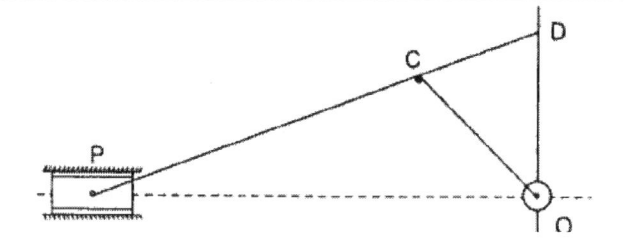

IES-5. Ans. (c)

IES-6. Ans. (c)

IES-7. Ans. (a)

T=40 N-m $\therefore F_T = \dfrac{40}{0.1} = 400$ N

Friction force = 400 sin 30 × 0.08 = 16 N

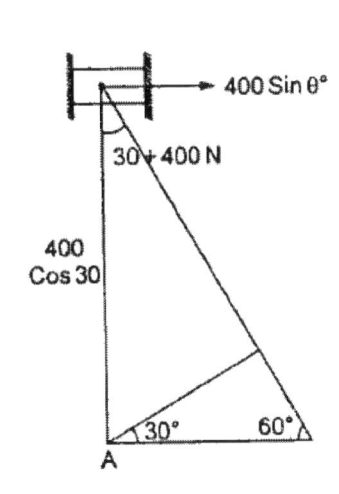

IES-8. Ans. (b)

IES-9. Ans. (d)

Klein's construction	→	Relative acceleration of linkages
Kennedy's theorem	→	eous centers in linkages
D' Alembert's principle	→	forces principle in linkages
Grubler's rule	→	Mobility of linkages

IES-10. Ans. (b)

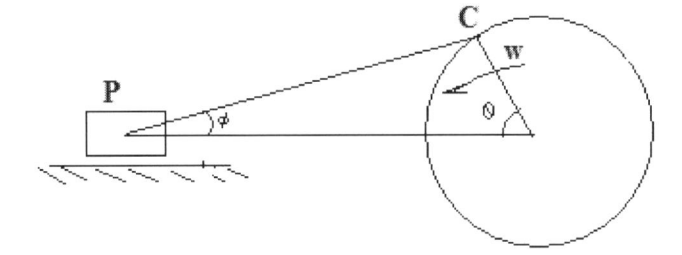

At $\theta = 90^0$, $\omega_{PC} = 0 \because \omega_{PC} = \dfrac{\omega\cos\theta}{\sqrt{(n^2 - \sin^2\theta)}}$

IES-11. Ans. (d) Angular acceleration of connecting rod $n = \dfrac{l}{r} = 1$ and $\theta = 45^0$

$\alpha_c = \dfrac{-\omega^2 \sin\theta(n^2 - 1)}{(n^2 - \sin^2\theta)^{3/2}} = 0$ [as n = 1]

IES-12. Ans. (c)

IES-13. Ans. (d)

IES-14. Ans. (b)

IES-15. Ans. (c)

IES-16. Ans. (b)

IES-17. Ans. (a)

Balancing of single and multi-cylinder engines
Chapter 6

IES-18. Ans. (d) Secondary force only involves ratio of length of connecting rod and crank radius and is equal to $m\omega^2 r \dfrac{\cos 2\theta}{n}$. If n increases, value of secondary force will decrease.

IES-19. Ans. (b) Maximum Secondary force
$$= \dfrac{m\omega^2 r}{n} = 1.2 \times \left(\dfrac{2\pi N}{60}\right)^2 \times \dfrac{r}{(l/r)} = 1.2 \times \left(\dfrac{2\pi \times 900}{60}\right)^2 \times \dfrac{0.050^2}{0.2} = 133 N$$

IES-20. Ans. (a)

IES-23. Ans. (a) The radial engine has resulted in a total resultant primary force of fixed magnitude viz. and directed along the first crank. Thus this fixed magnitude force —rotates‖ along with the first crank. Such a resultant force can therefore readily be balanced out by an appropriate mass kept on the crank. Therefore it is possible to get complete balance of the primary forces.
The inertia forces reveals that for even number of cylinders (n > 2) i.e., for four, six, eight etc. cylinders the secondary forces are also completely balanced out.

IES-24. Ans. (d) The inertia forces reveals that for even number of cylinders (n > 2) i.e., for four, six, eight etc. cylinders the secondary forces are also completely balanced out.

IES-25. Ans. (b)
IES-26. Ans. (b)
IES-27. Ans. (a)

Previous 20-Years IAS Answers

IAS-1. Ans. (a)

IAS-2. Ans. (a)
IAS-3. Ans. (b)
IAS-4. Ans. (a) When relative velocity V_{AB} will be zero. Or $V_{AB} = AB$. $\omega_{AB} = AB$.
$$\dfrac{\omega \cos\theta_2}{\sqrt{(n^2 - \sin^2 \theta_2)}} = 0 \text{ Or } \theta_2 = 90°$$

IAS-5. Ans. (d) Angular acceleration of connecting rod $n = \dfrac{l}{r} = 1$ and $\theta = 45°$
$$\alpha_c = \dfrac{-\omega^2 \sin\theta (n^2 - 1)}{(n^2 - \sin^2 \theta)^{3/2}} = 0 \text{ [as n = 1]}$$

IAS-6. Ans. (d)
IAS-7. ans. (b)
IAS-8. Ans. (a)
IAS-9. Ans. (a)
IAS-10. Ans. (d)
IAS-11. Ans. (b)
IAS-12. Ans. (a)
IAS-13. Ans. (a)
IAS-14. Ans. (b)

Linear Vibration Analysis of Mechanical Systems
Chapter 7

7. Linear Vibration Analysis of Mechanical Systems

Theory at a glance (IES, GATE & PSU)

Vibrations
A body is said to vibrate if has a to-and-fro motion
When any elastic body such as spring, shaft or beam is displaced from the equilibrium position by application of external forces and then release if commences cyclic motion. Such cyclic motion due to elastic deformation under the action of external forces is known as vibration.

DEFINITIONS

(i) **Free (Natural) Vibrations:** Elastic vibrations in which there are no friction and external forces after the initial release of the body are known as free or natural vibrations.

(ii) **Damped Vibrations:** When the energy of a vibrating system is gradually dissipated by friction and other resistances, the vibrations are said to be damped. The vibrations gradually cease and the system rests in its equilibrium position.

(iii) **Forced Vibrations:** When a repeated force continuously acts on a system, the vibrations are said to be force. The frequency of the vibrations is that of the applied force and is independent of their own natural frequency of vibrations.

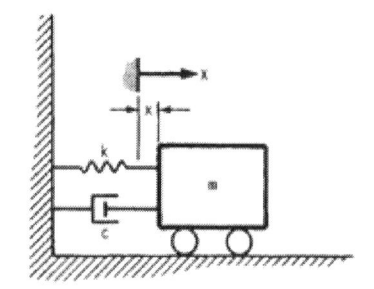

Fig. Representation of a single-degree-of-freedom system

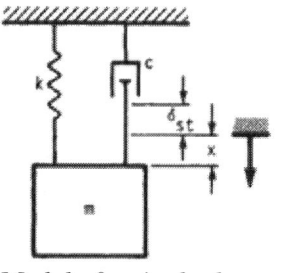

Fig. Model of a single-degree-of-freedom system showing the static deflection due to weight

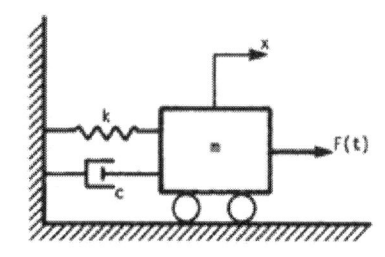

Fig. Oscillating force F(t) applied to the mass

Time period: It is the time taken by motion to repeat itself or it is the time require to complete one cycle.

$$t_P = \frac{2\pi}{\omega} \text{ sec.}$$

Frequency: The no. of cycle per unit time is known as frequency.

$$f = \frac{1}{t_P} = \frac{\omega}{2\pi} \text{ Hz.}$$

Linear Vibration Analysis of Mechanical Systems

Chapter 7

Stiffness of spring (K):
F = Kδ
F = force applied on spring N
δ = deflection of spring

Damping Coefficient (C):
$$C = \frac{F}{V} \text{ N-sec/m}$$
F = Force applied on damper
V = velocity of viscous fluid m/sec.

Resonance: When the frequency of external excitation force acting on a body is equal to the natural frequency of a vibrating body, the body starts vibrating with excessively large amplitude. Such state is known as resonance.

TYPES OF VIBRATIONS

Consider a vibrating body, e.g., a rod, shaft or spring. Figure. Shows a massless shaft end of which is fixed and the other end carrying a heavy disc. The system can execute the following types of vibrations.

(i) Longitudinal Vibrations: If the shaft is elongated and shortened so that same moves up and down resulting in tensile and compressive stresses in the shaft, the vibrations are said to be longitudinal. The different particles of the body move parallel to the axis of the body [figure (a)].

(ii) Transverse Vibrations: When the shaft is bent alternately [Figure (b)] and tensile and compressive stress due to bending result, the vibrations are said ti be transverse. The particles of the body move approximately perpendicular to its axis.

(iii) Torsional Vibrations: When the shaft is twisted and untwisted alternately and torsional shear stresses are induced, the vibrations are known as torsional vibrations. The particles of the body in a circle about the axis of the shaft [Figure.(c)].

S.H.M

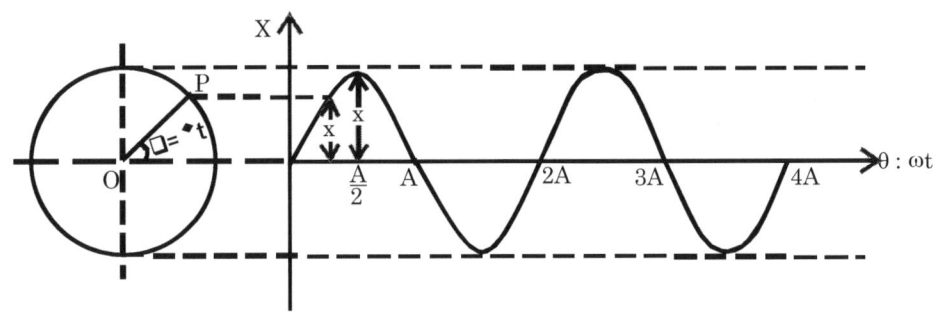

Let x = displacement of a point from mean position after time 't'.
X = Max. Displacement of point from mean position
$$x = X \sin \theta = X \sin \omega t$$

$$\dot{x} = \omega \times \cos \omega t$$

$$\ddot{x} = -\omega^2 \times \sin \omega t$$

$$\therefore \ddot{x} = -\omega^2 x$$

or $\boxed{\ddot{x} + \omega^2 x = 0}$ Equation of S.H.M

Linear Vibration Analysis of Mechanical Systems
Chapter 7

Equivalent Springs

1. **Springs in series:**

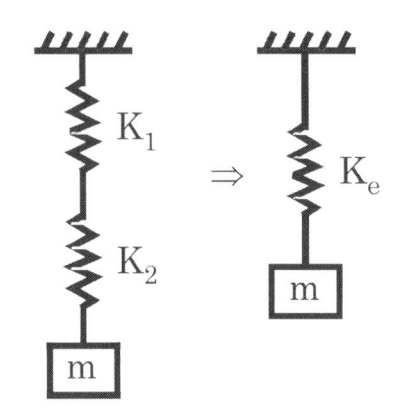

$\delta = \delta_1 + \delta_2$
and $mg = m_1g = m_2g$.

$\therefore \quad \dfrac{mg}{K_e} = \dfrac{m_1g}{K_1} + \dfrac{m_2g}{K_2}$

$\therefore \quad \boxed{\dfrac{1}{K_e} = \dfrac{1}{K_1} + \dfrac{1}{K_2}}$

2. **Springs in parallel:**

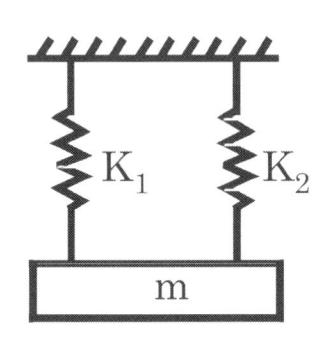

$mg = m_1g + m_2g$ and $\delta = \delta_1 = \delta_2$

$\therefore \quad K_e\delta = K_1\delta_1 + K_2\delta_2$

$\therefore \quad \boxed{K_e = K_1 + K_2}$

Natural Frequency of Free Longitudinal Vibrations

The natural frequency of the free longitudinal Vibrations may be determined by the following three methods:
1. Equilibrium Method
2. Energy method
3. Rayleigh's method

1. Equilibrium Method

Linear Vibration Analysis of Mechanical Systems
Chapter 7

Consider a constraint (i.e. spring) of negligible mass in an unstrained position, as shown in Figure (a).

Let   s = Stiffiness of constraint. It is the force required to produce unit displacement in the direction of vibration. It is usually expressed in N/m,
m = Mass of the body suspended from the constraint in kg,
W = Weight of the body in newtons = m.g,
δ = Static deflection of the spring in metres due to weight W newtons, and
x = Displacement given to the body by the external force, in metres.

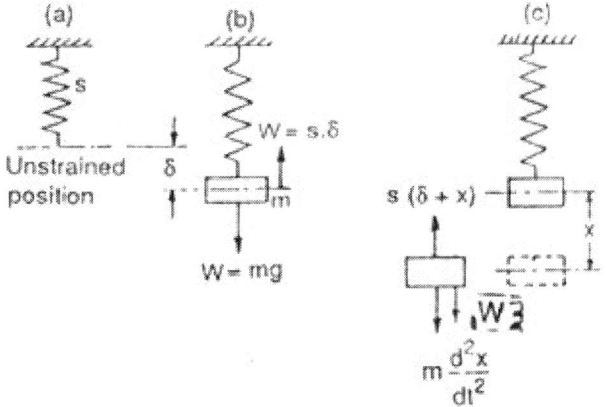

Fig. Natural frequency of free longitudinal vibrations

In the equilibrium position, as shown in Figure (b), the gravitational pull W= m.g, is balanced by a force of spring, such that W= s. δ.

Since the mass is now displaced from its equilibrium position by a distance x, as shown in Figure(c), and is then released, therefore after time t,

Restoring force $= W - s(\delta + x) = W - s.\delta - s.x$
$= s.\delta - s.\delta - s.x = -s.x$ ...($W = s.\delta$)   ... (i)
... (Taking upward force as nega5tive)

And Accelerating force = Mass × Acceleration
$= m \times \dfrac{d^2x}{dt^2}$ ... (Taking downward force as positive)

Equating equations (i) and (ii), the equation of motion of the body of mass m after time is

$m \times \dfrac{d^2x}{dt^2} = -s.x$ or $m \times \dfrac{d^2x}{dt^2} + s.x = 0$

$\therefore \dfrac{d^2x}{dt^2} + \dfrac{s}{m} \times x = 0$

We know that the fundamental equation of harmonic motion is

$$\dfrac{d^2x}{dt^2} + \omega^2.x = 0 \qquad \ldots(iv)$$

Comparing equations (iii) and (iv), we have

$$\omega = \sqrt{\dfrac{s}{m}}$$

$\therefore$ Time period, $t_p = \dfrac{2\pi}{\omega} = 2\pi\sqrt{\dfrac{m}{s}}$

And natural frequency, $f_n = \dfrac{1}{t_p} = \dfrac{1}{2\pi}\sqrt{\dfrac{s}{m}} = \dfrac{1}{2\pi}\sqrt{\dfrac{g}{\delta}}$ ...($m.g = s.\delta$)

Taking the value of g as 9.81 m/s² and δ in metres,

Linear Vibration Analysis of Mechanical Systems

Chapter 7

$$f_n = \frac{1}{2\pi}\sqrt{\frac{9.81}{\delta}} = \frac{0.4985}{\sqrt{\delta}} \text{ Hz}$$

Note: The value of static deflection δ may be found out from the given conditions of the problem. For longitudinal vibrations, it may be obtained by the relation,

$$\frac{\text{Stress}}{\text{Strain}} = E \text{ or } \frac{W}{A} \times \frac{1}{\delta} = E \text{ or } \delta = \frac{W \cdot t}{E \cdot A}$$

Where
δ = Static deflection i.e. extension of the conatraint,
W = load attached to the free end of constraint,
l = length of the constraint,
E = Young's modulus for the constraint, and
A = Cross-sectional area of the constraint.

2. Energy method

Energy Method:

K. E of mass = $-\frac{1}{2} m \dot{x}^2$

P. E of spring = $\frac{1}{2} K x^2$

$$F = Kx$$
$$mg = Kx$$
$$\boxed{\frac{K}{m} = \frac{g}{x}}$$

Total energy= constant
K. E + P. E = constant

$$\frac{d(K.E + P.E)}{dt} = 0$$

$$\frac{d\left(\frac{1}{2} m \dot{x}^2 + \frac{1}{2} K x^2\right)}{dt} = 0 \quad \Rightarrow \quad \boxed{m\ddot{x} + Kx = 0}$$

Torsional Stiffness (K_t) :

$$K_t = \frac{T}{\theta} = \frac{G I_P}{l}$$

G = modulus of rigidity of shaft material N/m²
I_P = Polar moment of inertia of shaft
 = $\frac{\pi}{32} d^4$
l = Length of the shaft
d = diameter of the shaft

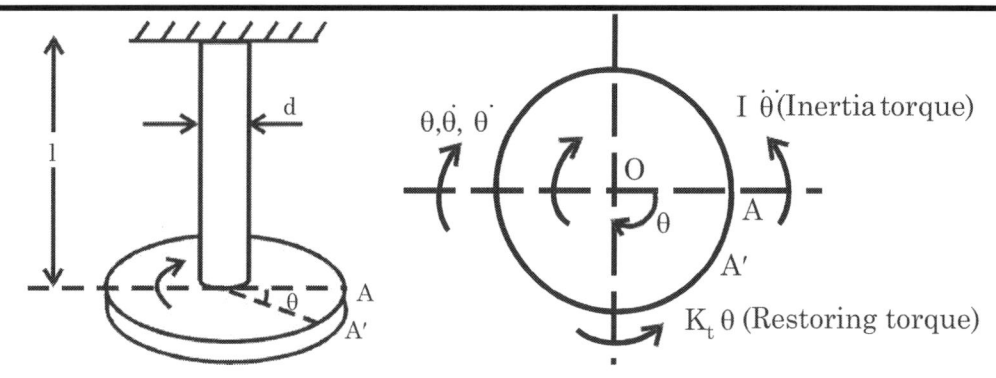

Linear Vibration Analysis of Mechanical Systems
Chapter 7

Σ (Inertia torque + External torque) = 0

$$I\ddot{\theta} + K_t \cdot \theta = 0$$

Q. $K_1 = 2400$ N/m, $K_2 = 1600$ N/m, $K_3 = 3600$ N/m $K_4 = K_5 = 500$ N/m. Find mass m such that system have natural frequency of 10 Hz.

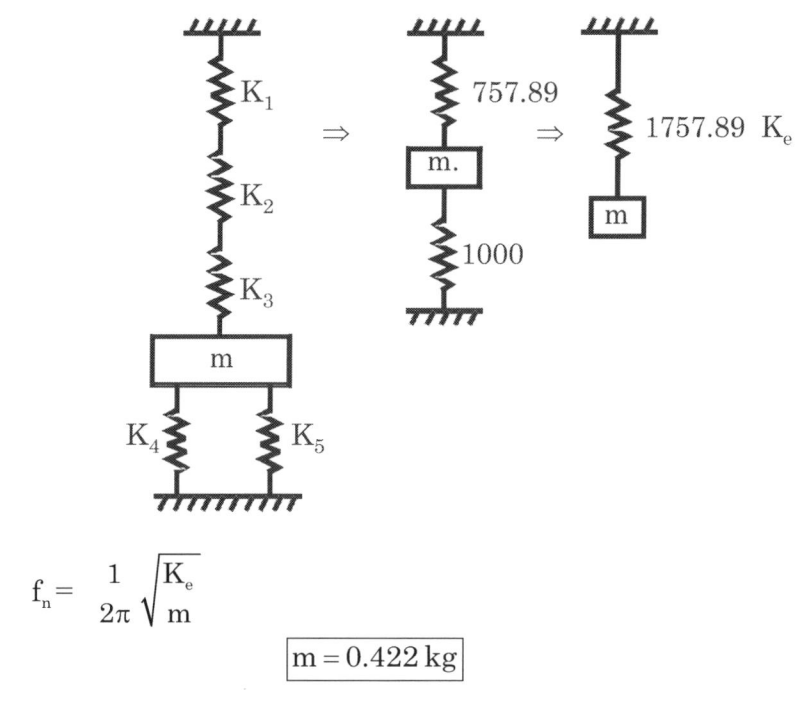

$$f_n = \frac{1}{2\pi}\sqrt{\frac{K_e}{m}}$$

$$\boxed{m = 0.422 \text{ kg}}$$

Q. Determine natural frequency.

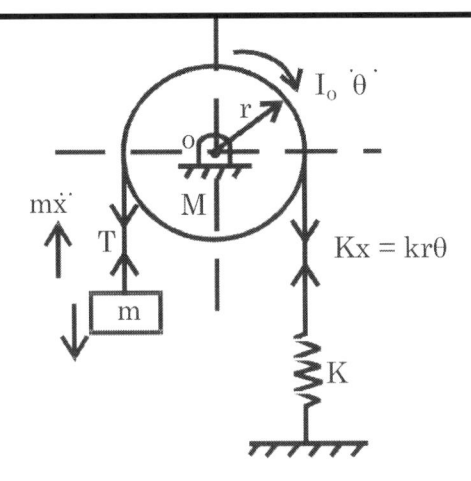

Linear Vibration Analysis of Mechanical Systems
Chapter 7

Angular displacement of pulley = θ

Angular velocity of pulley = $\dot{\theta}$

Angular Acceleration of pulley = $\ddot{\theta}$

Linear displacement of mass = $x = r\theta$

Linear velocity of mass = $\dot{x} = r\dot{\theta}$

Linear Acceleration of mass = $\ddot{x} = r\ddot{\theta}$

Equilibrium Method:

Consider linear motion of mass 'm'.

Σ(Inertia force + External force) = 0

$$m\ddot{x} + T = 0$$

$$T = -m\ddot{x}$$

Consider rotary motion of pulley.

Σ (Inertia force + External torque) = 0

$$I_0 \ddot{\theta} + Kr\theta \cdot r - T \cdot r = 0$$

$$I_0 \ddot{\theta} + Kr^2\theta + m\ddot{x}\, r = 0$$

$$I_0 \ddot{\theta} + Kr^2\theta + mr^2\ddot{\theta} = 0$$

$$(I_0 + mr^2)\ddot{\theta} + Kr^2\theta = 0$$

$$\ddot{\theta} + \frac{Kr^2}{I_0 + mr^2}\theta = 0 \qquad \boxed{I_0 = \frac{1}{2}Mr^2}$$

$$\ddot{\theta} + \frac{Kr^2}{\dfrac{Mr^2}{2} + mr^2}\theta = 0$$

$$\ddot{\theta} + \frac{K}{\dfrac{M}{2} + m}\theta = 0$$

$$\omega_n^2 = \ddot{\theta} + \frac{K}{\dfrac{M}{2} + m}$$

Linear Vibration Analysis of Mechanical Systems

Chapter 7

$$\omega_n = \sqrt{\dfrac{K}{\dfrac{M}{2} + m}}$$

$$\boxed{f_n = \dfrac{1}{2\pi}\sqrt{\dfrac{K}{\dfrac{M}{2} + m}}}$$

Energy Method:

K. E of the mass $= \dfrac{1}{2}m\dot{x}^2 = \dfrac{1}{2}mr^2\dot{\theta}^2$

K. E of pulley $= \dfrac{1}{2}I_0\dot{\theta}^2$

P. E of spring $= \dfrac{1}{2}Kx^2 = \dfrac{1}{2}Kr^2\theta^2$

Total energy $= \dfrac{1}{2}mr^2\dot{\theta}^2 + \dfrac{1}{2}I_0\dot{\theta}^2 + \dfrac{1}{2}Kr^2\theta^2$

$\dfrac{d(\text{Total energy})}{dt} = 0$

$\dfrac{1}{2}mr^2 \cdot 2\dot{\theta}\ddot{\theta} + \dfrac{1}{2}I_0 \cdot 2\dot{\theta}\ddot{\theta} + \dfrac{1}{2}Kr^2 \cdot 2\dot{\theta}\theta = 0$

$mr^2\ddot{\theta} + I_0\ddot{\theta} + Kr^2\theta = 0$

$\left(\dfrac{Mr^2}{2} + mr^2\right)\ddot{\theta} + Kr^2\theta = 0$

$\left(\dfrac{M}{2} + m\right)\ddot{\theta} + K\theta = 0$

$\ddot{\theta} + \dfrac{K}{\dfrac{M}{2} + m}\theta = 0$

$W_n = \sqrt{\dfrac{K}{\dfrac{M}{2} + m}}$

$$\boxed{f_n = \dfrac{1}{2\pi}\sqrt{\dfrac{K}{\dfrac{M}{2} + m}}}$$

Q. Let angular display of pulley $= \theta$. Linear display of mass $= x_1 = r\theta$

Difference of spring $\dot{x}_2 = R\theta$.

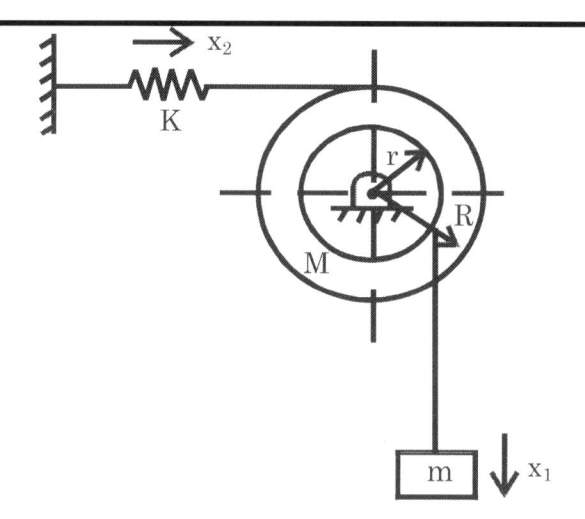

Linear Vibration Analysis of Mechanical Systems
Chapter 7

Now

K. E of mass = $\dfrac{1}{2} m \dot{x}^2 = \dfrac{1}{2} m r^2 \dot{\theta}^2$

K. E of pulley = $\dfrac{1}{2} I \dot{\theta}^2 = \dfrac{1}{2} M R^2 \dot{\theta}^2$

P. E of spring = $\dfrac{1}{2} K x^2 = \dfrac{1}{2} K R^2 \theta^2$

$\dfrac{d(\text{Total energy})}{dt} = 0$

$\dfrac{1}{2} m r^2 \cdot 2\dot{\theta}\ddot{\theta} + \dfrac{1}{2} M R^2 \cdot 2\dot{\theta}\ddot{\theta} + \dfrac{1}{2} K R^2 \cdot 2\theta\dot{\theta} = 0$

$\left(\dfrac{M R^2}{2} + m r^2\right) \ddot{\theta} + K R^2 \theta = 0$

$$\boxed{W_n = \sqrt{\dfrac{K R^2}{\dfrac{M R^2}{2} + m r^2}}}$$

Q. Let angular Display pulley = θ
Linear display of pulley = $x = r\theta$
Linear display of spring = x.

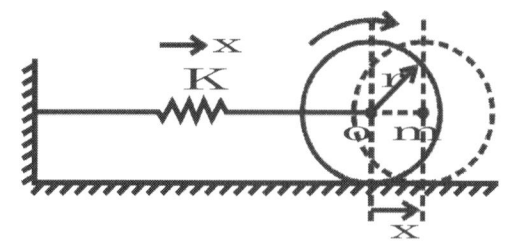

In this case pulley has both linear as well as angular display hence it has 2 K. E.

K. E of linear pulley = $\dfrac{1}{2} m \dot{x}^2 = \dfrac{1}{2} m r^2 \dot{\theta}^2$

K. E of rotary pulley = $\dfrac{1}{2} I \omega^2 = \dfrac{1}{2} I \dot{\theta}^2 = \dfrac{1}{4} m r^2 \dot{\theta}^2$

P. E of the spring = $\dfrac{1}{2} K x^2 = \dfrac{1}{2} K r^2 \theta^2$

Linear Vibration Analysis of Mechanical Systems

Now $\dfrac{d(T \cdot E)}{dt} = 0$

$\left(\dfrac{3mr^2}{2}\right)\ddot{\theta}^2 + Kr^2\theta = 0$

$\ddot{\theta} + \dfrac{2K}{3m}\theta = 0$

$\therefore \quad \boxed{\omega_n = \sqrt{\dfrac{2K}{3m}}}$

Q. Find the natural frequency of oscillation of liquid column when tube is slightly displaced.

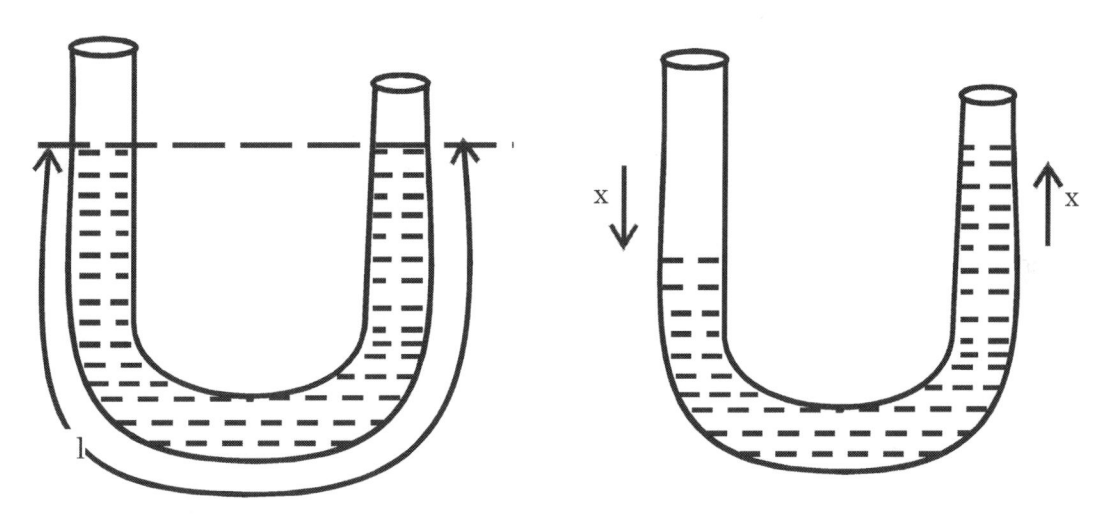

The liquid column is displaced from equilibrium position through a distance x.

Therefore each particle in liquid column has a velocity $\dot{x}$ at that instant.

$\therefore \quad$ K. E of fluid $= \dfrac{1}{2} m \dot{x}^2$

$\qquad \qquad \qquad \quad = \dfrac{1}{2} \rho Al \dot{x}^2$

$\rho \rightarrow$ density

A = cross area of column.

The liquid of mass ρAx is raised through a distance x. Therefore the P. E of display.

P. E $= \rho Ax \cdot g \cdot x$

Now $\dfrac{d(T \cdot E)}{dt} = 0$

$\ddot{x} + \dfrac{2g}{l} x = 0$

$\therefore \quad \boxed{\omega_n = \sqrt{\dfrac{2g}{l}}}$

Damped Free Vibration

Linear Vibration Analysis of Mechanical Systems

Chapter 7

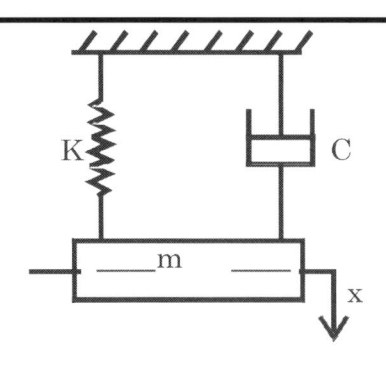

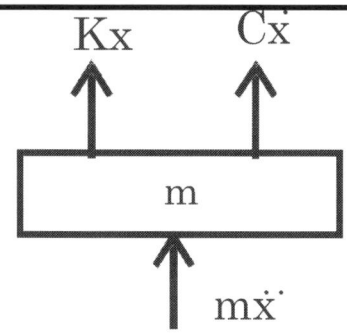

C = damping coefficient N.s/m

Σ (Intertia force + External force) = 0

$$m\ddot{x} + C\dot{x} + Kx = 0$$

Above equation is a linear differential equation of the second order and its solution can be written in the form

$x = e^{St}$

$\dot{x} = S \cdot e^{St}$

$\ddot{x} = S^2 e^{St}$

∴ Above equation will be

$mS^2 + CS + K = 0$

$$S_{1,2} = \frac{-C \pm \sqrt{C^2 - 4Km}}{2m}$$

$$= \frac{-C}{2m} \pm \sqrt{\left(\frac{C}{2m}\right)^2 - \frac{K}{m}}$$

The general solution to the different equation.

$x = Ae^{S_1 t} + Be^{S_2 t}$

A and B are constant.

Critical damping coefficient (C_C) : the critical coefficient $\dot{C}_C$ is that value of damping coefficient at which

$\sqrt{\left(\frac{C}{2m}\right)^2 - \frac{K}{m}} = 0$

$\left(\frac{C}{2m}\right)^2 = \frac{K}{m}$

$\frac{C_C}{2m} = \sqrt{\frac{K}{m}} = \omega_n$

$$\boxed{C_C = 2m\omega_n}$$

$$\boxed{C_C = 2\sqrt{Km}}$$

Damping Factor or Damping ratio (r):

$$r = \frac{C}{C_C}$$

General solution of differential equation:

$S_{1,2} = -r\omega_n \pm \sqrt{r^2 \omega_n^2 - \omega_n^2}$

Linear Vibration Analysis of Mechanical Systems
Chapter 7

$$= [-r \pm \sqrt{r^2 - 1}]\omega_n$$

$$S_1 = (-r + \sqrt{r^2 - 1})\omega_n$$

$$S_2 = (-r - \sqrt{r^2 - 1})\omega_n$$

1. over damped system (r > 1):

$$x = Ae^{(-r + \sqrt{r^2-1})\omega_n t} + Be^{(-r - \sqrt{r^2-1})\omega_n t}$$

The value of A and B can be determined from initial condition

At $t = 0$, $\quad x = X_o$

$t = 0$, $\quad \dot{x} = 0$

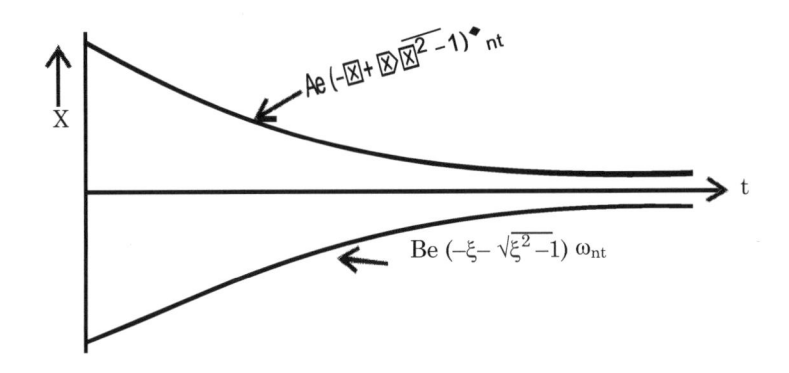

2. Critically damped system (r < 1):

$$S_1 = -\omega_n$$
$$S_2 = -\omega_n$$

$$\therefore \quad x = (A + Bt)e^{-\omega_n t}$$

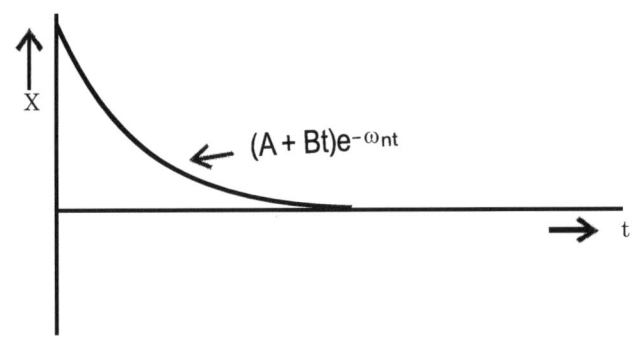

3. under damped system (r < 1):

$$S_1 = [-r + i\sqrt{1 - r^2}]\omega_n$$

$$S_2 = [-r - i\sqrt{1 - r^2}]\omega_n$$

$$x = Ae^{(-r + i\sqrt{1-r^2})\omega_n t} + Be^{(-r - i\sqrt{1-r^2})\omega_n t}$$

$$= e^{-rW_n t}[Ae^{iW_d t} + Be^{-iW_d t}]$$

$$\boxed{x = Xe^{-r\omega_n t} \sin(\omega_d t + \phi)}$$

X, ϕ are constant.

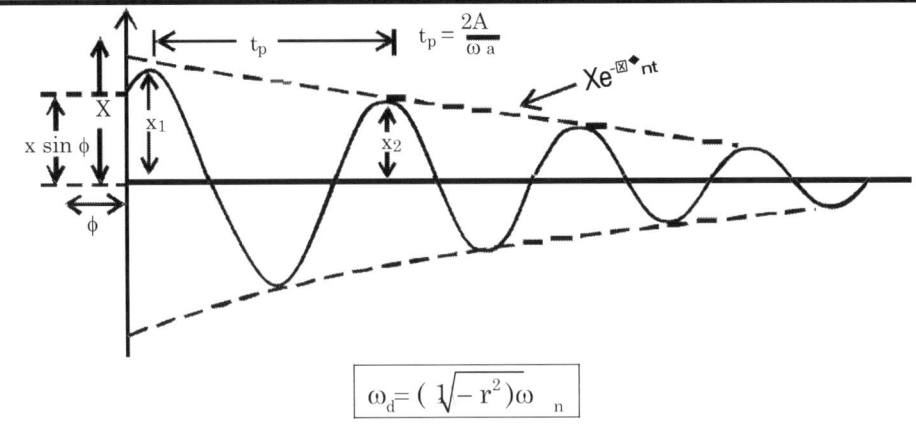

Linear Vibration Analysis of Mechanical Systems
Chapter 7

$$\omega_d = (\sqrt{1 - r^2})\omega_n$$

ω_d = natural frequency of damped vibration

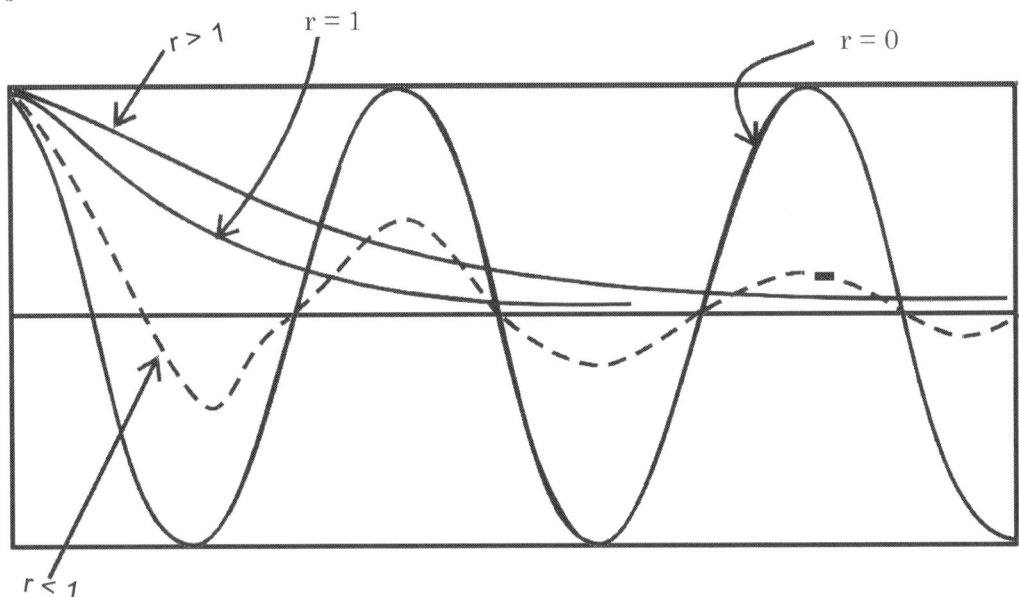

Logarithmic Decrement

The decrease in a amplitude from one cycle to the next depends on the extent of damping in the system. The successive peak amplitudes bear a certain specific relationship involving the damping of the system, leading us to the concept of —logarithmic decrement‖

or

It is defined as the logarithm of the ratio of any two success amplitudes on the same side of the mean position.

$$\delta = \log_e \left(\frac{X_0}{X_1}\right) = \log_e \left(\frac{X_1}{X_2}\right) = \ldots\ldots\ldots \log_e \left(\frac{X_{n-1}}{X_n}\right)$$

$$\therefore n\delta = \log_e \left(\frac{X_0}{X_1}\right) + \log_e \left(\frac{X_1}{X_2}\right) + \ldots\ldots \log_e \left(\frac{X_{n-1}}{X_n}\right)$$

$$\therefore \boxed{\delta = \frac{1}{n} \log_e \left(\frac{X_0}{X_n}\right)}$$

Now $X_1 = Xe^{-r\omega_n t_1} \sin(\omega_d t_1 + \phi)$

$X_2 = Xe^{-r\omega_n t_2} \sin(\omega_d t_2 + \phi)$

$\quad = Xe^{-r\omega_n t_2} \sin[\omega_d (t_1 + t_P) + \phi]$

$\quad = Xe^{-r\omega_n (t_1 + t_P)} \sin[\omega_d t_1 + \omega_d t_P + \phi]$

Linear Vibration Analysis of Mechanical Systems

Chapter 7

$$= Xe^{-r\omega_n(t_1+t_P)} \sin\left[\omega_d t_1 + \omega_d \times \frac{2\pi}{W_d} + \phi\right]$$

$$= Xe^{-r\omega_n(t_1+t_P)} \sin(\omega_d t_1 + \phi)$$

$$\delta = \log\left(\frac{X_1}{X_2}\right)$$

$$\frac{X_1}{X_2} = \frac{Xe^{-r\omega_n(t_1)} \sin(\omega_d t_1 + \phi)}{Xe^{-r\omega_n(t_1+t_P)} \sin(\omega_d t_1 + \phi)}$$

$$= e^{r W_n t_P}$$

$$\therefore \delta = r\omega_n t_P = r\omega_n \cdot \frac{2\pi}{W_d}$$

$$= r\omega_n \frac{2\pi}{(\sqrt{1-r^2})\omega_n}$$

$$\therefore \boxed{\delta = \frac{2\pi r}{\sqrt{1-r^2}}}$$

Damped Free Torsional vibration:

$$\boxed{I\ddot{\theta} + C_t \dot{\theta} + K_t \theta = 0}$$

All term are torque

Q. A spring mass dashpot system consists of spring of stiffness 400 N/m and the mass of 4 kg. The mass is displaced 20 mm beyond the equilibrium position and released. Find the equation of motion of the mass. If the damping coefficient of the dashpot is,
(i) 160 N-S/m
(ii) 80 N-S/m

Solution: M = 4 kg

K = 400 N/m

$$\omega_n = \sqrt{\frac{K}{m}} = 10 \text{ rad/sec.}$$

$$\xi_1 = \frac{C_1}{C_C} = \frac{160}{2\sqrt{4 \times 400}} = 2$$

$$\xi_2 = \frac{C_2}{C_C} = 1$$

1. Equation of motion for $\xi = 2$

$$x = Ae^{[-\xi + \sqrt{\xi^2+1}]\omega_n t} + Be^{[-\xi - \sqrt{\xi^2-1}]\omega_n t}$$

$$x = Ae^{-2.67t} + Be^{-37.32t}$$

$$\dot{x} = -2.67 Ae^{-2.67t} - 37.32 Be^{-37.32t}$$

Now at t = 0, x = 0.02 m

$$\therefore A + B = 0.02$$

at t = 0, $\dot{x} = 0$

$$\therefore -2.67A - 37.32B = 0$$

$$\therefore A = 0.0215$$

$$B = 1.54 \times 10^{-3}$$

Linear Vibration Analysis of Mechanical Systems
Chapter 7

$\therefore$ $\boxed{x = 0.215\, e^{-2.67t} - 1.54 \times 10^{-3}\, e^{-37.32t}}$

2. Equation of motion for $\xi = 1$

$x = (A + Bt)e^{-\omega_n t}$

$= (A + Bt)e^{-10t}$

$= Ae^{-10t} + Bte^{-10t}$

$\dot{x} = -10Ae^{-10t} - 10Bte^{-10t} + Be^{-10t}$

Substituting at, t = 0, x = 0.02
A = 0.02

And at t = 0, $\dot{x} = 0$
$0 = -10A + B$
B = 0.2

$\therefore$ $\boxed{x = 0.02 e^{-10t} + 0.2t e^{-10t}}$

Q. The system shown in fig. has a spring stiffness of 15 kN/m, a viscous damper having damping coefficient 1500 N-S/m and a mass of 10 kg. The mass is displaced by 0.01m and released with a velocity of 10 m/s in the direction of return motion

Find (i) An expression for the displacement x of the mass interms of times 't' and

(ii) the displacement of mass after $\dfrac{1}{100}$ sec.

Solution: m = 10 kg, K = 15 × 10³ N/m
C = 1500, N-S/m

At t = 0, x = 0.01 m

At t = 0, $\dot{x} = -10$ m/s

Now $W_n = \sqrt{\dfrac{K}{m}} = 38.72$ rad/sec.

$r = \dfrac{C}{C_C} = \dfrac{C}{2mW_n} = 1.936$

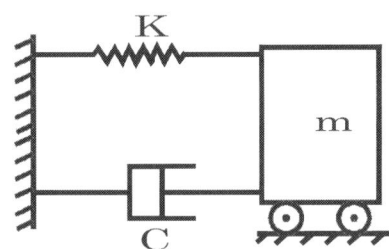

Expression for displacement:

Since r > 1

$x = Ae^{[-r + \sqrt{r^2 - 1}]\omega_n t} + Be^{[-r - \sqrt{r^2 - 1}]\omega_n t}$

$= Ae^{(-10.77)t} + Be^{(-139.14)t}$

$\dot{x} = -10.77\, Ae^{-10.77t} - 139.14\, Be^{-139.14t}$

Substitution t = 0 and x = 0.01
$\therefore$ A + B = 0.01 ... (i)

Substitution t = 0 and $\dot{x} = -10$
$-10.77\, A - 139.14\, B = -10$

$$\therefore A = -0.0670$$
$$B = 0.0770$$
$$\therefore \boxed{x = 0.0670e^{-(10.77)t} + 0.077e^{-(139.14)t}}$$

At $t = \dfrac{1}{100}$

$$\boxed{x = -0.041 \text{ m}}$$

Negative sign indicates the displacement is on opposite side of mean position.

Forced Damped Vibrations:

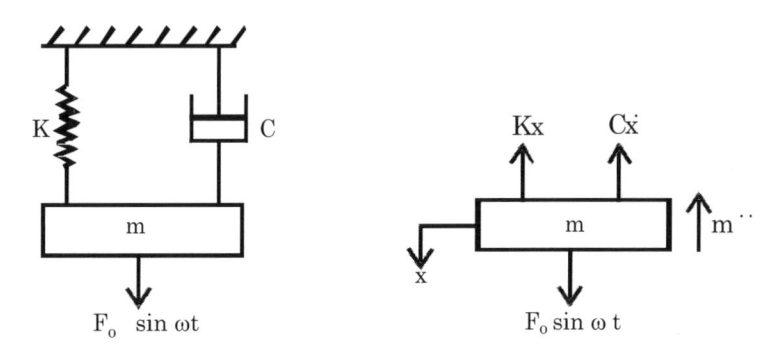

Σ (Inertia force + External force) $= 0$

$$m\ddot{x} + c\dot{x} + Kx = F_o \sin \omega t$$

The equation is a linear, second order differential equation. The solution of this equation consist two parts.

$$x = x_C + x_P$$

x_C = Complementary function

x_P = Particular integral

The complementary function is obtained by considering no forcing function i.e.

$m\ddot{x} + c\dot{x} + Kx = 0$ which is same as previous (3) cases

Now $x_P = X \sin(\omega t - \phi)$

X = amplitude of steady state vibration

ϕ = Angle by which the displacement vector lags force vector (phase angle).

ω = Angular frequency of external exciting hammons force rad/sec.

Then

$$X = \dfrac{\dfrac{F_o}{K}}{\sqrt{\left[1 - \left(\dfrac{\omega}{\omega_n}\right)^2\right]^2 + \left[2\xi \dfrac{\omega}{\omega_n}\right]^2}}$$

Where $\dfrac{F_o}{K} = X_{st}$ = deflection due to force F_o or zero frequency deflection or static deflection.

$$\phi = \tan^{-1}\left[\dfrac{2\xi \dfrac{\omega}{\omega_n}}{1 - \left(\dfrac{\omega}{\omega_n}\right)^2}\right]$$

Linear Vibration Analysis of Mechanical Systems
Chapter 7

∴ Solution $x = x_C + x_P$

$$x_P = \frac{F_o \sin(\omega t - \phi)}{k\sqrt{\left[1 - \left(\frac{\omega}{\omega_n}\right)^2\right]^2 + \left[2\xi \left(\frac{\omega}{\omega_n}\right)\right]^2}}$$

Magnification Factor: It is defined as the ratio of the amplitude of steady state vibration 'X' to the zero frequency deflection X_{st}.

$$M.F = \frac{X}{X_{St}}$$

$$M.F = \frac{1}{\sqrt{\left[1 - \left(\frac{\omega}{\omega_n}\right)^2\right]^2 + \left[2\xi \frac{\omega}{\omega_n}\right]^2}}$$

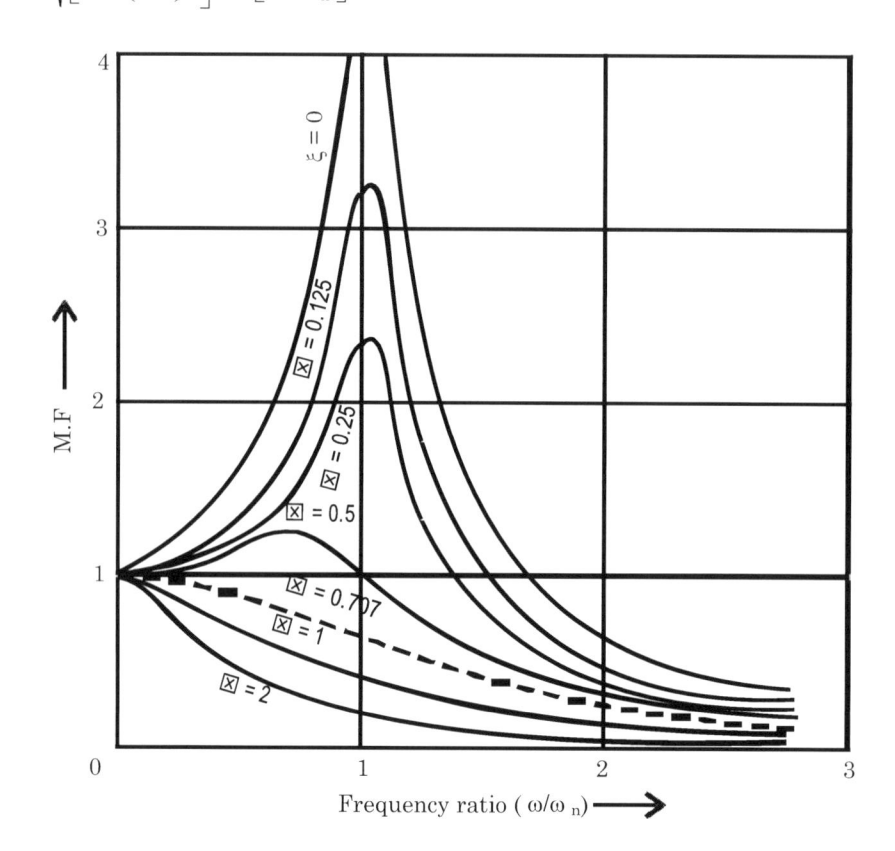

Frequency ratio (ω/ω_n) →

Observation made from frequency response curve:

1. The M.F is maximum when $\left(\frac{\omega}{\omega_n}\right) = 1$. the condition is known as resonance.
2. As the ξ decreases, the maximum value of M.F increases.
3. When there is no. damping ($\xi = 0$), the M.F reaches infinity at $\left(\frac{\omega}{\omega_n}\right) = 1$.
4. At zero frequency of excitation (i.e. when $\omega = 0$) the M.F is unity for all value of ξ. In other words, damping does not have any effect on M.F at zero frequency of excitation.
5. At very high frequency of excitation, the M.F tends to zero.
6. For ξ more than 0.707, the maximum M.F is below unity.

Linear Vibration Analysis of Mechanical Systems
Chapter 7

Phase angle (ϕ) Vs frequency ratio $\left(\dfrac{\omega}{\omega_n}\right)$

[Graph: Phase angle (ϕ) on y-axis (0° to 180°) vs frequency ratio (ω/ω_n) on x-axis (0 to 3), showing curves for $\xi = 0, 0.25, 0.5, 0.707, 1, 2$. All curves pass through 90° at $\omega/\omega_n = 1$.]

1. The phase angle varies from 0° at low frequency ratio to 180° at very high frequency ratio.
2. At resonance frequency ($\omega = \omega_n$) the phase angle is 90° and damping does not have any effect on phase angle.
3. At frequency ratio $\left(\dfrac{\omega}{\omega_n}\right)$ less than unity, higher the damping factor, higher is the phase angle, whereas at frequency $\left(\dfrac{\omega}{\omega_n}\right)$ greater than unity higher the damping factor lower is the phase angle.
4. The variation in phase angle is because of damping. If there is no damping ($\xi = 0$) the phase angle is either 0° or 180° and at resonance the phase angle suddenly change from 0° to 180°.

Q. A spring mass damper system has a mass of 80 kg suspended from a spring having stiffness of 1000 N/m and a viscous damper with a damping coefficient of 80 N-s/m. If the mass is subjected to a periodic disturbing force of 50 N at undamped natural frequency, determine

(i) The undamped natural frequency (w_n).
(ii) The damped natural frequency (w_d)
(iii) The amplitude of forced vibration of mass.
(iv) The phase difference.

Solution: m = 80 kg
K = 1000 N/m

Linear Vibration Analysis of Mechanical Systems
Chapter 7

$C = 80$ N-sec/m
$F_o = 50$ N

$$\omega_n = \sqrt{\frac{K}{m}} = 3.53 \text{ rad/sec.}$$

$$\xi = \frac{C}{C_C} = \frac{C}{2m\omega_n} = 0.1416$$

$$w_d = (\sqrt{1-\xi^2})\omega_n = 3.49 \text{ rad/sec.}$$

$$X = \frac{\frac{F_o}{K}}{\sqrt{1-\left(\frac{\omega}{\omega_n}\right)^2 + \left[2\xi\frac{\omega}{\omega_n}\right]^2}} \qquad [\text{here } \omega = \omega_n]$$

$= 0.1765$ m.

$$\phi = \tan^{-1}\left[\frac{2\xi\frac{\omega}{\omega_n}}{1-\left(\frac{\omega}{\omega_n}\right)^2}\right]$$

$\phi = 90°$

Forced variation due to rotating unbalance:

In this case

$$\boxed{F_o = m_o e\omega^2}$$

m_o = unbalanced mass
E = eccentricity of the unbalanced mass.
ω = angular velocity of rotation of unbalanced mass.

Transmissibility: (T_R)

The ratio of the force transmitted (F_T) to the force applied (F_0 is known as the isolation factor or transmissibility ratio of the spring support.

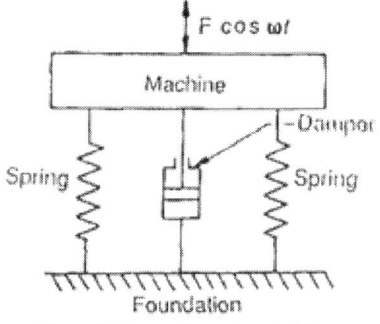

Fig. Vibration isolation

$$\text{Force transmissibility} = \frac{\text{Force transmitted to the foundation}}{\text{Force impressed upon the system}}$$

$$\boxed{T_R = \frac{F_T}{F_o}}$$

Linear Vibration Analysis of Mechanical Systems
Chapter 7

$$T_R = \frac{\sqrt{1 + \left(2\xi\frac{\omega}{\omega_n}\right)^2}}{\sqrt{\left[1 - \left(\frac{\omega}{\omega_n}\right)^2\right]^2 + \left[2\xi\frac{\omega}{\omega_n}\right]^2}}$$

$$\text{Motion transmissiblity} = \frac{X}{Y}$$

X = Absolute amplitude of the mass (body)
Y = Amplitude of the base excitation.
Frequency ratio $\left(\frac{\omega}{\omega_n}\right) \to$

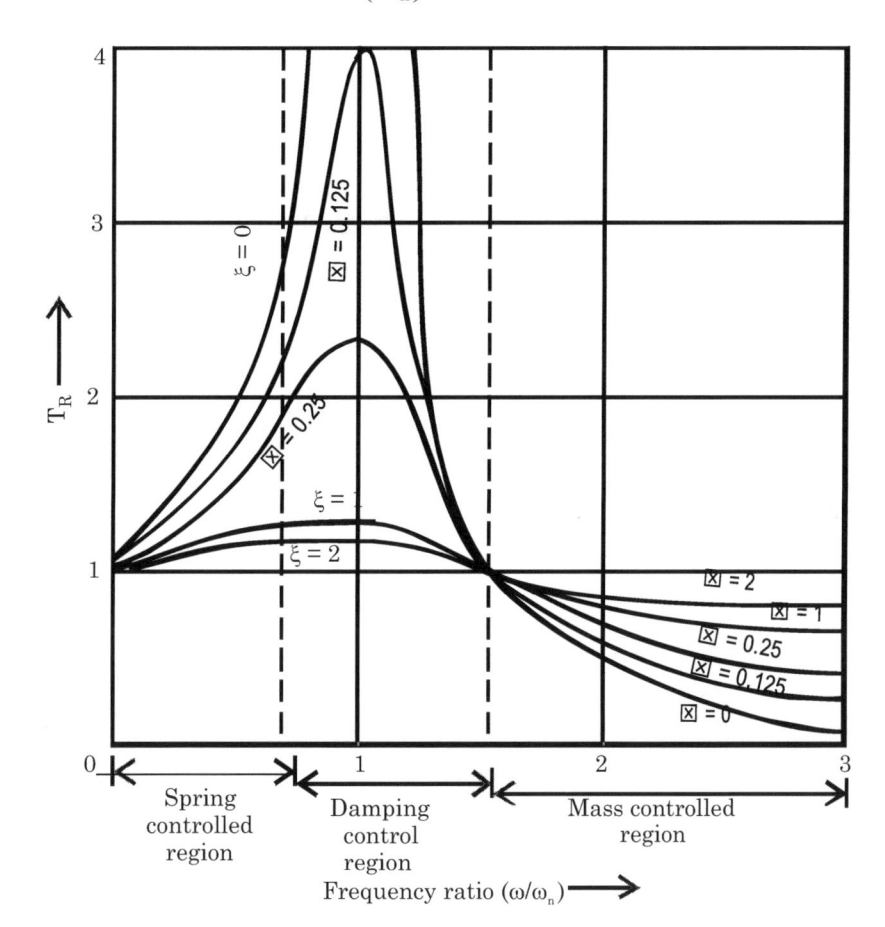

[Graph of T_R vs frequency ratio (ω/ω_n) showing curves for $\xi = 0$, $\xi = 0.125$, $\xi = 0.25$, $\xi = 1$, $\xi = 2$. Regions marked: Spring controlled region, Damping control region, Mass controlled region.]

1. All the curves start from the unit value of transmissibility and pass through the unit value of transmissibility at frequency ratio $\left(\frac{\omega}{\omega_n}\right) = \sqrt{2}$.

2. The transmissibility tends to zero as the frequency ratio $\left(\frac{\omega}{\omega_n}\right)$ tend to infinity.

3. When $\left(\frac{\omega}{\omega_n}\right) < \sqrt{2}$, the transmissibility is greater than one. This means the transmitted force is always greater than the impressed exciting force.

4. When $\frac{\omega}{\omega_n} = 1$, the transmissibility is maximum.

Linear Vibration Analysis of Mechanical Systems
Chapter 7

5. In order to have low value of transmissibility, the operation of vibrating system generally kept $\left|\dfrac{\omega}{\omega_n}\right| > \sqrt{2}$.

Q. Find the natural frequency of vibration for the system shown in fig. Neglect the mass of cantilever beam. Also, find the natural frequency of vibration when: [IAS – 1997]
(i) $K = \infty$
(ii) $I = \infty$

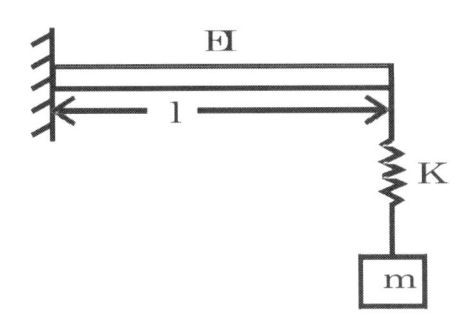

Solution: The deflection of cantilever beam due to weight mg.

$$\delta = \dfrac{mgl^3}{3EI}$$

∴ Stiffness of cantilever beam is

$$K_1 = \dfrac{mg}{\delta} = \dfrac{3EI}{l^3}$$

Then the system is converted into

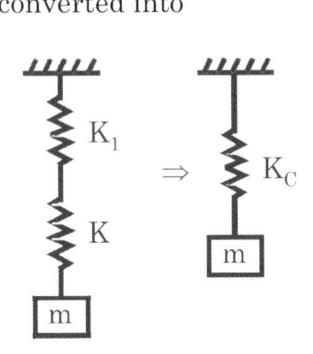

$$\dfrac{1}{K_e} = \dfrac{1}{K_1} + \dfrac{1}{K}$$

$$K_e = \dfrac{3EIK}{KL^3 + 3EI}$$

$$\therefore f_n = \dfrac{1}{2\pi}\sqrt{\dfrac{K_e}{m}} = \dfrac{1}{2\pi}\sqrt{\dfrac{3EIK}{m(KL^3 + 3EI)}} \; Hz$$

(i) When $K = \infty$

$$\therefore \dfrac{1}{K_e} = \dfrac{1}{K_1} + \dfrac{1}{K} \Rightarrow K_e = K_1 = \dfrac{3EI}{l^3}$$

$$f_n = \dfrac{1}{2\pi}\sqrt{\dfrac{3EI}{ml^3}}$$

(ii) When $I = \infty$

# Linear Vibration Analysis of Mechanical Systems
Chapter 7

$$\frac{1}{K_e} = \frac{1}{K_1} + \frac{1}{K} = \frac{L^3}{3E.\infty} + \frac{1}{K}$$

$$\therefore K_e = K$$

$$\therefore f_n = \frac{1}{2\pi}\sqrt{\frac{K}{m}}$$

Q. A m/c of mass 500 kg. It is supported on helical springs which deflect by 5 mm due to the weight of the m/c. The m/c has rotating unbalanced equal to 250 kg mm. The speed of the m/c is 1200 rpm. Determine the dynamic amplitude. The damping factor of the viscous damper is 0.4.

Now this m/c is mounted on a larger concrete block of mass 1200 kg. The stiffness of the spring is changed such that the static deflection is still the same with the same viscous damper as in earlier case. Determine the change in the dynamic amplitude. **[IES-2007]**

Solution: m = 500 kg

$$\delta = 5 \times 10^{-3}$$

$$m_o e = 250 \times 10^{-3}$$

$$N = 1200 \text{ rpm}, \omega = 125.66 \text{ rad/sec.}$$

$$\xi = 0.4$$

Now $K = \dfrac{m_g}{\delta} = 9.81 \times 10^5$ N/m

$$\omega_n = \sqrt{\frac{K}{m}} = 44.29 \text{ rad/sec.}$$

$$\therefore X = \frac{\dfrac{F_o}{K}}{\sqrt{\left[1-\left(\dfrac{\omega}{\omega_n}\right)^2\right]^2 + \left[2\xi\dfrac{\omega}{\omega_n}\right]^2}}$$

$$= \frac{\dfrac{250 \times 10^{-3} \times 125.66^2}{9.81 \times 10^5}}{\sqrt{\left[1-\left(\dfrac{125.66}{44.29}\right)^2\right]^2 + \left[2 \times 0.4 \times \dfrac{125.66}{44.29}\right]^2}}$$

$$= 0.543 \times 10^{-3} \text{ m.}$$

Now if m = 500 + 1200 = 1700 kg.

$$\therefore K = 33.354 \times 10^5 \text{ N/m.}$$

$$\omega_n = \sqrt{\frac{33.354 \times 10^5}{1700}} = 44.29 \text{ rad/s.}$$

$$X_{new} = \frac{\dfrac{F_o}{K}}{\sqrt{\left[1-\left(\dfrac{\omega}{\omega_n}\right)^2\right]^2 + \left[2\xi\dfrac{\omega}{\omega_n}\right]^2}}$$

$$= 0.16 \times 10^{-3} \text{ m.}$$

$\therefore$ Change in dynamic amplitude = $X - X_{new}$

$$= 0.383 \times 10^{-3} \text{ m.}$$

Linear Vibration Analysis of Mechanical Systems

Chapter 7

Objective Questions (IES, IAS, GATE)

Previous 20-Years GATE Questions

Natural frequency of free longitudinal vibration

GATE-1. A simple pendulum of length 5 m, with a bob of mass 1 kg, is in simple harmonic motion as it passes through its mean position, the bob has a speed of 5 m/s. The net force on the bob at the mean position is [GATE-2005]
(a) zero (b) 2.5 N (c) 5 N (d) 25N

GATE-2. A mass m attached to a light spring oscillates with a period of 2 sec. If the mass is increased by 2 kg, the period increases by 1sec. The value of m is
(a) 1 kg (b) 1.6 kg (c) 2 kg (d) 2.4kg **[GATE-1994]**

GATE-3. The natural frequency of a spring-mass system on earth is ω_n. The natural frequency of this system on the moon $(g_{moon} = g_{earth}/6)$ is   [GATE-2010]
(a) ω_n (b) $0.408\omega_n$ (c) $0.204\omega_n$ (d) $0.167\omega_n$

GATE-4. Consider the system of two wagons shown in Figure. The natural frequencies of this system are   [GATE-1999]

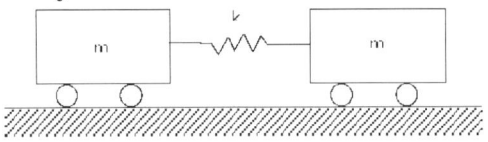

(a) $0, \dfrac{\sqrt{2k}}{m}$ (b) $\dfrac{\sqrt{k}}{m}, \dfrac{\sqrt{2k}}{m}$ (c) $\dfrac{\sqrt{k}}{m}, \dfrac{\sqrt{k}}{2m}$ (d) $0, \dfrac{\sqrt{k}}{2m}$

GATE-5. The differential equation governing the vibrating system is   [GATE-2006]

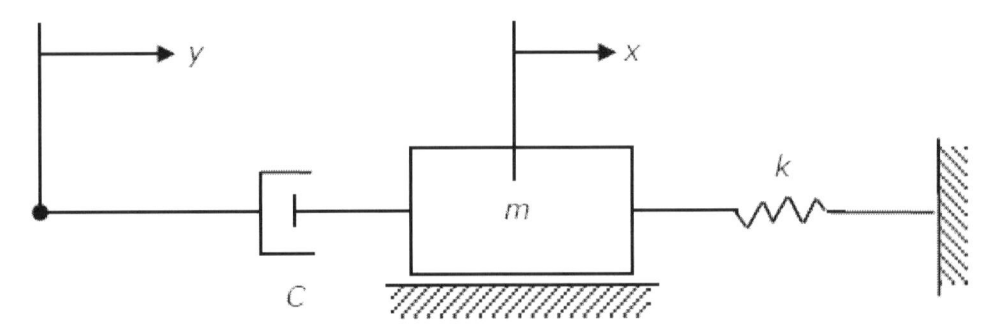

(a) $m\ddot{x} + c\dot{x} + k(x-y) = 0$
(b) $m(\ddot{x}-\ddot{y}) + c(\dot{x}-\dot{y}) + kx = 0$
(c) $m\ddot{x} + c(\dot{x}-\dot{y}) + kx = 0$
(d) $m(\ddot{x}-\ddot{y}) + c(\dot{x}-\dot{y}) + k(x-y) = 0$

GATE-6. The natural frequency of the spring mass system shown in the figure is closest to   [GATE-2008]

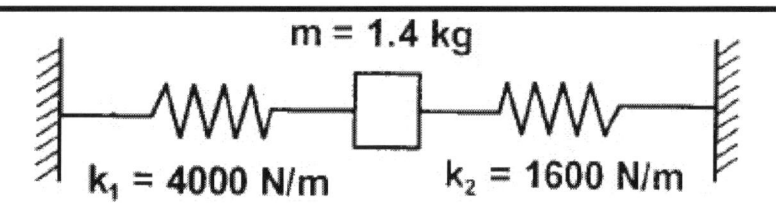

Linear Vibration Analysis of Mechanical Systems
Chapter 7

(A) 8 Hz (B) 10 Hz (C) 12 Hz (D) 14 Hz

GATE-7. A uniform rigid rod of mass m = 1 kg and length L = 1 m is hinged at its centre and laterally supported at one end by a spring of constant k = 300 N/m. The natural frequency (ω_n in rad/s is **[GATE-2008]**
(A) 10 (B) 20 (C) 30 (D) 40

GATE-8. Consider the arrangement shown in the figure below where J is the combined polar mass moment of inertia of the disc and the shafts. K₁, K₂, K₃ are the torsional stiffness of the respective shafts. The natural frequency of torsional oscillation of the disc is given by **[GATE-2003]**

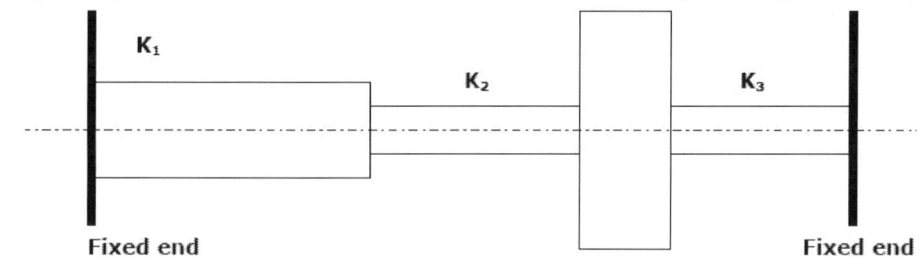

(a) $\sqrt{\dfrac{K_1 + K_2 + K_3}{J}}$
(b) $\sqrt{\dfrac{K_1 K_2 + K_2 K_3 + K_3 K_1}{J(K_1 + K_2)}}$
(c) $\sqrt{\dfrac{K_1 K_2 K_3}{J(K_1 K_2 + K_2 K_3 + K_3 K_1)}}$
(d) $\sqrt{\dfrac{K_1 K_2 + K_2 K_3 + K_3 K_1}{J(K_2 + K_3)}}$

GATE-10. As shown in Figure, a mass of 100 kg is held between two springs. The natural frequency of vibration of the system, in cycles/s, is

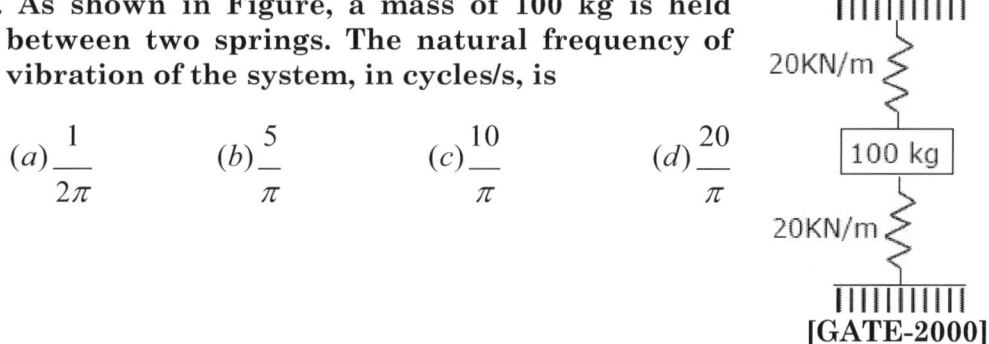

(a) $\dfrac{1}{2\pi}$ (b) $\dfrac{5}{\pi}$ (c) $\dfrac{10}{\pi}$ (d) $\dfrac{20}{\pi}$

[GATE-2000]

Data for Q. 11 - 12 are given below. Solve the problems and choose correct answers. A uniform rigid slender bar of mass 10 kg, hinged at the left end is suspended with the help of spring and damper arrangement as shown in the figure where K = 2 kN/m, C = 500 Ns/m and the stiffness of the torsional spring k_θ is 1 kN/m/rad. Ignore the hinge dimensions.

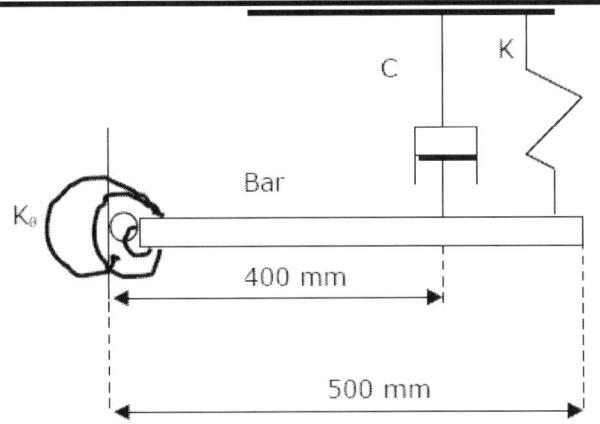

Linear Vibration Analysis of Mechanical Systems
Chapter 7

GATE-11. The un-damped natural frequency of oscillations of the bar about the hinge point is [GATE-2003]
(a) 42.43 rad/s (b) 30 rad/s (c) 17.32 rad/s (d) 14.14 rad/s

GATE-12. The damping coefficient in the vibration equation is given by [GATE-2003]
(a) 500 Nms/rad (b) 500 N/(m/s) (c) 80 Nms/rad (d) 80 N/(m/s)

GATE-13. In the figure shown, the spring deflects by δ to position A (the equilibrium position) when a mass m is kept on it. During free vibration, the mass is at position B at some instant. The change in potential energy of the spring-mass system from position A to position B is [GATE-2001]

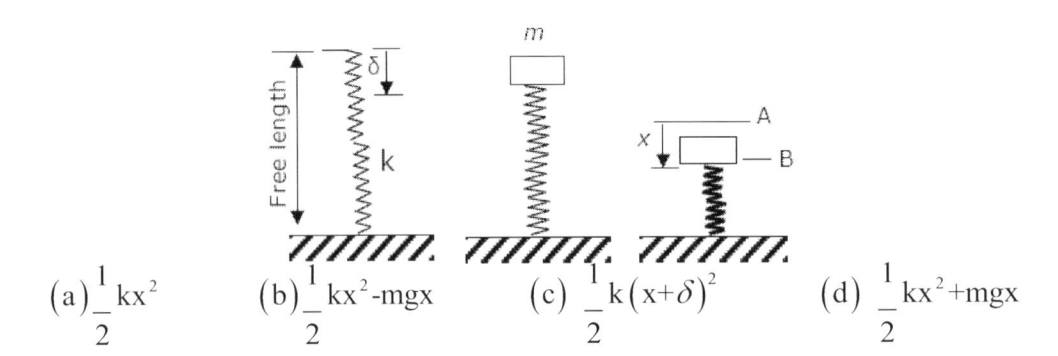

(a) $\dfrac{1}{2}kx^2$ (b) $\dfrac{1}{2}kx^2 - mgx$ (c) $\dfrac{1}{2}k(x+\delta)^2$ (d) $\dfrac{1}{2}kx^2 + mgx$

GATE-14. A mass of 1 kg is suspended by means of 3 springs as shown in figure. The spring constants K_1, K_2 and K_3 are respectively 1 kN/m, 3kN/m and 2 kN/m. The natural frequency of the system is approximately

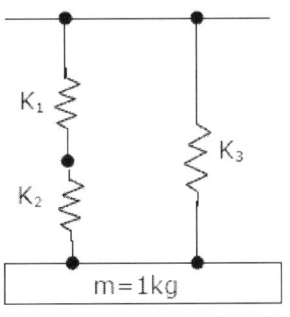

[GATE-1996]
(a) 46.90 Hz (b) 52.44 Hz (c) 60.55 Hz (d) 77.46 Hz

Linear Vibration Analysis of Mechanical Systems

Chapter 7

GATE-15. The assembly shown in the figure is composed of two mass less rods of length 'l' with two particles, each of mass m. The natural frequency of this assembly for small oscillations is

(a) $\sqrt{g/l}$

(b) $\sqrt{2g/(l\cos\alpha)}$

(c) $\sqrt{g/(l\cos\alpha)}$

(d) $\sqrt{(g\cos\alpha)/l}$

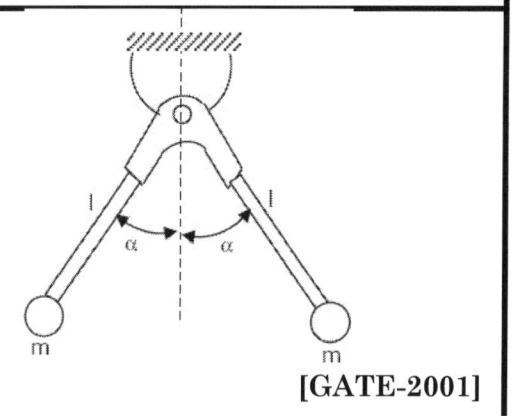

[GATE-2001]

Natural frequency of free transverse vibration

GATE-16. A cantilever beam of negligible weight is carrying a mass M at its free end, and is also resting on an elastic support of stiffness k_1 as shown in the figure below.
If k_2 represents the bending stiffness of the beam, the natural frequency (rad/s) of the system is [GATE-1993]

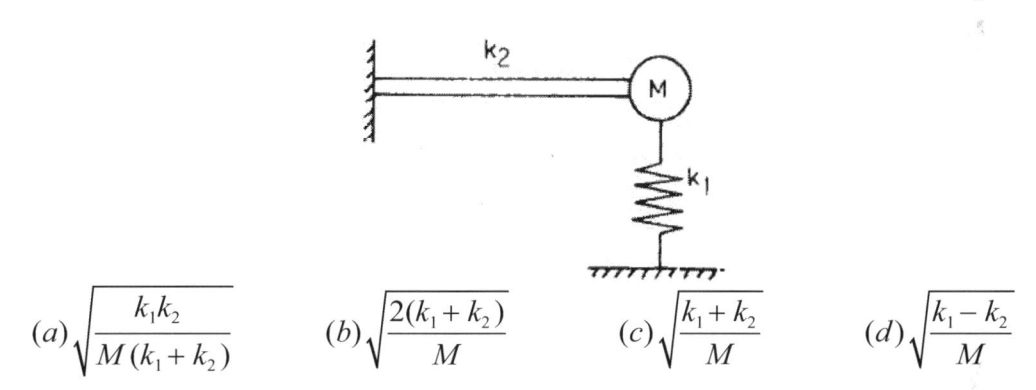

(a) $\sqrt{\dfrac{k_1 k_2}{M(k_1+k_2)}}$
(b) $\sqrt{\dfrac{2(k_1+k_2)}{M}}$
(c) $\sqrt{\dfrac{k_1+k_2}{M}}$
(d) $\sqrt{\dfrac{k_1-k_2}{M}}$

Effect of Inertia on the longitudinal and transverse vibration

GATE-17. If the length of the cantilever beam is halved, then natural frequency of the mass M at the end of this cantilever beam of negligible mass is increased by a factor of [GATE-2002]

(a) 2   (b) 4   (c) $\sqrt{8}$ (d) 8

Rayleigh's method (accurate result)

GATE-18. There are four samples P, Q, R and S, with natural frequencies 64, 96, 128 and 256 Hz, respectively. They are mounted on test setups for conducting vibration experiments. If a loud pure note of frequency 144 Hz is produced by some instrument, which of the samples will show the most perceptible induced vibration? [GATE-2005]

(a) P   (b) Q   (c) R   (d) S

Linear Vibration Analysis of Mechanical Systems

Chapter 7

Damping factor

GATE-19. A machine of 250 kg mass is supported on springs of total stiffness 100 kN/m. Machine has an unbalanced rotating force of 350 N at speed of 3600 rpm. Assuming a damping factor of 0.15, the value of transmissibility ratio is [GATE-2006]
(a) 0.0531 (b) 0.9922 (c) 0.0162 (d) 0.0028

GATE-20. The natural frequency of an undamped vibrating system is 100 rad/s A damper with a damping factor of 0.8 is introduced into the system, The frequency of vibration of the damped system, m rad/s, is [GATE-2000]
(a) 60 (b) 75 (c) 80 (d) 100

GATE-21. A mass M, of 20 kg is attached to the free end of a steel cantilever beam of length 1000 mm having a cross-section of 25 x 25 mm. Assume the mass of the cantilever to be negligible and E_{steel} = 200 GPa. If the lateral vibration of this system is critically damped using a viscous damper, then damping constant of the damper is
(a) 1250 Ns/m (b) 625 Ns/m
(c) 312.50 Ns/m (d) 156.25 Ns/m

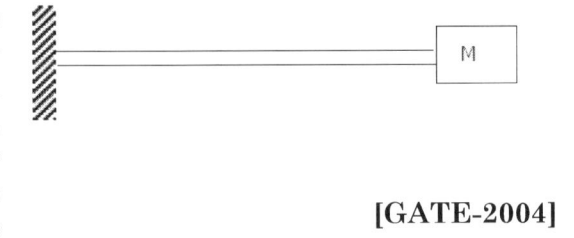

[GATE-2004]

GATE-22. In a spring-mass system, the mass is 0.1 kg and the stiffness of the spring is 1 kN/m. By introducing a damper, the frequency of oscillation is found to be 90% of the original value. What is the damping coefficient of the damper? [GATE-2005]
(a) 1.2 N.s/m (b) 3.4 N.s/m (c) 8.7 N.s/m (d) 12.0 N.s/m

GATE-23. A mass m attached to a spring is subjected to a harmonic force as shown in figure. The amplitude of the forced motion is observed to be 50 mm. The value of m (in kg) is[GATE-2010]

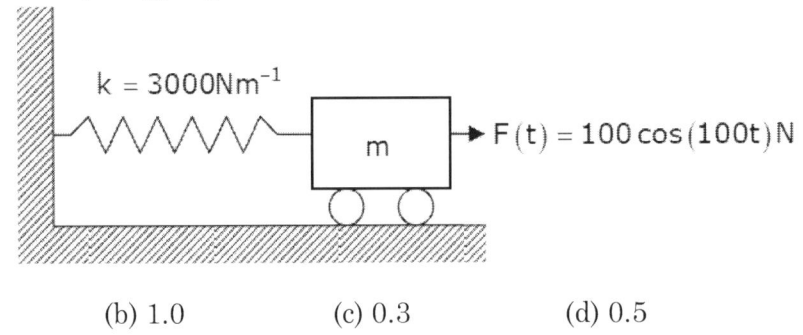

(a) 0.1 (b) 1.0 (c) 0.3 (d) 0.5

Magnification factor or Dynamic magnifier

Statement for Linked Answer Questions 24 & 25:
A vibratory system consists of a mass 12.5 kg, a spring of stiffness 1000 N/m, and a dashpot with damping coefficient of 15 Ns/m.

GATE-24. The value of critical damping of the system is [GATE-2006]
(a) 0.223 Ns/m (b) 17.88 Ns/m (c) 71.4 Ns/m (d) 223.6 Ns/m

Linear Vibration Analysis of Mechanical Systems

Chapter 7

GATE-25. The value of logarithmic decrement is [GATE-2006]
 (a) 1.35 (b) 1.32 (c) 0.68 (d) 0.42

GATE-26(i). A vibrating machine is isolated from the floor using springs. If the ratio of excitation frequency of vibration of machine to the natural frequency of the isolation system is equal to 0.5, then transmissibility of ratio of isolation is [GATE-2004]
 (a) $\dfrac{1}{2}$ (b) $\dfrac{3}{4}$ (c) $\dfrac{4}{3}$ (d))2

GATE-27. A vehicle suspension system consists of a spring and a damper. The stiffness of the spring is 3.6 kN/m constant of the damper is 400 Ns/m. If the mass is 50 kg, then the damping factor (D) and damped natural frequency (f_n), respectively, are [GATE -2009]
 (a) 0.471 and 1.19 H_z (b) 0.471 and 7.48 H_z
 (c) 0.666 and 1.35 H_z (d) 0.666 and 8.50 H_z

Previous 20-Years IES Questions

IES-1. Match List-I (Property) with List-II (System) and select the correct answer using the code given below the Lists: [IES-2006]

List-I	List - II
A. Resonance	1. Closed-loop control system
B. On-off control	2. Free vibrations
C. Natural frequency	3. Excessively large amplitude
D. Feedback signal	4. Mechanical brake

	A	B	C	D			A	B	C	D
(a)	1	2	4	3	(b)		3	4	2	1
(c)	1	4	2	3	(d)		3	2	4	1

IES-2. A reciprocating engine, running at 80 rad/s, is supported on springs. The static deflection of the spring is 1 mm. Take g = 10 rn/s². When the engine runs, what will be the frequency of vibrations of the system? [IES-2009]
 (a) 80 rad/s (b) 90 rad/s (c) 100 rad/s (d) 160 rad/s

IES-3. The static deflection of a shaft under a flywheel is 4 mm. Take g = 10m/s². What is the critical speed in rad/s? [IES-2009]
 (b) 50 (b) 20 (c) 10 (d) 5

IES-4. A rod of uniform diameter is suspended from one of its ends in vertical plane. The mass of the rod is 'm' and length' ℓ ', the natural frequency of this rod in Hz for small amplitude is [IES-2002]
 (a) $\dfrac{1}{2\pi}\sqrt{\dfrac{g}{l}}$ (b) $\dfrac{1}{2\pi}\sqrt{\dfrac{g}{3l}}$ (c) $\dfrac{1}{2\pi}\sqrt{\dfrac{2g}{3l}}$ (d) $\dfrac{1}{2\pi}\sqrt{\dfrac{3g}{2l}}$

IES-5. The equation of free vibrations of a system is $\ddot{x}+36\pi^2 x=0$. Its natural frequency is [IES-1995]
 (a) 6 Hz (b) 3π Hz (c) 3 Hz (d) 6π Hz.

Linear Vibration Analysis of Mechanical Systems
Chapter 7

IES-6. If air resistance is neglected, while it is executing small oscillations the acceleration of the bob of a simple pendulum at the mid-point of its swing will be [IES-1997]
(a) zero
(b) a minimum but not equal to zero
(c) a maximum
(d) not determinable unless the length of the pendulum and the mass of the bob are known

IES-7. A simple spring mass vibrating system has a natural frequency of N. If the spring stiffness is halved and the mass is doubled, then the natural frequency will become [IES-1993]
(a) N/2 (b) 2N (c) 4N (d) 8N

IES-8.

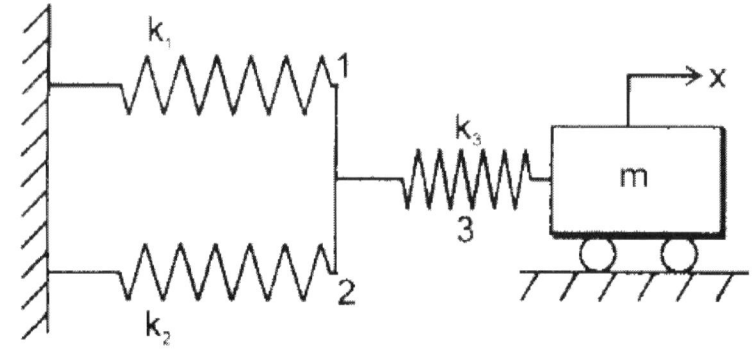

Which one of the following is the correct value of the natural frequency (ω_n) of the system given above? [IES-2005]

(a) $\left[\dfrac{1}{\left\{\dfrac{1}{(k_1+k_2)}+\dfrac{1}{k_3}\right\}m}\right]^{1/2}$
(b) $\left(\dfrac{3k}{m}\right)^{1/2}$
(c) $\left(\dfrac{k}{3m}\right)^{1/2}$
(d) $\left[\dfrac{k_3+\left(\dfrac{1}{\dfrac{1}{k_1}+\dfrac{1}{k_2}}\right)}{m}\right]^{1/2}$

IES-9. A mass M vibrates on a frictionless platform between two sets of springs having individual spring constant as shown in the figure below. What is the combined spring constant of the system? [IES-2009]

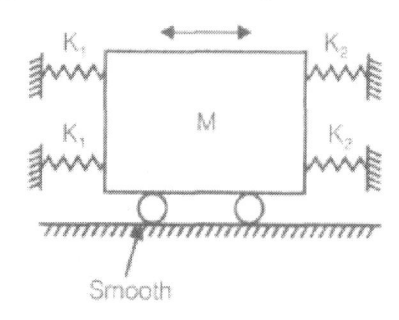

(a) K_1+K_2
(b) $2(K_1+K_2)$
(c) $\dfrac{K_1K_2}{K_1+K_2}$
(d) $\dfrac{2.(K_1K_2)}{K_1+K_2}$

Linear Vibration Analysis of Mechanical Systems
Chapter 7

IES10. The figure above shows the schematic of an automobile having a mass of 900 kg and the suspension spring constant of 81×10^4 N/m. If it travels at a speed of 72 km/hr on a rough road with periodic waviness as shown, what is the forcing frequency of the road on the wheel?

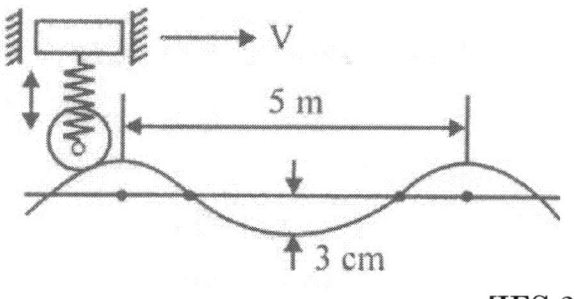

[IES-2008]

(a) 10 Hz   (b) 4 Hz   (c) 1·5 Hz   (d) 20 Hz

IES-11. For the system shown in the given figure the moment of inertia of the weight W and the ball about the pivot point is I_o. The natural frequency of the system is given by   [IES-1993]

$$f_n = \frac{1}{2\pi}\sqrt{\frac{Ka^2 - Wb}{I_o}}$$

The system will vibrate when

(a) $b < \dfrac{Ka^2}{W}$ (b) $b = \dfrac{Ka^2}{W}$ (c) $b > \dfrac{Ka^2}{W}$ (d) $a = 0$

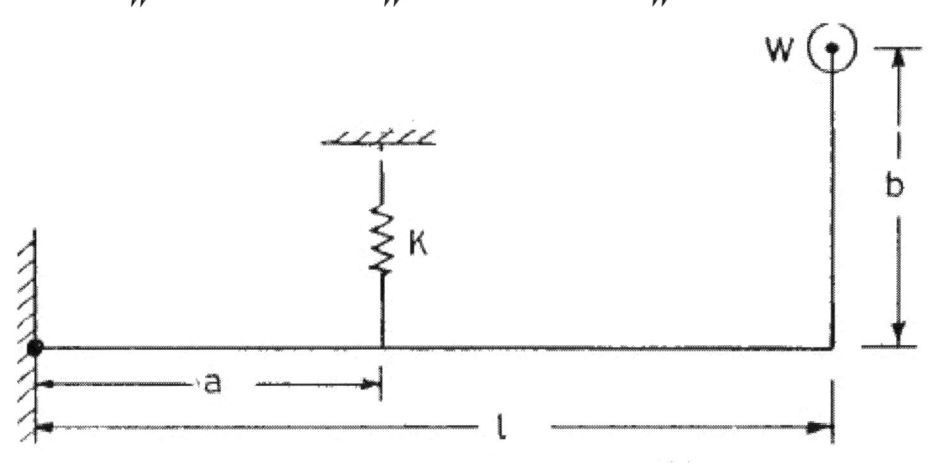

IES-12. For the spring-mass system shown in the given figure, the frequency of oscillations of the block along the axis of the springs is

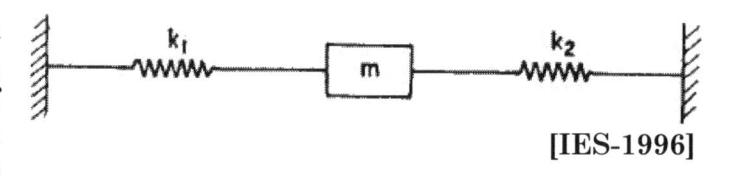

[IES-1996]

(a) $\dfrac{1}{2\pi}\sqrt{\dfrac{k_1 - k_2}{m}}$ (b) $\dfrac{1}{2\pi}\sqrt{\dfrac{k_1 k_2}{(k_1 + k_2)m}}$ (c) $\dfrac{1}{2\pi}\sqrt{\dfrac{k_1 + k_2}{m}}$ (d) $\dfrac{1}{2\pi}\sqrt{\dfrac{m}{(k_1 + k_2)}}$

Linear Vibration Analysis of Mechanical Systems

Chapter 7

IES-13. For the spring-mass system shown in the figure 1, the frequency of vibration is N. What will be the frequency when one more similar spring is added in series, as shown in figure 2?
(a) N/2
(b) N/$\sqrt{2}$
(c) $\sqrt{2}$/N
(d) 2N.

[IES-1995]

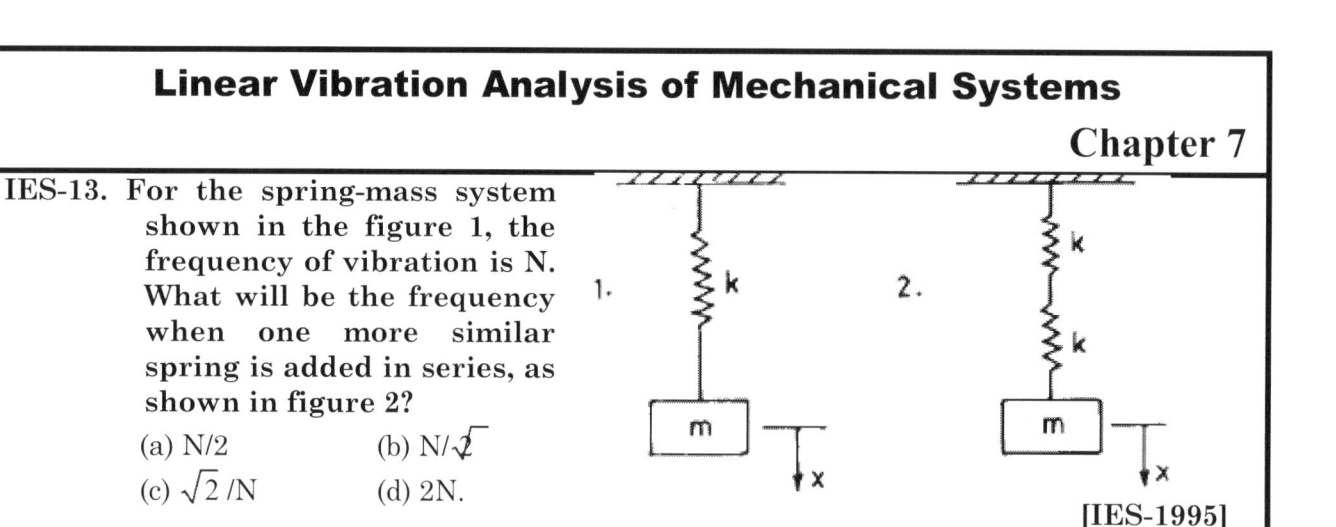

IES-14. Match List I (Applications) with List II (Features of vibration) and select the correct answer using the codes given below the Lists: [IES-2000]

List I
A. Vibration damper
B. Shock absorber
C. Frahm tachometer
D. Oscillator

List II
1. Frequency of free vibration
2. Forced vibration
3. Damping of vibration
4. Transverse vibration
5. Absorption of vibration

Code:	A	B	C	D		A	B	C	D
(a)	5	3	2	1	(b)	3	1	4	2
(c)	5	3	4	1	(d)	3	4	2	5

Natural frequency of free transverse vibration

IES-15. The natural frequency of transverse vibration of a massless beam of length L having a mass m attached at its midspan is given by (EI is the flexural rigidity of the beam) [IES-2001]

(a) $\left(\dfrac{mL^3}{48EI}\right)^{\frac{1}{2}}$ rad/s
(b) $\left(\dfrac{48mL^3}{EI}\right)^{\frac{1}{2}}$ rad/s
(c) $\left(\dfrac{48EI}{mL^3}\right)^{\frac{1}{2}}$ rad/s
(d) $\left(\dfrac{3EI}{mL^3}\right)^{\frac{1}{2}}$ rad/s

IES-16. A system is shown in the following figure. The bar AB is assumed to be rigid and weightless.
The natural frequency of vibration of the system is given by

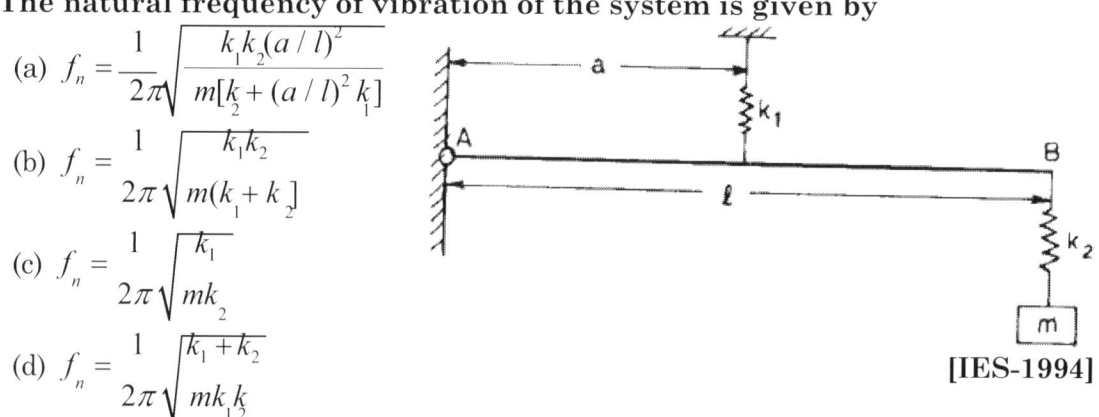

(a) $f_n = \dfrac{1}{2\pi}\sqrt{\dfrac{k_1 k_2 (a/l)^2}{m[k_2 + (a/l)^2 k_1]}}$

(b) $f_n = \dfrac{1}{2\pi}\sqrt{\dfrac{k_1 k_2}{m(k_1 + k_2)}}$

(c) $f_n = \dfrac{1}{2\pi}\sqrt{\dfrac{k_1}{mk_2}}$

(d) $f_n = \dfrac{1}{2\pi}\sqrt{\dfrac{k_1 + k_2}{mk_1 k_2}}$

[IES-1994]

Effect of Inertia on the longitudinal and transverse vibration

Linear Vibration Analysis of Mechanical Systems
Chapter 7

IES-17. A uniform bar, fixed at one end carries a heavy concentrated mass at the other end. The system is executing longitudinal vibrations. The inertia of the bar may be taken into account by which one of the following portions of the mass of the bar at the free end? [IES 2007]

(a) $\dfrac{5}{384}$ (b) $\dfrac{1}{48}$ (b) $\dfrac{33}{140}$ (d) $\dfrac{1}{3}$

IES-18. If a mass 'm' oscillates on a spring having a mass m_s and stiffness 'k', then the natural frequency of the system is given by [IES-1998]

(a) $\sqrt{\dfrac{k}{m + \dfrac{m_s}{3}}}$ (b) $\sqrt{\dfrac{k}{\dfrac{m}{3} + m_s}}$ (c) $\sqrt{\dfrac{3k}{m + m_s}}$ (d) $\sqrt{\dfrac{k}{m + m_s}}$

Rayleigh's method (accurate result)

IES-19.

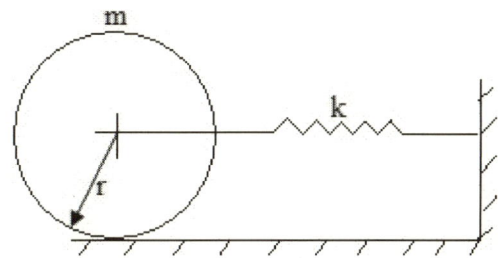

A rolling disc of radius 'r' and mass 'm' is connected to one end of a linear spring of stiffness 'k', as shown in the above figure. The natural frequency of oscillation is given by which one of the following? [IES 2007]

(a) $\omega = \sqrt{\dfrac{2k}{3m}}$ (b) $\omega = \sqrt{\dfrac{k}{m}}$ (c) $\omega = \sqrt{\dfrac{k}{2m}}$ (d) $\omega = \sqrt{\dfrac{2k}{m}}$

IES-20. The value of the natural frequency obtained by Rayleigh's method
(a) is always greater than the actual fundamental frequency [IES-1999]
(b) is always less than the actual fundamental frequency
(c) depends upon the initial deflection curve chose and may be greater than or less than the actual fundamental frequency
(d) is independent of the initial deflection curve chosen

IES-21. Which of the following methods can be used to determine the damping of machine element? [IES-1995]
1. Logarithmic method   2. Band-width method
3. Rayleigh method   4. Hozer method
Select the correct answer using the codes given below:
Codes:
(a) 1 and 3   (b) 1 and 2   (c) 3 and 4   (d) 1, 3 and 4.

Frequency of free damped vibration

IES-22. A system has viscous damped output. There is no steady-state lag if input is [IES-2001]
(a) unit step displacement   (b) step velocity
(c) harmonic   (d) step velocity with error-rate damping

Linear Vibration Analysis of Mechanical Systems

Chapter 7

Damping factor

IES-23. A motion is aperiodic at what value of the damping factor? [IES 2007]
(a) 1.0 or above (b) 0.5 (c) 0.3 (d) 0.866

IES-24. The equation of motion for a damped viscous vibration is $3\ddot{x} + 9\dot{x} + 27x = 0$
The damping factor is [IES-2000]
(a) 0.25 (b) 0.50 (c) 0.75 (d) 1.00

IES-25. The equation of motion for a single degree of freedom system [IES-1996]
$4\ddot{x} + 9\dot{x} + 16x = 0$
The damping ratio of the system is

(a) $\dfrac{9}{128}$
(b) $\dfrac{9}{16}$
(c) $\dfrac{9}{8\sqrt{2}}$
(d) $\dfrac{9}{8}$

IES-26. A mass of 1 kg is attached to the end of a spring with stiffness 0.7 N/mm. The critical damping coefficient of this system is [IES-1994]
(a) 1.40 Ns/m (b) 18.522 Ns/m (c) 52.92 Ns/m (d) 529.20 Ns/m

Logarithmic Decrement

IES-27. A damped free vibration is expressed by the general equation
$x = Xe^{-\xi\omega_n t}\sin\left(\sqrt{1-\xi^2}\,\omega_n t + \phi\right)$
which is shown graphically below:
The envelope A has the equation: [IES-1997]

(a) Xe^{-t}
(b) $X\sin\left(\sqrt{1-\xi^2}\,\omega t\right)_n$
(c) $e^{-\xi\omega_n t}$
(d) $Xe^{-\xi\omega_n t}$

IES-28. The amplitude versus time curve of a damped-free vibration is shown in the figure. Curve labelled 'A' is [IES-1998]

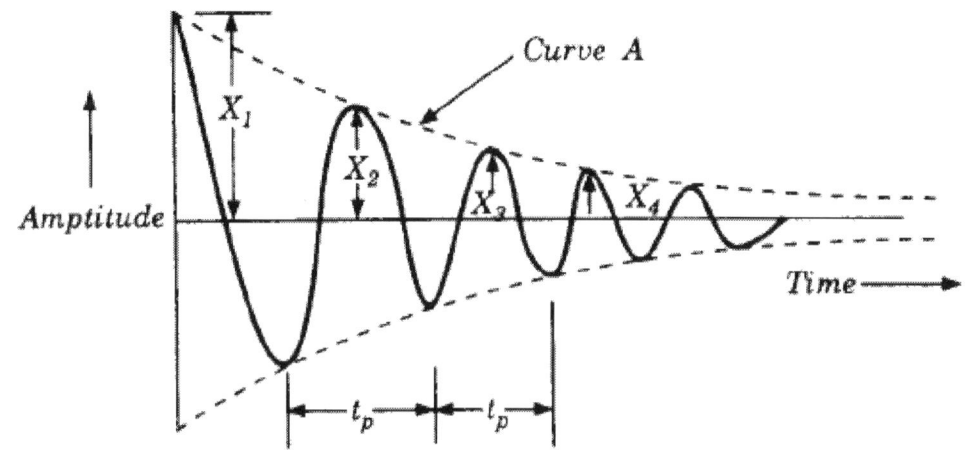

(a) a logarithmic decrement curve
(b) an exponentially decreasing curve
(c) a hyperbolic curve
(d) a linear curve

Linear Vibration Analysis of Mechanical Systems
Chapter 7

Frequency of under damped forced vibration

IES-29. With symbols having the usual meanings, the single degree of freedom system, $m\ddot{x} + c\dot{x} + kx = F\sin\omega t$ represents [IES-1993]
(a) free vibration with damping
(b) free vibration without damping
(c) forced vibration with damping
(d) forced vibration without damping

IES-30. The given figure depicts a vector diagram of forces and displacements in the case of Forced Damped Vibration. If vector A represents the forcing function P = P₀sin ω t, vector B the displacement y = Y sin ωt, and φ the phase angle between them, then the vectors C and D represent respectively
(a) the force of inertia and the force of damping
(b) the elastic force and the damping force
(c) the damping force and the inertia force
(d) the damping force and the elastic force

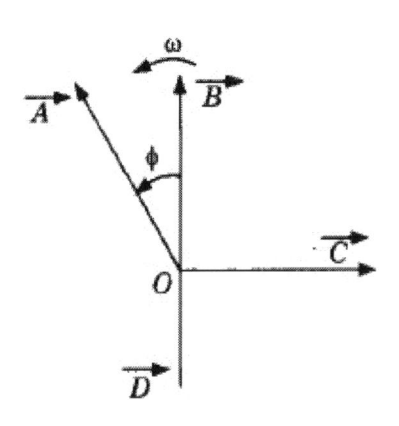

[IES-1997]

IES-31. In a forced vibration with viscous damping, maximum amplitude occurs when forced frequency is [IES-1999]
(a) Equal to natural frequency
(b) Slightly less than natural frequency
(c) Slightly greater than natural frequency
(d) Zero

IES-32. When the mass of a critically damped single degree of freedom system is deflected from its equilibrium position and released, it will
(a) return to equilibrium position without oscillation [IES-1996]
(b) Oscillate with increasing time period
(c) Oscillate with decreasing amplitude
(d) Oscillate with constant amplitude.

IES-33. Under logarithmic decrement, the amplitude of successive vibrations are
(a) Constant   (b) in arithmetic progression [IES-1992]
(c) In geometric progression   (d) in logarithmic progression

Statement for Linked Answer Questions 34 & 35:
A vibratory system consists of a mass 12.5 kg, a spring of stiffness 1000 N/m, and a dashpot with damping coefficient of 15 Ns/m.

IES-34. Match List-l with List-ll and select the correct answer using the code given below the lists: [IES-2009]

List-l	List-ll
A. Node point	1. Balancing of reciprocating masses
B. Critical damping	2. Torsional vibration of shafts
C. Magnification factor	3. Forced vibration of spring-mass system
D. Hammer blow	4. Damped vibration

   A B C D
(a) 1 4 3 2 (b) 2 4 3 1
(c) 1 3 4 2 (d) 2 3 4 1

Linear Vibration Analysis of Mechanical Systems
Chapter 7

Vibration Isolation and Transmissibility

IEA-35. If $\omega/\omega_n = \sqrt{2}$, where ω is the frequency of excitation and ω_n is the natural frequency of vibrations, then the transmissibility of vibrations will be
(a) 0.5 (b) 1.0 (c) 1.5 (d) 2.0 [IES-1995]

IES-36. Match List I (force transmissibility) with List II (frequency ratio) and select the correct answer using the codes given below the Lists: [IES-1994]

List I
A. 1
B. Less than 1
C. Greater than 1
D. Tending to zero

List II
1. $\dfrac{\omega}{\omega_n} > \sqrt{2}$
2. $\dfrac{\omega}{\omega_n} = \sqrt{2}$
3. $\dfrac{\omega}{\omega_n} \gg \sqrt{2}$
4. $\dfrac{\omega}{\omega_n} < \sqrt{2}$

Code: A B C D   A B C D
(a)   1 2 3 4   (b) 2 1 4 3
(c)   2 1 3 4   (d) 1 2 4 3

IES-37. When a shaking force is transmitted through the spring, damping becomes detrimental when the ratio of its frequency to the natural frequency is greater than [IES-1996]
(a) 0.25 (b) 0.50 (c) 1.00 (d) $\sqrt{2}$

IES-39. When a vehicle travels on a rough road whose undulations can be assumed to he sinusoidal, the resonant conditions of the base excited vibrations are determined by the [IES-2001]
(a) mass of the vehicle, stiffness of the suspension spring, speed of the vehicle, wavelength of the roughness curve
(b) speed of the vehicle only
(c) speed of the vehicle and the stiffness of the suspension spring
(d) amplitude of the undulations

IES-40. Given figure shows a flexible shaft of negligible mass of torsional stiffness K coupled to a viscous damper having a coefficient of viscous damping c. If at any instant the left and right ends of this shaft have angular displacements θ_1 and θ_2 respectively, then the transfer function, θ_2/θ_1 of the system is

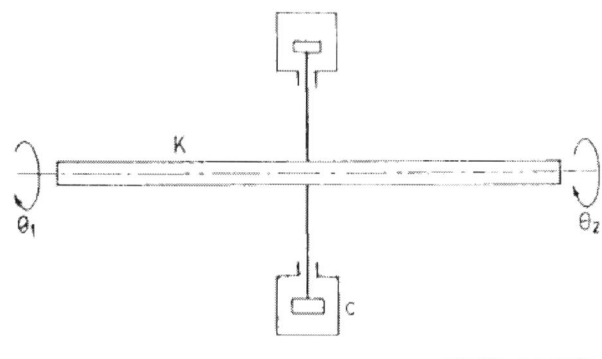

[IES-1995]

Linear Vibration Analysis of Mechanical Systems
Chapter 7

(a) $\dfrac{K}{K+c}$ (b) $\dfrac{1}{1+\dfrac{c}{K}s}$ (c) $\dfrac{1}{1+\dfrac{K}{c}s}$ (d) $1+\dfrac{K}{c}s$

IES-41. Consider the following statements: [IES-2008]
1. one way of improving vibration isolation is to decrease the mass of the vibrating object.
2. For effective isolation, the natural frequency of the system should be far less than the exciting frequency.

Which of the statements given above is/are correct?
(a) 1 only (b) 2 only
(c) Both 1 and 2 (d) Neither 1 nor 2

Torsional Vibration

IES-42. During torsional vibration of a shaft, the node is characterized by the
(a) maximum angular velocity (b) maximum angular displacement
(c) maximum angular acceleration (d) zero angular displacement [IES-2001]

IES-43. In a multi-rotor system of torsional vibration maximum number of nodes that can occur is [IES-1999]
(a) two (b) equal to the number of rotor plus one
(c) equal to the number of rotors (d) equal to the number of rotors minus one

IES-44. The above figure shows two rotors connected by an elastic shaft undergoing torsional vibration. The rotor (1) has a mass of 2 kg and a diameter of 60 cm, while the rotor (2) has a mass of 1 kg and a diameter of 20 cm. what is the distance l at which the node of vibration of torsional vibration occurs? [IES-2009]

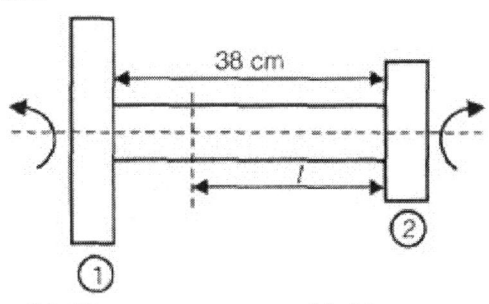

(a) 36 cm (b) 30 cm (c) 22 cm (d) 18

Torsionally equivalent shaft

IES-45. Two heavy rotating masses are connected by shafts of length l_1, l_2 and l_3 and the corresponding diameters are d_1, d_2 and d_3. This system is reduced to a torsionally equivalent system having uniform diameter d_1 of the shaft. The equivalent length of the shaft is equal to [IES-1997]

(a) $l_1 + l_2 + l_3$

(b) $\dfrac{l_1 + l_2 + l_3}{3}$

(c) $l_1 + l_2\left(\dfrac{d_1}{d_2}\right)^3 + l_3\left(\dfrac{d_1}{d_3}\right)^3$

(d) $l_1 + l_2\left(\dfrac{d_1}{d_2}\right)^4 + l_3\left(\dfrac{d_1}{d_3}\right)^4$

Linear Vibration Analysis of Mechanical Systems

Chapter 7

IES-46. Two heavy rotating masses are connected by shafts of lengths l_1, l_2 and l_3 and the corresponding diameters are d_1, d_2 and d_3. This system is reduced to a torsionally equivalent system having uniform diameter "d_1" of the shaft. The equivalent length of the shaft is [IES-1994]

(a) $\dfrac{l_1 + l_2 + l_3}{3}$

(b) $l_1 + l_2 \left(\dfrac{d_1}{d_2}\right)^3 + l_3 \left(\dfrac{d_1}{d_3}\right)^3$

(c) $l_1 + l_2 \left(\dfrac{d_1}{d_2}\right)^4 + l_3 \left(\dfrac{d_1}{d_3}\right)^4$

(d) $l_1 + l_2 + l_3$

Previous 20-Years IAS Questions

Natural frequency of free longitudinal vibration

IAS-1. Consider the following statements: [IAS-2002]

1. SHM is characteristic of all oscillating motions, where restoring force exists.
2. In SHM, the motion is of uniform velocity.
3. Frequency in SHM is equal to number of oscillations.
4. Frequency is number of complete cycles per unit time.

Which of the above statements are correct?
(a) 1, 2 and 3   (b) 1 and 4   (c) 1, 2 and 4   (d) 2, 3 and 4

IAS-2. Assertion (A): In a simple harmonic motion, the potential energy reaches its maximum value twice during each cycle. [IAS-2000]
Reason(R): Velocity becomes zero twice during each cycle.
(a) Both A and R are individually true and R is the correct explanation of A
(b) Both A and R are individually true but R is not the correct explanation of A
(c) A is true but R is false
(d) A is false but R is true

IAS-3. A disc of mass 'm' and radius 'r' is attached to a spring of stiffness 'k' During its motion, the disc rolls on the ground. When released from some stretched position, the centre of the disc will execute harmonic motion with a time period of [IAS 1994]

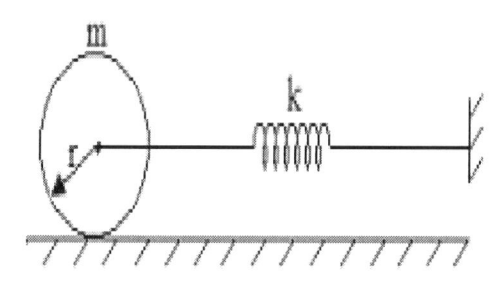

(a) $2\pi\sqrt{\dfrac{m}{ak}}$

(b) $2\pi\sqrt{\dfrac{m}{k}}$

(c) $2\pi\sqrt{\dfrac{3m}{2k}}$

(d) $2\pi\sqrt{\dfrac{2m}{k}}$

IAS-4. Consider the following statements: [IAS-1996]
The period of oscillation of the fluid column in a U-tube depends upon the
1. diameter of U-tube
2. length of the fluid column
3. acceleration due to gravity
Of these statements:
(a) 1, 2 and 3 are correct   (b) 1 and 3 are correct

Linear Vibration Analysis of Mechanical Systems

Chapter 7

(c) 1 and 2 are correct (d) 2 and 3 are correct

IAS-5. Consider the following statements: [IAS-1999]
1. Periodic time is the time for one complete revolution.
2. The acceleration is directed towards the centre of suspension.
3. The acceleration in proportional to distance from mean position.
Of these statements:
(a) 1, 2 and 3 are correct. (b) 2, 3 and 4 are correct
(c) 1, 3 and 4 correct (d) 1, 2 and 4 are correct

IAS-6. Two vibratory systems are shown in the given figures. The ratio of the natural frequency of longitudinal vibration of the second system to that of the first is
(a) 4 (b) 2 (c) 0.5 (d) 0.25

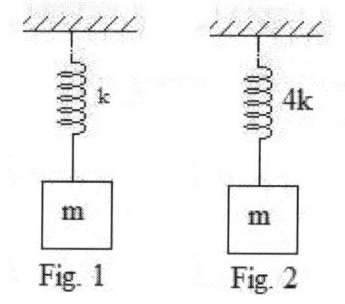

[IAS-1998]

IAS-7. A machine mounted on a single coil spring has a period of free vibration of T. If the spring is cut into four equal parts and placed in parallel and the machine is mounted on them, then the period of free vibration of the new system will become. [IAS-1995]
(a) 16T (b) 4T (c) $\dfrac{T}{4}$ (d) $\dfrac{T}{16}$

IAS-8. For the vibratory system shown in the given figure, the natural frequency of vibration in rad./sec is
(a) 43.3 (b) 86.6
(c) 100 (d) 200

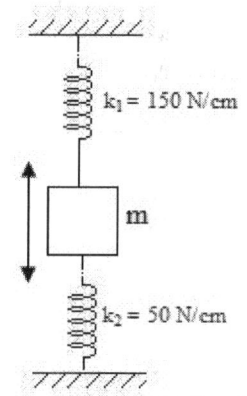

[IAS-1997]

Linear Vibration Analysis of Mechanical Systems
Chapter 7

IAS-9. The figure shows a rigid body oscillating about the pivot A. If J is mass moment of inertia of the body about the axis of rotation, its natural frequency for small oscillations is proportional to
(a) J
(b) J^2
(c) $\dfrac{1}{J}$
(d) $\dfrac{1}{\sqrt{J}}$

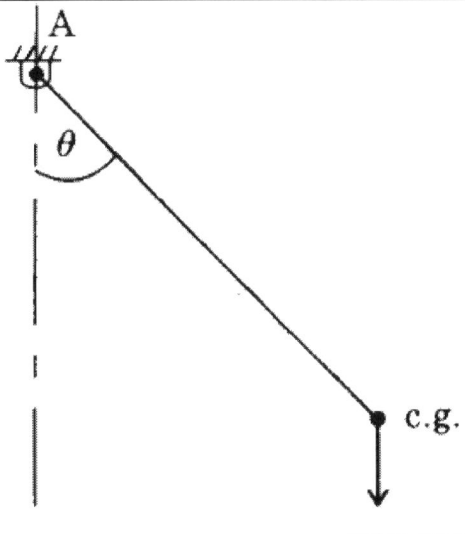

[IAS-2003]

IAS-10. A vibratory system is shown in the given figure. The flexural rigidity of the light cantilever beam is EI. The frequency of small vertical vibrations of mass m is [IAS-1997]

(a) $\dfrac{\sqrt{3EIk}}{(3EI+Kl^3)m}$
(b) $\dfrac{k}{m}$
(c) $\dfrac{\sqrt{kl^3+3EI}}{ml^3}$
(d) $\dfrac{\sqrt{kl^3-3EI}}{ml^3}$

IAS-11. A uniform cantilever beam undergoes transverse vibrations. The number of natural frequencies associated with the beam is [IAS-1998]
(a) 1
(b) 10
(c) 100
(d) infinite

IAS-12. A reed type tachometer uses the principle of
(a) torsional vibration
(b) longitudinal vibration
(c) transverse vibration
(d) damped free vibration

Effect of Inertia on the longitudinal and transverse vibration

IAS-13. In a simple spring mass vibrating system, the natural frequency ω_n of the system is (k is spring stiffness, m is mass and m_s, is spring mass) [IAS-2000]

(a) $\sqrt{\dfrac{K}{m - \dfrac{m_s}{3}}}$
(b) $\sqrt{\dfrac{K}{m + \dfrac{m_s}{3}}}$
(c) $\sqrt{\dfrac{K}{m + 3m_s}}$
(d) $\sqrt{\dfrac{K}{m - 3m_s}}$

Linear Vibration Analysis of Mechanical Systems

Chapter 7

Rayleigh's method (accurate result)

IAS-14. Consider the following methods: [IAS-2001]
1. Energy method 2. Equilibrium method 3. Rayleigh's method
Which of these methods can be used for determining the natural frequency of the free vibrations?
(a) 1 and 2 (b) 1, 2 and 3 (c) 1 and 3 (d) 2 and 3

IAS-15. Which one of the following pairs is correctly matched? [IAS-1995]
(a) Coulomb ------------Energy Principle
(b) Rayleigh ------------Dynamic Equilibrium
(c) D' Alembert --------Damping Force
(d) Fourier --------------Frequency domain analysis

IAS-16. Consider the following statements: [IAS-2003]
1. Critical or whirling speed of the shaft is the speed at which it tends to vibrate violently in the transverse direction.
2. To find the natural frequency of a shaft carrying several loads, the energy method gives accurate result.
3. Dunkerley's method gives approximate results of the natural frequency of a shaft carrying several loads.
Which of these statements is/are correct?
(a) 1 only (b) 2 and 3 (c) 1 and 3 (d) 1, 2 and 3

Frequency of free damped vibration

IAS-17. A viscous damping system with free vibrations will be critically damped if the damping factor is [IAS-2000]
(a) zero (b) less than one (c) equal to one (d) greater than one

IAS-18 The transmitted force through a mass-spring damper system will be greater than the transmitted through rigid supports for all values of damping factors, if the frequency ratio $\left(\dfrac{\omega}{\omega_n}\right)$ is [IAS-1999]

(a) more than $\sqrt{2}$ (b) less than $\sqrt{2}$
(c) equal to one (d) less than one

IAS-19. If a damping factor in a vibrating system is unity, then the system will
(a) have no vibrations (b) be highly damped [IAS-1996]
(c) be under damped (d) be critically damped

IAS-20. The figure shows a critically damped spring-mass system undergoing single degree of freedom vibrations. If m = 5 kg and k = 20 N/m, the value of viscous damping coefficient is
(a) 10 Ns/m (b) 20 Ns/m
(c) 4 Ns/m (d) 8 Ns/m

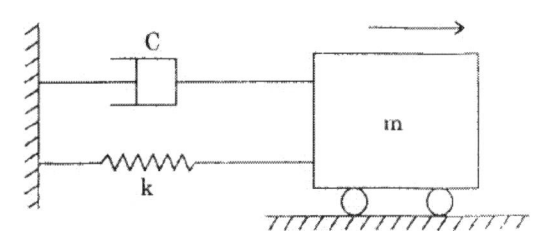

[IAS-2003]

Linear Vibration Analysis of Mechanical Systems

Chapter 7

IAS-21. A spring-mass suspension has a natural frequency of 40 rad/s. What is the damping ratio required if it is desired to reduce this frequency to 20 rad/s by adding a damper to it? [IAS-2004]
(a) $\dfrac{\sqrt{3}}{2}$ (b) $\dfrac{1}{2}$ (c) $\dfrac{1}{\sqrt{2}}$ (d) $\dfrac{1}{4}$

Logarithmic Decrement

IAS-22. The given figure shows vibrations of a mass 'M' isolated by means of springs and a damper. If an external force 'F' (=A sin ωt) acts on the mass and the damper is not used, then
(a) $\sqrt{\dfrac{k}{M}}$ (b) $\dfrac{1}{2}\sqrt{\dfrac{k}{M}}$
(c) $2\sqrt{\dfrac{k}{M}}$ (d) $\sqrt{\dfrac{k}{2M}}$

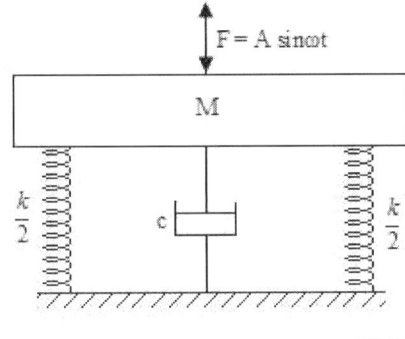

[IAS-1999]

IAS-23. For steady-state forced vibrations, the phase lag at resonance is [IAS-1996]
(a) 0^0 (b) 45^0 (c) 90^0 (d) 180^0

IAS-24. For a harmonically excited single degree of freedom viscous damped system, which one of the following is correct? [IAS-2007]
(a) Inertia force leads damping force by 90° while damping force leads spring force by 90°
(b) Spring force leads damping force by 90° while damping force leads inertia force by 180°
(c) Spring force and damping force are in phase, and inertia force leads them by 90°
(d) Spring force and inertia force are in phase, and damping force leads them by 90°

IAS-25. In a forced vibrations with viscous damping, maximum amplitude occurs when the forced frequency is [IAS-1999]
(a) equal to natural frequency (b) slightly less than natural frequency
(c) slightly greater than natural frequency (d) zero

IAS-27. The assumption of viscous damping in practical vibrating system is
(a) one of reality [IAS 1994]
(b) to make the resulting differential equation linear
(c) to make the resulting differential equatic1n non-liner
(d) to make the response of the mass linear with time

Magnification factor or Dynamic magnifier

IAS-30. In a system subjected to damped forced vibrations, the ratio of maximum displacement to the static deflection is known as [IAS-2003]
(a) Critical damping ratio (b) Damping factor
(c) Logarithmic decrement (d) Magnification factor

Linear Vibration Analysis of Mechanical Systems

Chapter 7

IAS-31. The ratio of the maximum dynamic displacement due to a dynamic force to the deflection due to the static force of the same magnitude is called the
(a) displacement ratio (b) deflection ratio [IAS 1994]
(c) force factor (d) magnification factor

IAS-32. Logarithmic decrement of a damped single degree of freedom system is δ. If the stiffness of the spring is doubled and the mass is made half, then the logcrithmic decrement of the new system will be equal to [IAS-1997]
(a) $\frac{1}{4}\delta$ (b) $\frac{1}{2}\delta$ (c) δ (d) 2δ

Vibration Isolation and Transmissibility

IAS-33. In a vibration isolation system, if $\frac{\omega}{\omega_n} > 1$, then what is the phase difference between the transmitted force and the disturbing force? [IAS-2007]
(a) 0° (b) 45° (c) 90° (d) 180°

IAS-34. For effective vibration isolation, the natural frequency w of the system must be (w is the forcing frequency) [IAS 1994]
(a) $\omega/4$ (b) ω (c) 4ω (d) 10ω

IAS-35. For a single degree of freedom viscous damped system, transmissibility is less than 1 if frequency ratio is [IAS-2007]
(a) Equal to 1 (b) < 1 (c) $< \sqrt{2}$ (d) $> \sqrt{2}$

IAS-36. Transmissibility is unity at two points. [IAS-2004]
Which one of the following is true for these two points?
(a) ω/ω_n is zero and $\sqrt{3}$ for all values of damping
(b) ω/ω_n is zero and $\sqrt{2}$ for all values of damping
(c) ω/ω_n is unity and 2 for all values of damping
(d) ω/ω_n is unity and $\sqrt{3}$ for all values of damping

IAS-37. Consider the following statements: [IAS-2003]
1. When frequency ratio is $< \sqrt{2}$, the force transmitted to the foundations is more than the exciting force.
2. When frequency ratio is $> \sqrt{2}$, the force transmitted to the foundations increases as the damping is decreased.
3. The analysis of base-excited vibrations is similar to that of forced vibrations.
Which of these statements are correct?
(a) 1 and 2 (b) 2 and 3 (c) 1 and 3 (d) 1, 2 and 3

IAS-38. Consider the following statements: [IAS-2001]
1. In forced vibrations, the body vibrates under the influence of an applied force.
2. In damped vibrations, amplitude reduces over every cycle of vibration.
3. In torsional vibrations, the disc moves parallel to the axis of shaft.
4. In transverse vibrations, the particles of the shaft moves approximately perpendicular to the axis of the shaft.

Linear Vibration Analysis of Mechanical Systems

Chapter 7

Which of these statements are correct?
(a) 1, 2 and 3 (b) 1, 3 and 4 (c) 2, 3 and 4 (d) 1, 2 and 4

IAS-39. A shaft, supported on two bearings at its ends, carries two flywheels 'L' apart. Mass moment of inertia of the two flywheels are I_a and I_b, I being the polar moment of inertia of cross-sectional area of the shaft. Distance I_a of the mode of torsional vibration of the shaft from flywheel I_a is given by
[IAS-1998]

(a) $l_a = \dfrac{LI_b}{I_a + I_b}$ (b) $l_a = \dfrac{LI_a}{I_a + I_b}$ (c) $l_a = \dfrac{LI_b}{I_a + I_b - I}$ (d) $l_a = \dfrac{LI_a}{I_a + I_b - I}$

IAS-40. Assertion (A): The longitudinal, transverse and torsional vibrations are simple harmonic. [IAS-1996]
Reason (R): The restoring force or couple is proportional velocity in the case of these vibrations.
(a) Both A and R are individually true and R is the correct explanation of A
(b) Both A and R are individually true but R is **not** the correct explanation of A
(c) A is true but R is false
(d) A is false but R is true

Answers with Explanation (Objective)

Previous 20-Years GATE Answers

GATE-1. Ans. (a) Force at mean position is zero.

GATE-2. Ans. (b) Period of oscillation = $\dfrac{\sqrt{\delta}}{4.99} \cdot \dfrac{\text{sec onds}}{\text{cycle}}$

∴ $\dfrac{\sqrt{\delta_1}}{4.99} = 2$ and $\dfrac{\sqrt{\delta_2}}{4.99} = 3$

Hence $\dfrac{\delta_1}{\delta_2} = \dfrac{4}{9}$

Now $mgk = \delta_1$ and $(m+2)gk = gk = \delta_2$

Where k is the stiffness of the spring.
∴ $9\delta_1 = 4\delta_2$
or $9\,mgk = 4(m+2)gk$
∴ $m = 1.6$ kg

GATE-3. Ans. (a) $\omega_n = \sqrt{\dfrac{k}{m}}$ neither mass nor stiffness depends on gravity. If you think about $\sqrt{\dfrac{g}{\delta}}$ then δ as g changes δ will also change by same factor.

GATE-4. Ans. (c)

GATE-5. Ans. (c) This is the differential equation governing the above vibrating system.

Linear Vibration Analysis of Mechanical Systems

Chapter 7

GATE-6. Ans. (b) $\dfrac{d^2y}{dx^2} + \dfrac{(K_1+K_2)}{m} y = 0$ Therefore $\omega_n = \dfrac{\pi}{2N} = \sqrt{\dfrac{K_1+K_2}{m}}$ or

$$N = \dfrac{1}{2\pi}\sqrt{\dfrac{4000+1600}{1.4}} = 10.06\,Hz$$

GATE-7. Ans. (a) A uniform rigid rod of mass m = 1 kg and length L = 1 m is hinged at its centre & laterally supported at one end by a spring of spring constant k = 300N/m. The natural frequency ω_n in rad/s is 10

GATE-8. Ans. (b) Equivalent stiffness $= \dfrac{K_1 K_2}{K_1+K_2} + K_3$

$$= \dfrac{K_1 K_2}{K_1+K_2} + K_3$$

Now natural frequency

$$= \sqrt{\dfrac{K_{eq}}{J}}$$

$$= \dfrac{\sqrt{K_1 K_2 + K_1 K_3 + K_2 K_3}}{J(K_1+K_2)}$$

GATE-10. Ans. (c)
$S = S_1 + S_2 = 20 + 20$
$= 40\ \text{kN/m} = 40{,}000\ \text{N/m}$

∴ Natural frequency of vibration of the system,

$$f_n = \dfrac{1}{2\pi}\sqrt{\dfrac{S}{m}} = \dfrac{1}{2\pi}\sqrt{\dfrac{40\times1000}{100}} = \dfrac{20}{2\pi} = \dfrac{10}{\pi}$$

$S_1 = 20\ KN/$
$m = 100\,kg$
$S_2 = 20$

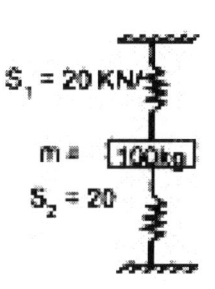

GATE-11. Ans. (a)

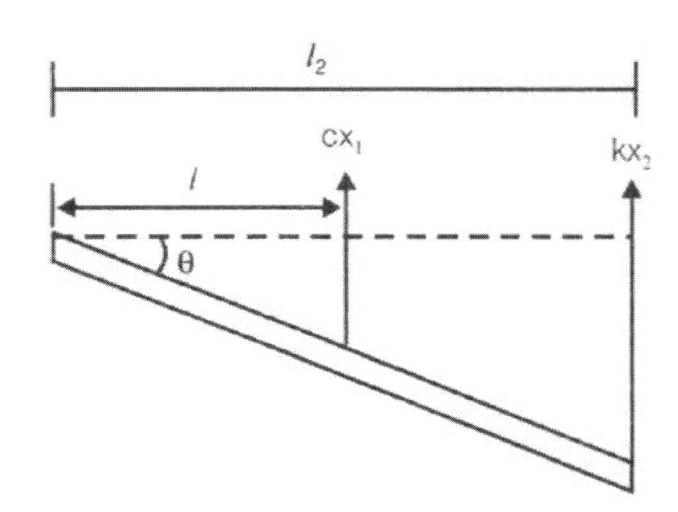

For small deflection, after equilibrium

Now, $\theta = \dfrac{x_1}{0.4} = \dfrac{x_2}{0.5}$

∴ $x_1 = 0.4\theta$

and $x_2 = 0.5\theta$

Moment of inertia

Linear Vibration Analysis of Mechanical Systems
Chapter 7

$$= \frac{ml^2}{3} = \frac{10 \times (0.5)^2}{3} = 0.833 \text{ kg-m}^2$$

$$c\dot{x}_1 l_1 + kx_2 l_2 + k_\theta \cdot \theta + I\ddot{\theta} = 0$$

$$\Rightarrow Cl_1^2 \dot{\theta} + kl_2^2 + k_\theta \cdot \theta + I\ddot{\theta} = 0$$

$$\Rightarrow 500 \times (0.4)^2 \dot{\theta} + (2000 \times (0.5)^2 + 1000)\theta + 0.833\ddot{\theta} = 0$$

$$\Rightarrow 0.833\ddot{\theta} + 80\dot{\theta} + 1500\theta \quad \ldots\ldots (1)$$

equation can be written as:
$$I\ddot{\theta} + C_T\dot{\theta} + k_T\theta = 0$$

$$\therefore I = 0.833, C_T = 80, k_T = 1500$$

Now natural frequency, $\omega_n = \sqrt{\frac{k_T}{I}} = \sqrt{\frac{1500}{0.833}} = 42.43$ rad/s

GATE-12. Ans. (c) Equivalent damping coefficient = C_T = 80 Nms/rad

GATE-13. Ans. (b)

Explanation: Potential energy at A = mg (l - δ)

Total energy at B = mg [l – (δ + x)] + $\frac{1}{2}kx^2$

∴ Change in energy = mgl - mg(δ + x) + $\frac{1}{2}kx^2$ - mgl + mgδ.

$$= \frac{1}{2}kx^2 - mgx \cdot \delta$$

GATE-14. Ans. (b)

Explanation. We know, $\frac{1}{K_s} = \frac{1}{K_1} + \frac{1}{K_2^t} = \frac{1}{1} + \frac{1}{3} = \frac{4}{3}$ kN/m

Combined stiffness = $K_s + K_3 = \frac{3}{4} + 2 = \frac{11}{4}$ kN/m

$$\therefore f = \frac{1}{2\pi}\sqrt{\frac{11 \times 10^3}{4 \times 1}} = 52.44 \text{ Hz}$$

GATE-15. Ans. (d) Net restoring torque when displaced by a small angle θ,

$$\tau = mg \cos(\alpha - \theta)l - mg(\alpha + \theta)l = 2mgl \cos\alpha \sin\theta$$

For very small θ, $\sin\theta \approx \theta$

∴ $\tau = 2mgl \cos\alpha \, \theta$ (restorative)

Now, $I\frac{d^2\theta}{dt^2} + 2mgl \cos\alpha \, \theta = 0$

But $I = 2ml^2$

∴ $2ml^2 \frac{d^2\theta}{dt^2} + 2mgl \cos\alpha \, \theta = 0$

or $\frac{d^2\theta}{dt^2} + \frac{g \cos\alpha}{l} \theta = 0$

∴ $\omega_n = \sqrt{\frac{g \cos\alpha}{l}}$

GATE-16. Ans. (c)

Natural frequency, $f = \frac{1}{2\pi}\sqrt{\frac{k}{M}}$

Linear Vibration Analysis of Mechanical Systems

Chapter 7

where $\quad k = k_1 + k_2 \quad$ and $\quad f = \dfrac{1}{2\pi}\sqrt{\dfrac{(k_1+k_2)}{M}}$

GATE-17. Ans. (c)
GATE-18. Ans. (a)
GATE-19. Ans. (c)

$$\omega = \dfrac{2\pi \times 3600}{60} = 377 \text{ rad/s}$$

Natural frequency

$$\omega_n = \sqrt{\dfrac{k}{m}} = \sqrt{\dfrac{100 \times 1000}{250}} = 20 \text{ rad/s}$$

Now, $\quad r = \dfrac{\omega}{\omega_n} = \dfrac{377}{20} = 18.85$

Transmissibility ratio

$$(TR) = \dfrac{\sqrt{1+(2\xi r)^2}}{\sqrt{(1-r^2)^2+(2\xi r)^2}} = \dfrac{\sqrt{1+(2 \times 0.15 \times 18.85)^2}}{\sqrt{1-(18.85)^2)+(2 \times 0.15 \times 18.85)^2}} = 0.0162$$

GATE-20. Ans. (a) $\xi < 1$, hence it is under damped vibration case.

∴ Frequency of the system, $\omega_d = 1\sqrt{\xi^2}.\omega_n$

$$= \sqrt{-0.64} \times 100 = 60$$

GATE-21. Ans. (a)

$$\delta = \dfrac{wl^3}{3EI} = \dfrac{mgl^3}{3E\dfrac{a^4}{12}} = \dfrac{4mgl^3}{Ea^4}$$

$$\omega_n = \sqrt{\dfrac{s}{m}} \times \sqrt{\dfrac{g}{\delta}} = \dfrac{a^2}{2}\sqrt{\dfrac{E}{ml^3}} = \dfrac{(0.025)^2}{2}\sqrt{\dfrac{200 \times 10^9}{20 \times 1^3}} = 31.25 \text{ cycle/s}$$

Therefore $c_c = 2m\omega_n = 2 \times 20 \times 31.25 \text{ Ns/m} = 1250 \text{ Ns/m}$

GATE-22. Ans. (c)

Given $\quad \omega_d = 0.9\omega_n$

We know that $\quad \omega_d = \omega_n\sqrt{1-\xi^2}$

$\Rightarrow \quad 0.9\omega_n = \omega_n\sqrt{1-\xi^2}$

∴ $\quad \xi = 0.436$

Now $\quad \xi = \dfrac{c}{2\sqrt{km}}$

∴ $\quad c = 2 \times 0.436 \times \sqrt{1000 \times 0.1} = 8.71 \text{ –s/m}$

GATE-23. Ans. (a)

Linear Vibration Analysis of Mechanical Systems
Chapter 7

$F_0 = 100$, $\omega = 100$, $K = 3000$, $X = 50mm$

$\omega_n = ?$ $m = ?$

$$X = \frac{F_0/K}{\sqrt{\left[1-\left(\frac{\omega}{\omega_n}\right)^2\right]^2 + \left[2\xi\frac{\omega}{\omega_n}\right]^2}} \quad Here\ \xi = 0$$

$$0.050 = \frac{100/3000}{\sqrt{\left[1-\left(\frac{100}{\omega_n}\right)^2\right]^2}}$$

$$0.050\left[1-\left(\frac{100}{\omega_n}\right)^2\right] = \frac{1}{30}$$

$$\therefore 1 - \left(\frac{100}{\omega_n}\right)^2 = 0.66$$

$$\therefore \omega_n = 173.2; \quad \omega_n = \sqrt{\frac{K}{m}} \text{ and } m = 0.1 kg$$

GATE-24. Ans. (d)

For critical damping, $C_c = 2m \times \sqrt{\frac{S}{M}} = 223.6\ Ns/m$

GATE-25. Ans. (d)

Logarithmic decrement, $\delta = \frac{2\pi \times C}{\sqrt{C_c^2 - C^2}} = 0.42$

GATE-26(i). Ans. (c)

Transmissibility ratio, $\xi = \dfrac{\sqrt{1+\left(\dfrac{2c\omega}{\omega_n C_c}\right)^2}}{\sqrt{\left(\dfrac{2c\omega}{\omega_n C_c}\right)^2 + \left(1-\dfrac{\omega^2}{\omega_n^2}\right)^2}}$

If $c = 0$, then $\xi\ \dfrac{1}{\left(1-\dfrac{\omega^2}{\omega_n^2}\right)^2} = \dfrac{1}{1-(0.5)^2} = \dfrac{4}{3}$

GATE-27. Ans. (a) Given $K = 3600$ N/m; $c = 400$ Ns/m; $m = 50$ kg $\omega_n = \sqrt{\dfrac{k}{m}} = 2\pi N$

$$\xi = \frac{C}{C_c} = \frac{C}{2m\omega_n} = \frac{C}{2m\sqrt{\dfrac{k}{m}}} = \frac{C}{2\sqrt{km}}$$

Linear Vibration Analysis of Mechanical Systems

Chapter 7

$$\omega_d = \omega_n \sqrt{1-\xi^2}$$

Previous 20-Years IES Answers

IES-1. Ans. (b)

IES-2. Ans. (c) Frequency of vibration of system

$$\omega = \sqrt{\frac{g}{\delta}} = \sqrt{\frac{10}{1\times 10^{-3}}} = 100 \text{ rad/sec}.$$

IES-3. Ans. (a) Critical speed $= \sqrt{\dfrac{g}{\delta}} = \sqrt{\dfrac{10}{4\times 10^{-3}}} = 50$ rad/sec

IES-4. Ans. (c)

IES-5. Ans. (c) $\omega = \sqrt{36\pi^2}$ and $f = \dfrac{\omega}{2\pi}$

IES-6. Ans. (a)

IES-7. Ans. (a) Natural frequency of vibration $f_n \propto \sqrt{\dfrac{k}{m}}$ In new system

$$f_n \propto \sqrt{\frac{k/2}{2m}} = \frac{1}{2}\sqrt{\frac{k}{m}} = \frac{N}{2}$$

IES-8. Ans. (a)

$$\omega_n = \sqrt{\frac{K_e}{m}};\ \text{Equivalent stiffness } \frac{1}{(k_e)} = \frac{1}{K_3} + \frac{1}{k_1+k_2};\ \omega_n = \left[\frac{1}{\left\{\dfrac{1}{k_1+k_2}+\dfrac{1}{k_3}\right\}m}\right]^{1/2}$$

IES-9. Ans. (b) K_1 and K_1 are in parallel and K_2 and K_2 are in parallel

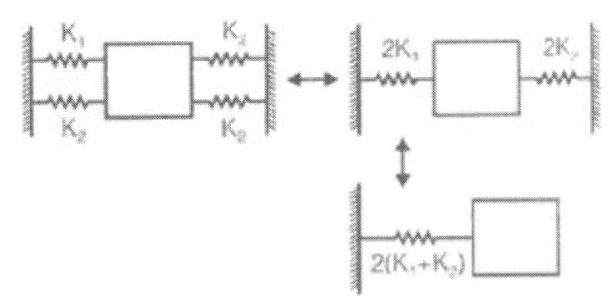

∴ Equivalent spring constant is $2(K_1 + K_2)$

IES-10. Ans. (b) Speed = 72 km/hr = $72 \times \dfrac{5}{18}$ = 20 m/sec

Since Wavelength = λ = 5 m.
Forcing frequency of road on the wheel
$$= \frac{20}{5} = 4 \text{ Cycles/second} = 4\text{Hz}$$

IES-11. Ans. (a) For system to vibrate, f_n should be positive, which is possible when $b < \dfrac{Ka^2}{W}$

Linear Vibration Analysis of Mechanical Systems
Chapter 7

$S = S_1 + S_2 = 20 + 20$
$= 40$ kN/m $= 40,000$ N/m

∴ Natural frequency of vibration of the system,

$$f_n = \frac{1}{2\pi}\sqrt{\frac{S}{m}} = \frac{1}{2\pi}\sqrt{\frac{40 \times 1000}{100}} = \frac{20}{2\pi} = \frac{10}{\pi}$$

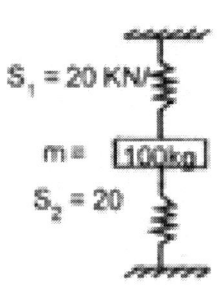

$S_1 = 20$ KN
$m = 100$ kg
$S_2 = 20$

IES-12. Ans. (c)

IES-13. Ans. (b)

IES-14. Ans. (a)

IES-15. Ans. (c)

IES-16. Ans. (a)

IES-17. Ans. (d)

IES-18. Ans. (a)

IES-19. Ans. (a) Energy method $\frac{d}{dt}\left[\frac{1}{2}m\left(\frac{dx}{dt}\right)^2 + \frac{1}{2}I\left\{\frac{1}{r}\left(\frac{dx}{dt}\right)\right\}^2 + \frac{1}{2}kx^2\right] = 0$

where $I = mk^2$

or $\frac{d}{dt}\left[\frac{3m}{2}\left(\frac{dx}{dt}\right)^2 + kx^2\right] = 0$ 　　　 or $\frac{3m}{2}\cdot\frac{d^2x}{dt^2} + kx = 0$

or $\frac{d^2x}{dt^2} + \left(\frac{2k}{3m}\right)x = 0$ 　　　 or $\omega^2 = \frac{2k}{3m}$

IES-20. Ans. (d)
IES-21. Ans. (a)

IES-22. Ans. (d)
IES-23. Ans. (a)
IES-24. Ans. (b)
IES-25. Ans. (b) $\omega_n = \sqrt{\frac{16}{4}} = 2$; $2\xi\omega_n = \frac{9}{4}$; $\xi = \frac{9}{4\times 4} = \frac{9}{16}$

IES-26. Ans. (c) For critical damping, $\xi = \frac{c}{2m\omega_n}$, $c = 2\times 1 \times \sqrt{\frac{s}{m}} = 2\sqrt{\frac{700}{1}} = 52.92$ Ns/m

IES-27. Ans. (d)
IES-28. Ans. (b)
IES-29. Ans. (c) Since the equation involves $c\dot{x}$ and $F\sin\omega t$, It means it is case of forced vibrations with damping.
IES-30. Ans. (c) Inertia force is in phase with displacement but opposite in direction to acceleration, and damping force lags displacement by 90°.
IES-31. Ans. (b)
IES-32. Ans. (a)

Linear Vibration Analysis of Mechanical Systems

Chapter 7

IES-33. Ans. (c)
IES-34. Ans. (b)
IES-35. Ans. (b) Transmissibility of vibration is 1 when $\omega/\omega_n = \sqrt{2}$
IES-36. Ans. (b)
IES-38. Ans. (d)
IES-39. Ans. (a)
IES-40. Ans. (b)
IES-41. Ans. (b) Both the statements are correct.
1. One way of improving vibration isolation is to decrease the mass of the vibrating object.
2. For effective isolation, the natural frequency of the system should be far less than the exciting frequency.

Transmissibility: which is the magnitude of the ratio of the force transmitted to the force applied.

$$T = \left[\frac{1 + (2\xi\omega/\omega_n)^2}{(1-\omega^2/\omega_n^2)^2 + (2\xi\omega/\omega_n)^2} \right]^{1/2}$$

For $\omega/\omega_n > \sqrt{2}$, transmissibility, although below unity, increases with an increase in damping, contrary to normal expectations. At higher frequencies, transmissibility goes to zero.

$\omega_n = \sqrt{\dfrac{K}{m}}$ if $m \downarrow$ then $\omega_n \uparrow$ and $\dfrac{\omega}{\omega_n} \downarrow$ and we want $\dfrac{\omega}{\omega_n}$ should be high. So —statement -1|| is wrong

IES-42. Ans. (d)
IES-43. Ans. (d)
IES-44. Ans. (a)

$$I_1 l_1 = I_2 l_2$$
$$l_1(2 \times 60^2) = l_2(1 \times 20^2)$$
$$18 l_1 = l_2$$

Given that
$$l_1 + l_2 = 38$$
$$19\, l_1 = 38 \text{ or } l_1 = 2 \text{ cm and } l_2 = 38 - 2 = 36 \text{ cm}$$

IES-45. Ans. (d)
IES-46. Ans. (c)

Previous 20-Years IAS Answers

IAS-1. Ans. (b)
IAS-2. Ans. (a) As total energy is constant when V = 0, P.E is maximum. And V = 0 becomes at both extreme ends.
IAS-3. Ans. (c)
IAS-4. Ans. (d)
IAS-5. Ans. (c)
IAS-6. Ans. (b) $n = \dfrac{1}{2\pi}\sqrt{\dfrac{K}{m}}$ $\dfrac{n_2}{n_1} = \sqrt{\dfrac{4k}{k}} = 2$

Linear Vibration Analysis of Mechanical Systems

Chapter 7

IAS-7. Ans. (c) Period of free vibration of a spring $T \propto \sqrt{\frac{1}{k}}$ (k = spring stiffness). When a spring is cut into 4 equal pieces, spring stiffness of each cut spring will be $4k$.
When four such springs are placed in parallel. Spring stiffness of combination will be $4 \times (4k) = 16k$.

∴ now $T \propto \sqrt{\frac{1}{16k}}$ or $\frac{T}{4}$

IAS-8. Ans. (c) Equivalent $(K) = K_1 + K_2 = 200$ N/cm $= 20000$ N/m
Mass = 2 kg. Natural frequency $(\omega) = \sqrt{\frac{K}{m}} = \sqrt{\frac{20000}{2}} = 100$ rad/s

IAS-9. Ans. (d) Potential energy at $A = mg(l - \delta)$
Total energy at $B = mg[l - (\delta + x)] + \frac{1}{2}kx^2$

∴ Change in energy $= mgl - mg(\delta + x) + \frac{1}{2}kx^2 - mgl + mg\delta$

$= \frac{1}{2}kx^2 - mgx \cdot \delta$

IAS-10. Ans. (a)
IAS-11. Ans. (d)
IAS-12. Ans. (c)
IAS-13. Ans. (b)
IAS-14. Ans. (b)
IAS-15. Ans. (d) Coulomb is concerned with damping force, Rayleigh with energy principle, D' Alembert with dynamic equilibrium, and Fourier with frequency domain analysis. Thus the correctly matched pair is (d).
Dunkerley's method (Approximate result)
IAS-16. Ans. (c)
IAS-17. Ans. (c)
IAS-18. Ans. (b)
IAS-19. Ans. (d)
IAS-20. Ans. (b) Critical dampling co-efficient $= 2m\omega_d$

$= 2 \times m \times \sqrt{\frac{5}{m}} = 2\sqrt{5m} = 2\sqrt{20 \times 5} = 20$ Ns/m

IAS-21. Ans. (a) $W_d = W_n\sqrt{1-\varepsilon^2}$ or $20 = 40\sqrt{1-\varepsilon^2}$ or $\varepsilon = \frac{\sqrt{3}}{2}$

IAS-22. Ans. (a) As damper is not used, $c = 0$, $m\frac{d^2x}{dt^2} + \left(\frac{k}{2} + \frac{k}{2}\right)x = 0$ gives $\omega = \sqrt{\frac{K}{m}}$

IAS-23. Ans. (c)

IAS-24. Ans. (a) $x = A\cos(\omega t - \phi)$

Linear Vibration Analysis of Mechanical Systems

Chapter 7

$$\frac{dx}{dt} = -\omega A \sin(\omega t - \phi) = \omega A \cos[90 + (\omega t - \phi)]$$

$$\frac{d^2x}{dt^2} = -\omega^2 A \cos(\omega t - \phi) = \omega^2 A \cos[180 + (\omega t - \phi)]$$

$$m \times \frac{d^2x}{dt^2} + c\frac{dx}{dt} + sx = F\cos(\omega t - \phi)$$

IAS-25. Ans. (b)

IAS-27. Ans. (a)
IAS-30. Ans. (d)

IAS-31. Ans. (d)

IAS-32. Ans. (c) Logarithmic decrement $(\delta) = \ln\left(\frac{x_n}{x_{n+1}}\right) = \frac{2\pi c}{\sqrt{c_c^2 - c^2}}$ $c_c = 2m\omega_n = 2m\sqrt{\frac{s}{m}} = 2\sqrt{sm}$

$$\delta = \frac{2\pi c}{\sqrt{4sm - c^2}}$$ if s ↑ to double and m ↓ to half so   sm = constant and δ remains the same.

IAS-33. Ans. (d)

IAS-34. Ans. (a)

IAS-35. Ans. (d)

IAS-36. Ans. (b)
IAS-37. Ans. (c)

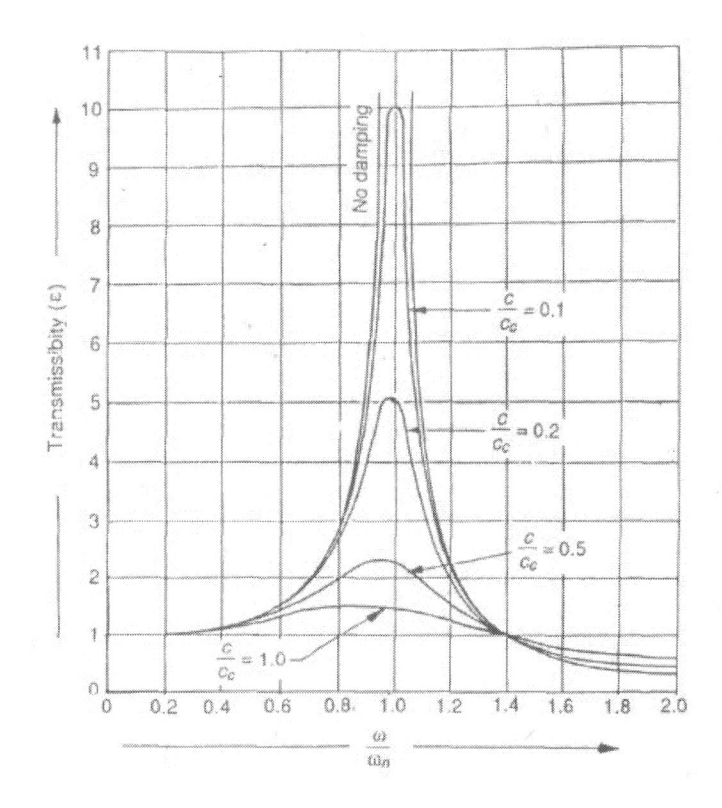

IAS-38. Ans. (d) 3 is false. In torsional vibrations, the disc moves in a circle about the axis of the shaft.

Linear Vibration Analysis of Mechanical Systems
Chapter 7

IAS-39. Ans. (c)

IAS-40. Ans. (c) The restoring force or couple is proportional to displacement from the mean position.

Critical speeds or whirling of Shaft

Chapter 8

8. Critical speeds or whirling of Shaft

Theory at a glance (IES, GATE & PSU)

Critical Speed

Critical speeds are associated with uncontrolled large deflections, which occur when inertial loading on a slightly deflected shaft exceeds the restorative ability of the shaft to resist. Shafts must operate well away from such speeds. Rayleigh's equation

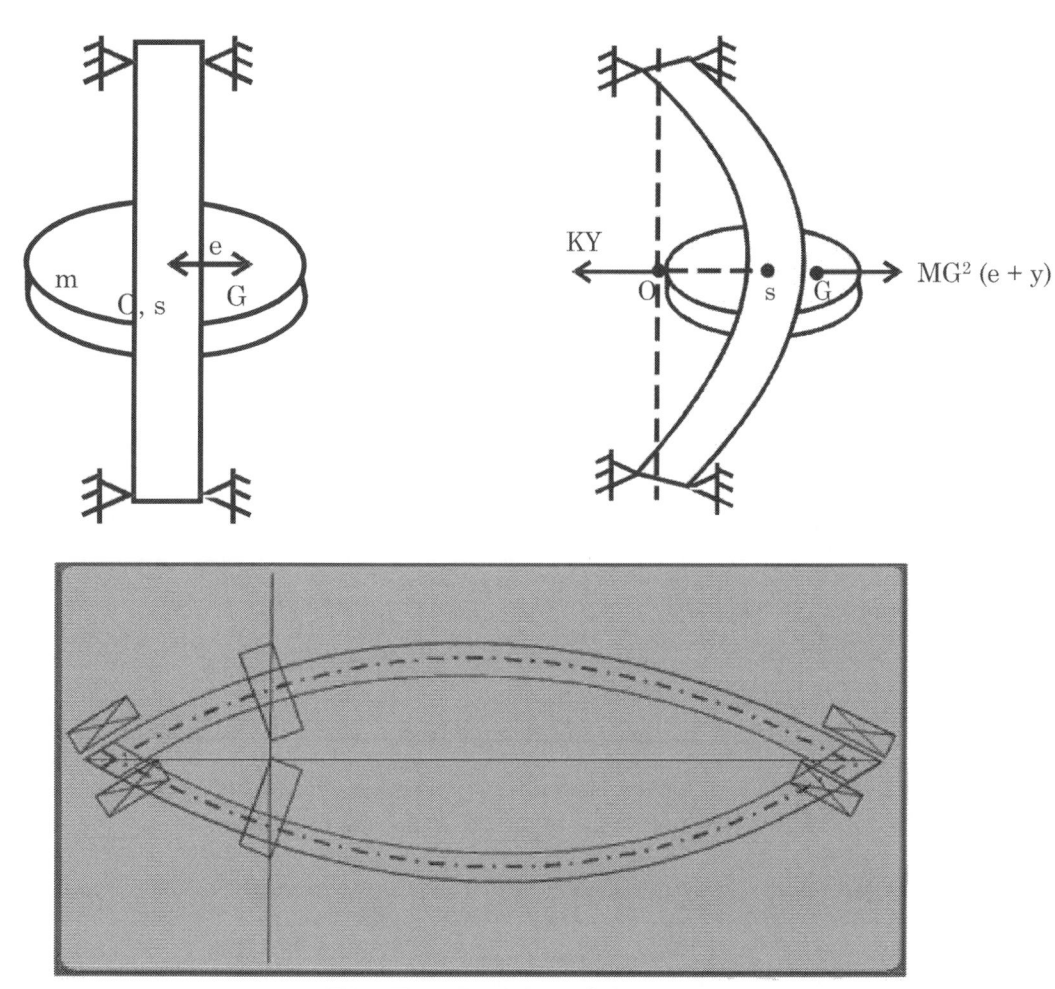

Fig. Two Position of the rotor

Critical Speed of Shafts

Critical speeds or whirling of Shaft

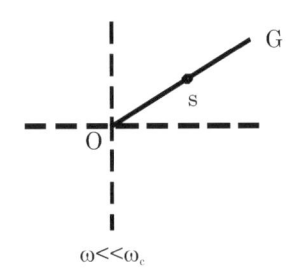

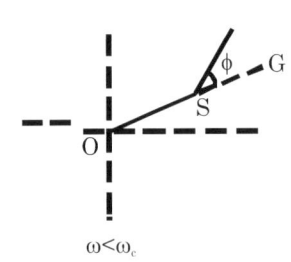

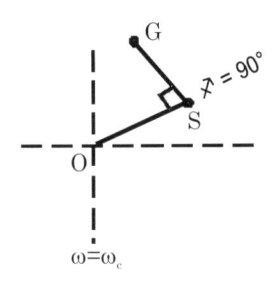

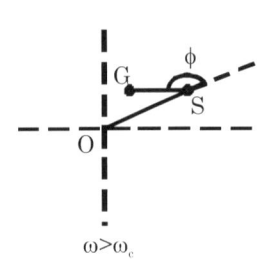

Chapter 8

The speed at which the shafts starts to vibrate violently in the direction perpendicular to the axis of the shaft is **known as critical speed or whirling speed.**

1. Critical speed of shaft carrying single rotor (without damping):

$$y = \frac{\left(\dfrac{\omega}{\omega_c}\right)^2 e}{1 - \left(\dfrac{\omega}{\omega_c}\right)^2}$$

y = deflection of geometric centre due to C.F
e = eccentricity of the rotor
ω_c = critical speed of shaft.

$$\omega_c = \sqrt{\frac{K}{m}} = \sqrt{\frac{g}{\delta}}$$

2. Critical speed of shaft carrying single rotor (with damping).

$$y = \frac{\left(\dfrac{\omega}{\omega_c}\right)^2 e}{\sqrt{\left[1 - \left(\dfrac{\omega}{\omega_c}\right)^2\right]^2 + \left[2\xi\dfrac{\omega}{\omega_c}\right]^2}}$$

Let 0 = point of intersection of bearing centre line with the rotor.
S = geometric centre of the rotor.
G = centre of gravity of the rotor.
ϕ = phase angle between e and y.

$$\phi = \tan^{-1}\left[\frac{2\xi\left(\dfrac{\omega}{\omega_c}\right)}{1 - \left(\dfrac{\omega}{\omega_c}\right)^2}\right]$$

$\omega \ll \omega_c$

$\omega < \omega_c$

$\omega = \omega_c$, $\phi = 90°$

$\omega > \omega_c$

Critical speeds or whirling of Shaft

Chapter 8

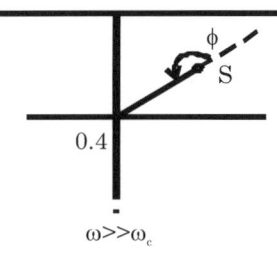

$\omega \gg \omega_c$

Multi Rotor

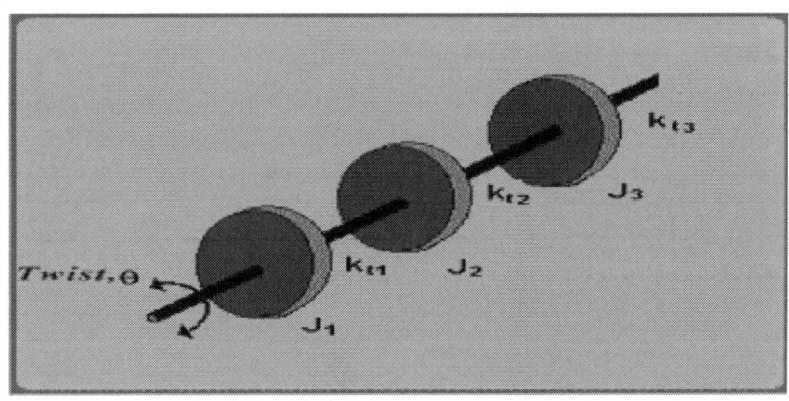

Fig. Typical multi-rotor system

1. Rayleigh's Method:

$$w_C = \sqrt{\frac{g \, \Sigma m_i y_i}{\Sigma m_i y_i^2}}$$

2. Dunkerley's Method:

$$\frac{1}{\omega_C^2} = \frac{1}{(\omega_{C_1})^2} + \frac{1}{(\omega_{C_2})^2} + \ldots \frac{1}{(\omega_{C_5})^2}$$

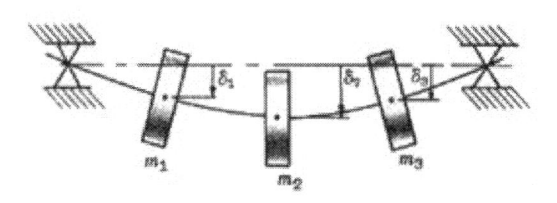

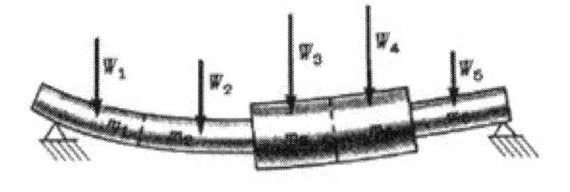

Objective Questions (IES, IAS, GATE)

Previous 20-Years GATE Questions

GATE-1. An automotive engine weighing 240 kg is supported on four springs with linear characteristics. Each of the front two springs have a stiffness of 16

Critical speeds or whirling of Shaft

Chapter 8

MN/m while the stiffness of each rear spring is 32 MN/m. The engine speed (in rpm), at which resonance is likely to occur, is [GATE -2009]
(a) 6040 (b) 3020 (c) 1424 (d) 955

GATE-2. For lightly damped heavy rotor systems, resonance occurs when the forcing ω is equal to [GATE-1992]
(a) $2\omega_{cr}$ (b) $\sqrt{2}\omega_{cr}$ (c) ω_{cr} (d) $\frac{1}{2}\omega_{cr}$

Where ω_{cr} is the critical speed

GATE-3. A flexible rotor-shaft system comprises of a 10 kg rotor disc placed in the middle of a mass-less shaft of diameter 30 mm and length 500 mm between bearings (shaft is being taken mass-less as the equivalent mass of the shaft is included in the rotor mass) mounted at the ends. The bearings are assumed to simulate simply supported boundary conditions. The shaft is made of steel for which the value of E is 2.1×10^{11} Pa. What is the critical speed of rotation of the shaft? [GATE-2003]
(a) 60 Hz (b) 90 Hz (c) 135 Hz (d) 180 Hz

Previous 20-Years IES Questions

IES-1. Which one of the following causes the whirling of shafts? [IES 2007]
(a) Non-homogeneity of shaft material (b) Misalignment of bearings
(c) Fluctuation of speed (d) Internal damping

IES-2. Critical speed of a shaft with a disc supported in between is equal to the natural frequency of the system in [IES-1993]
(a) Transverse vibrations
(b) Torsional vibrations
(c) Longitudinal vibrations
(d) Longitudinal vibrations provided the shaft is vertical.

IES-3. Rotating shafts tend of vibrate violently at whirling speeds because
(a) the shafts are rotating at very high speeds [IES-1993]
(b) Bearing centre line coincides with the shaft axis
(c) The system is unbalanced
(d) Resonance is caused due to the heavy weight of the rotor

IES-4. A shaft carries a weight W at the centre. The CG of the weight is displaced by an amount e from the axis of the rotation. If y is the additional displacement of the CG from the axis of rotation due to the centrifugal force, then the ratio of y to e (where ω_c, is the critical speed of shaft and w is the angular speed of shaft) is given by [IES-2001]

(a) $\dfrac{1}{\left[\dfrac{\omega_c}{\omega}\right]^2+1}$ (b) $\dfrac{\pm e}{\left[\dfrac{\omega_c}{\omega}\right]^2-1}$ (c) $\left[\dfrac{\omega_c}{\omega}\right]^2+1$ (d) $\dfrac{\omega}{\left[\dfrac{\omega_c}{\omega}\right]^2-1}$

IES-5. The critical speed of a rotating shaft depends upon [IES-1996]
(a) Mass (b) stiffness (c) mass and stiffness (d) mass, stiffness and eccentricity.

Critical speeds or whirling of Shaft

Chapter 8

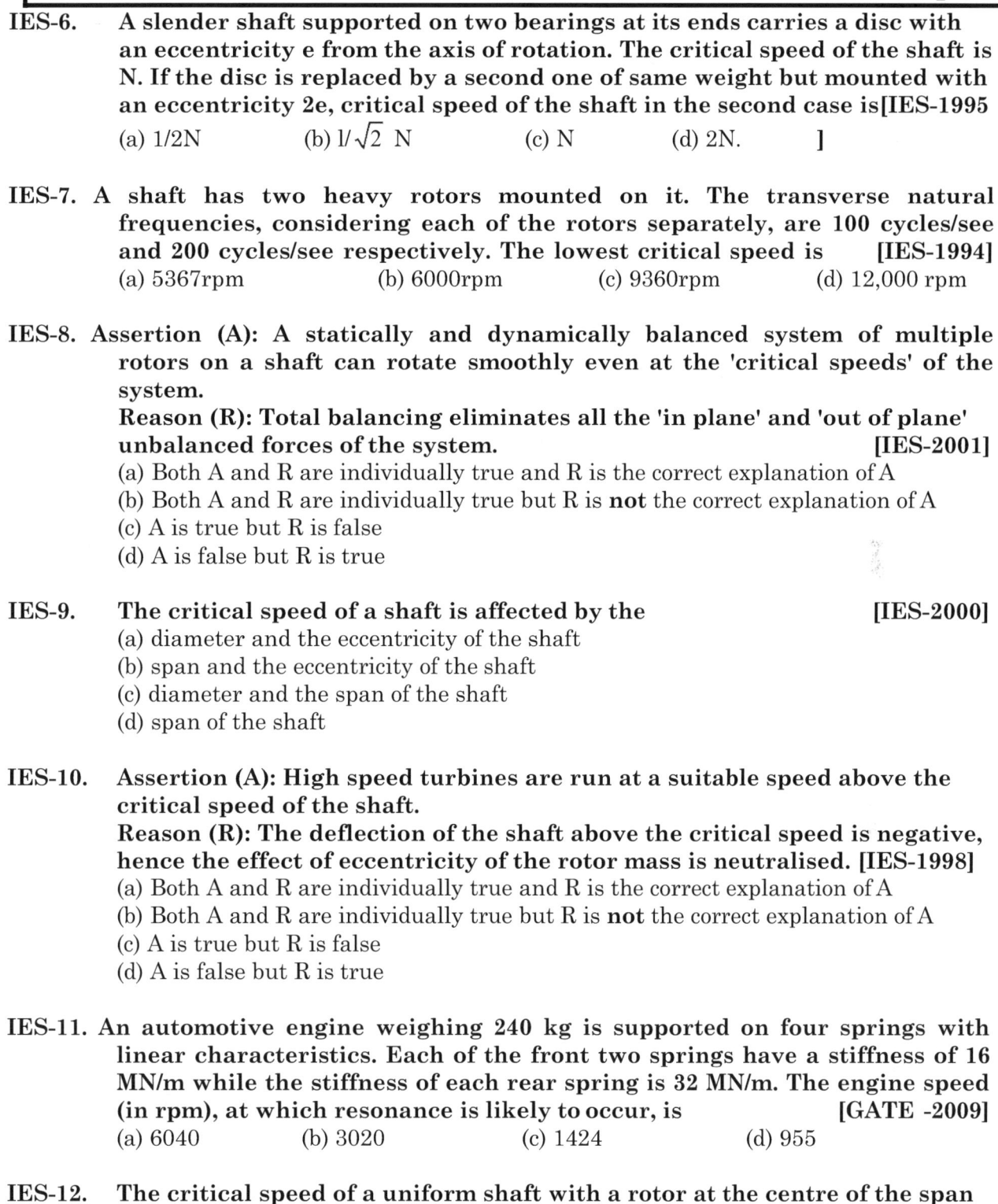

IES-6. A slender shaft supported on two bearings at its ends carries a disc with an eccentricity e from the axis of rotation. The critical speed of the shaft is N. If the disc is replaced by a second one of same weight but mounted with an eccentricity 2e, critical speed of the shaft in the second case is [IES-1995]
(a) 1/2N (b) 1/$\sqrt{2}$ N (c) N (d) 2N.]

IES-7. A shaft has two heavy rotors mounted on it. The transverse natural frequencies, considering each of the rotors separately, are 100 cycles/sec and 200 cycles/sec respectively. The lowest critical speed is [IES-1994]
(a) 5367rpm (b) 6000rpm (c) 9360rpm (d) 12,000 rpm

IES-8. Assertion (A): A statically and dynamically balanced system of multiple rotors on a shaft can rotate smoothly even at the 'critical speeds' of the system.
Reason (R): Total balancing eliminates all the 'in plane' and 'out of plane' unbalanced forces of the system. [IES-2001]
(a) Both A and R are individually true and R is the correct explanation of A
(b) Both A and R are individually true but R is **not** the correct explanation of A
(c) A is true but R is false
(d) A is false but R is true

IES-9. The critical speed of a shaft is affected by the [IES-2000]
(a) diameter and the eccentricity of the shaft
(b) span and the eccentricity of the shaft
(c) diameter and the span of the shaft
(d) span of the shaft

IES-10. Assertion (A): High speed turbines are run at a suitable speed above the critical speed of the shaft.
Reason (R): The deflection of the shaft above the critical speed is negative, hence the effect of eccentricity of the rotor mass is neutralised. [IES-1998]
(a) Both A and R are individually true and R is the correct explanation of A
(b) Both A and R are individually true but R is **not** the correct explanation of A
(c) A is true but R is false
(d) A is false but R is true

IES-11. An automotive engine weighing 240 kg is supported on four springs with linear characteristics. Each of the front two springs have a stiffness of 16 MN/m while the stiffness of each rear spring is 32 MN/m. The engine speed (in rpm), at which resonance is likely to occur, is [GATE -2009]
(a) 6040 (b) 3020 (c) 1424 (d) 955

IES-12. The critical speed of a uniform shaft with a rotor at the centre of the span can be reduced by [IES-1998]
(a) reducing the shaft length (b) reducing the rotor mass
(c) increasing the rotor mass (d) increasing the shaft diameter

IES-13. Assertion (A): The critical speed of an elastic shaft calculated by the Rayleigh's method is higher than the actual critical speed.
Reason (R): The higher critical speed is due to higher damping ratio.
(a) Both A and R are individually true and R is the correct explanation of A
(b) Both A and R are individually true but R is **not** the correct explanation of A

Critical speeds or whirling of Shaft

Chapter 8

(c) A is true but R is false [IES-2005]
(d) A is false but R is true

IES-14. A shaft of 50 mm diameter and 1 m length carries a disc which has mass eccentricity equal to 190 microns. The displacement of the shaft at a speed which is 90% of critical speed in microns is [IES-2002]
(a) 810 (b) 900 (c) 800 (d) 820

IES-15. The danger of breakage and vibration is maximum? [IES-1992]
(a) below critical speed (b) near critical speed
(c) above critical speed (d) none of the above.

IES-16. If a spring-mass-dashpot system is subjected to excitation by a constant harmonic force, then at resonance, its amplitude of vibration will be
(a) Infinity [IES-1999]
(b) Inversely proportional to damp in
(c) Directly proportional to damping
(d) Decreasing exponentially with time

IES-17. Match List-I with List-II and select the correct answer using the codes given below the lists: [IES-1998]

List-I
A. Node and mode
B. Equivalent inertia
C. Log decrement
D. Resonance

List-II
1. Geared vibration
2. Damped-free vibration
3. Forced vibration
4. Multi-rotor vibration

Code: A   B   C   D   A   B   C   D
(a)   1   4   3   2   (b)   4   1   2   3
(c)   1   4   2   3   (d)   4   1   3   2

Previous 20-Years IAS Questions

IAS-1. Whirling speed of a shaft coincides with the natural frequency of its
(a) longitudinal vibration (b) transverse vibration [IAS-1995]
(c) torsional vibration (d) coupled bending torsional vibration

IAS-2. Assertion (A): Every rotating shaft has whirling speeds [IAS 1994]
Reason (R): Eccentricity of rotors on rotating shafts is unavoidable.
(a) Both A and R are individually true and R is the correct explanation of A
(b) Both A and R are individually true but R is **not** the correct explanation of A
(c) A is true but R is false
(d) A is false but R is true

IAS-3. Whirling speed of shaft is the speed at which [IAS-2002]
(a) shaft tends to vibrate in longitudinal direction
(b) torsional vibration occur
(c) shaft tends to vibrate vigorously in transverse direction
(d) combination of transverse and longitudinal vibration occurs

IAS-4. The rotor of a turbine is generally rotated at

Critical speeds or whirling of Shaft

Chapter 8

(a) the critical speed [IAS-1999]
(b) a speed much below the critical speed
(c) 3 speed much above the critical speed
(d) a speed having no relation to critical speed

IAS-5. Consider the following statements [IAS 1994]
The critical speed of a shaft if affected by the
1. eccentricity of the shaft   2. span of the shaft   3. diameter of the shaft
Of these statements:
(a) 1 and 2 are correct   (b) 1 and 3 are correct
(c) 2 and 3 are correct   (d) 1, 2 and 3 are correct.

Answers with Explanation (Objective)

Previous 20-Years GATE Answers

GATE-1. Ans. (a) $K = K_1 + K_2 + K_3 + K_4$

$$f_n = \frac{1}{2\pi}\sqrt{\frac{k}{m}}$$

GATE-2. Ans. (c)

GATE-3. Ans. (b)
Here, $m = 10$ kg = mass of rotar
d = diameter of shaft = 30×10^5 m
l = length of shaft = 500×10^{-3} m
E for steel = 2.1×10^{11} N/m²

Δ = deflection of shaft = $\dfrac{mgl^3}{4gEI}$

$I = \dfrac{\pi}{64}d^4 = \dfrac{\pi}{64} \times (30 \times 10^{-3})^4$

Critical speeds or whirling of Shaft

Chapter 8

$$= 3.976 \times 10^{-8} \text{m}^4$$

$$\Delta = \frac{mgl^3}{48EI}$$

$$= \frac{10 \times 9.81 \times (500 \times 10^{-3})^3}{48 \times 2.1 \times 10^{11} \times 3.976 \times 10^{-8}}$$

$$= 3.06 \times 10-5 \text{ m}$$

$$\omega_n = \sqrt{\frac{g}{\Delta}} = \sqrt{\frac{9.81}{3.06 \times 10^{-5}}} = 566.24 \text{ rad/s}$$

$$f_n = \frac{\omega_n}{2\pi}$$

$$= \frac{566.24}{2 \times 3.142} = 90 \text{ Hz}.$$

Previous 20-Years IES Answers

IES-1. Ans. (a)
IES-2. Ans. (a)
IES-3. Ans. (d)
IES-4. Ans. (b)
IES-5. Ans. (c)
$$\omega_1 = \left(\frac{\pi}{l}\right)^2 \sqrt{\frac{EI}{m}} = \left(\frac{\pi}{l}\right)^2 \sqrt{\frac{gEI}{A\gamma}}$$

IES-6. Ans. (c)
$$\omega_1 = \left(\frac{\pi}{l}\right)^2 \sqrt{\frac{EI}{m}} = \left(\frac{\pi}{l}\right)^2 \sqrt{\frac{gEI}{A\gamma}}$$

IES-7. Ans. (a) $\dfrac{1}{f_n^2} = \dfrac{1}{f_1^2} + \dfrac{1}{f_2^2}$

IES-8. Ans. (d)
IES-9. Ans. (c) $\omega_1 = \left(\dfrac{\pi}{l}\right)^2 \sqrt{\dfrac{EI}{m}} = \left(\dfrac{\pi}{l}\right)^2 \sqrt{\dfrac{gEI}{A\gamma}}$

IES-10. Ans. (c)
IES-11. Ans. (a) $K = K_1 + K_2 + K_3 + K_4$

$$f_n = \frac{1}{2\pi}\sqrt{\frac{k}{m}}$$

IES-12. Ans. (c)
$$\omega_1 = \left(\frac{\pi}{l}\right)^2 \sqrt{\frac{EI}{m}} = \left(\frac{\pi}{l}\right)^2 \sqrt{\frac{gEI}{A\gamma}}$$

IES-13. Ans. (c)
IES-14. Ans. (a)
IES-15. Ans. (b)
IES-16. Ans. (a)
IES-17. Ans. (b)

Critical speeds or whirling of Shaft

Chapter 8

Previous 20-Years IAS Answers

IAS-1. Ans. (b)

IAS-2. Ans. (b) $\omega_1 = \left(\dfrac{\pi}{l}\right)^2 \sqrt{\dfrac{EI}{m}} = \left(\dfrac{\pi}{l}\right)^2 \sqrt{\dfrac{gEI}{A\gamma}}$

IAS-3. Ans. (c)
IAS-4. Ans. (c)
IAS-5. Ans. (c)

Gear Train and Gear Design

Chapter 9

Gear Train and Gear Design
(Same chapter is added to Machine design booklet)

Theory at a glance (GATE, IES, IAS & PSU)

Spur gear

Basic Purpose of Use of Gears

Gears are widely used in various mechanisms and devices to transmit power
And motion positively (without slip) between parallel, intersecting (axis) or
Non-intersecting non parallel shafts,
- Without change in the direction of rotation
- With change in the direction of rotation
- Without change of speed (of rotation)
- With change in speed at any desired ratio

Often some gearing system (rack – and – pinion) is also used to transform Rotary motion into linear motion and vice-versa.

- **A SPUR GEAR** is cylindrical in shape, with teeth on the outer circumference that are straight and parallel to the axis (hole). There are a number of variations of the basic spur gear, including pinion wire, stem pinions, rack and internal gears.

Fig.

Gear Train and Gear Design

Chapter 9

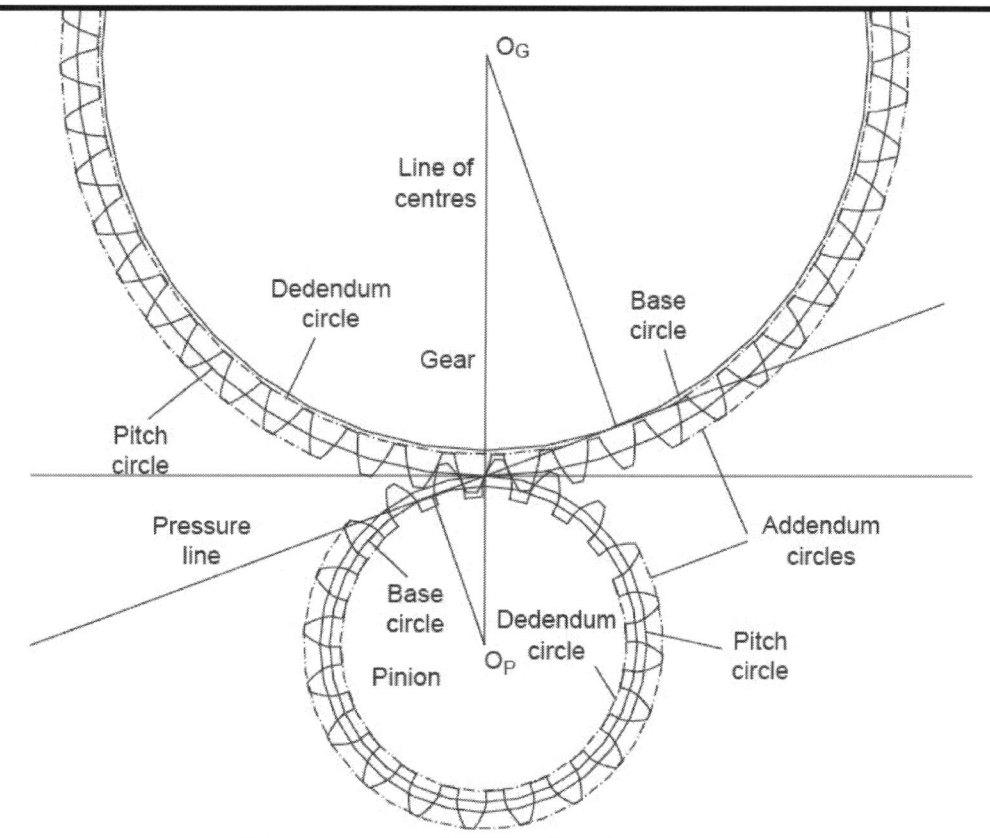

Figure- Layout of a pair of meshing spur gears

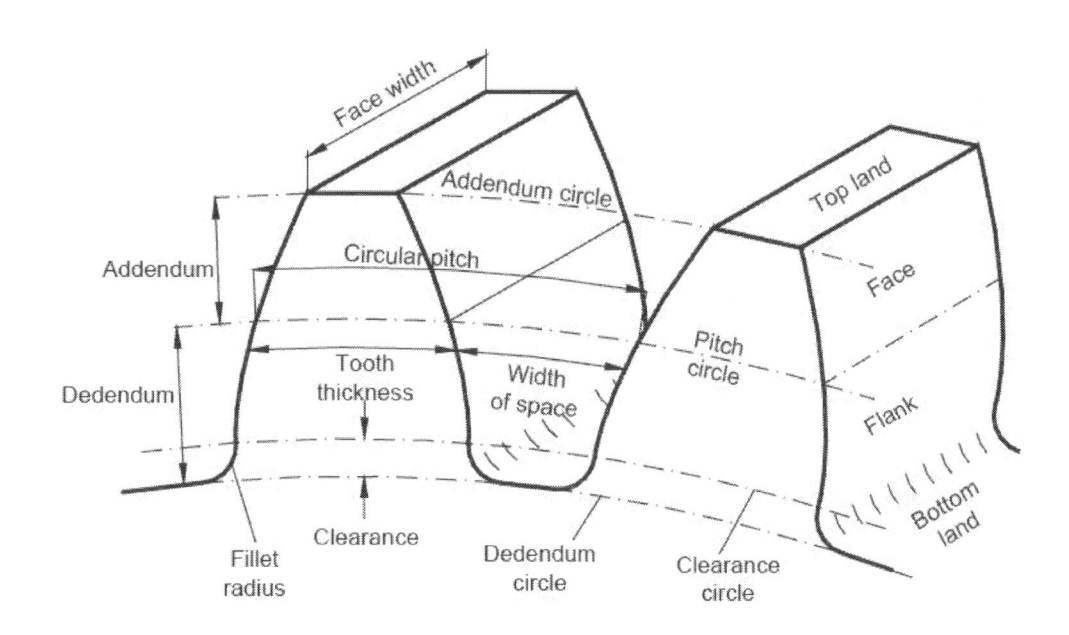

Figure- Spur gear schematic showing principle terminology

For a pair of meshing gears, the smaller gear is called the „pinion", the larger is called the „gear wheel" or simply the „gear".

Pitch circle

Gear Train and Gear Design

Chapter 9

This is a theoretical circle on which calculations are based. Its diameter is called the pitch diameter.

$$d = mT$$

Where d is the pitch diameter (mm); m is the module (mm); and T is the number of teeth.
Care must be taken to distinguish the module from the unit symbol for a meter.

Circular pitch

This is the distance from a point on one tooth to the corresponding point on the adjacent tooth measured along the pitch circle.

$$p = \pi m = \frac{\pi d}{T}$$

Where p is the circular pitch (mm); m the module; d the pitch diameter (mm); and T the Number of teeth.

Module.

This is the ratio of the pitch diameter to the number of teeth. The unit of the module should be millimeters (mm). The module is defined by the ratio of pitch diameter and number of teeth. Typically the height of a tooth is about 2.25 times the module. Various modules are illustrated in figure.

$$m = \frac{d}{T}$$

- **Addendum**, (a). This is the radial distance from the pitch circle to the outside of the tooth.

- **Dedendum**, (b). This is the radial distance from the pitch circle to the bottom land.

Clearance (C) is the amount by which the dedendum in a given gear exceeds the addendum of its mating gear.

Backlash

BACKLASH is the distance (spacing) between two "mating" gears measured at the back of the driver on the pitch circle. Backlash, which is purposely built in, is very important because it helps prevent noise, abnormal wear and excessive heat while providing space for lubrication of the gears.

- The backlash for spur gears depends upon (i) module and (ii) pitch line velocity.

- Factor affected by changing center distance is backlash.

Gear Train and Gear Design

Chapter 9

Fig.

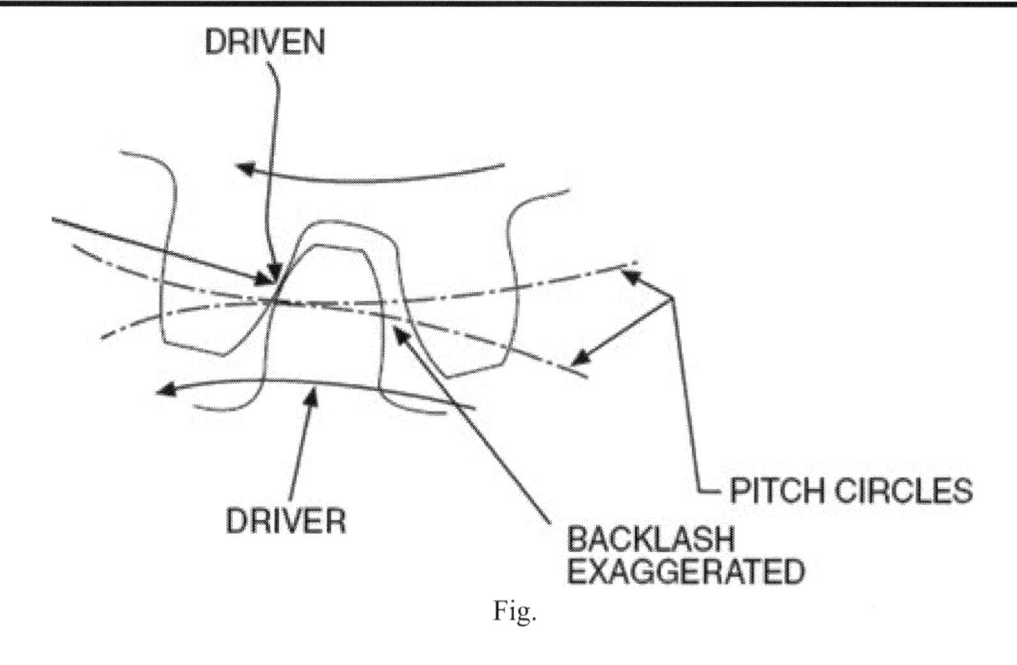

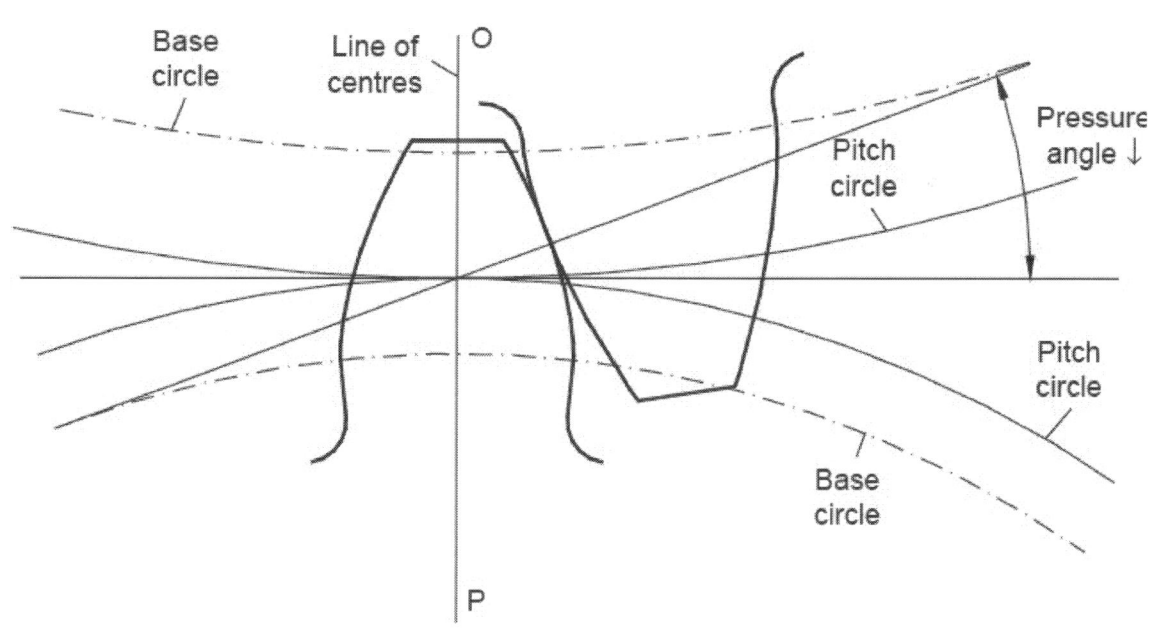

Figure- Schematic showing the pressure line and pressure angle

Gear Train and Gear Design

Chapter 9

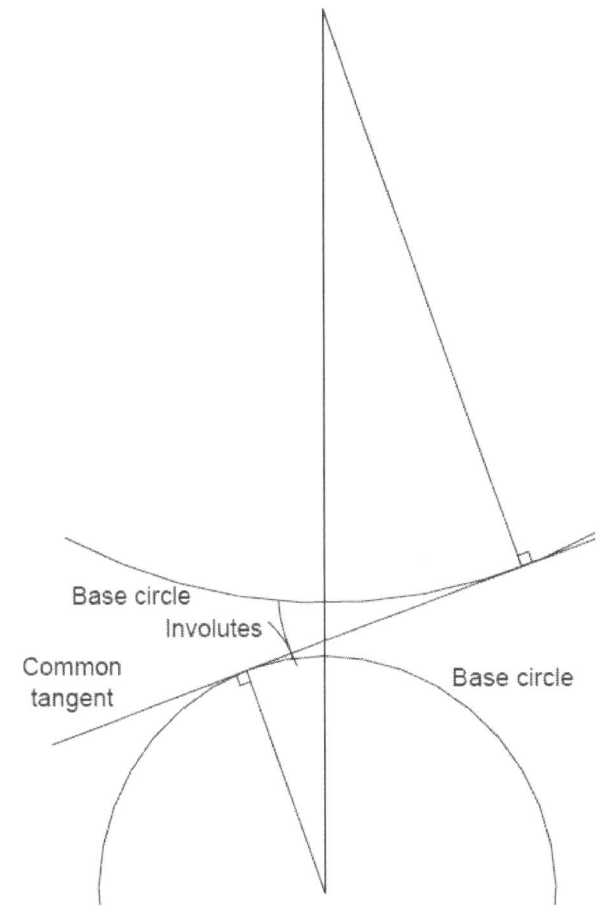

Figure- Schematic of the involute form

Pitch Circle and pitch point

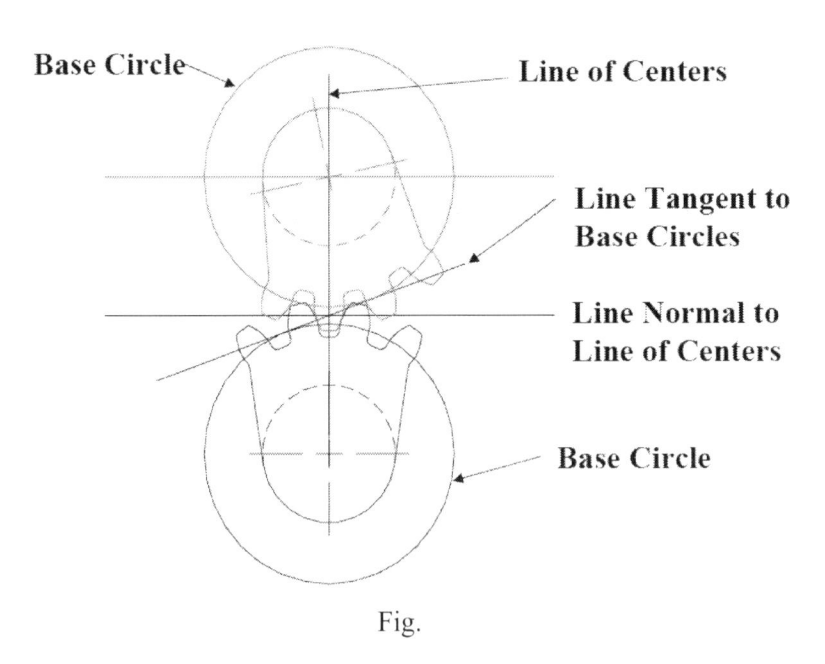

Fig.

Line of Action – Line tangent to both base circles

Gear Train and Gear Design

Chapter 9

Pitch Point – Intersection of the line of centers with the line of action

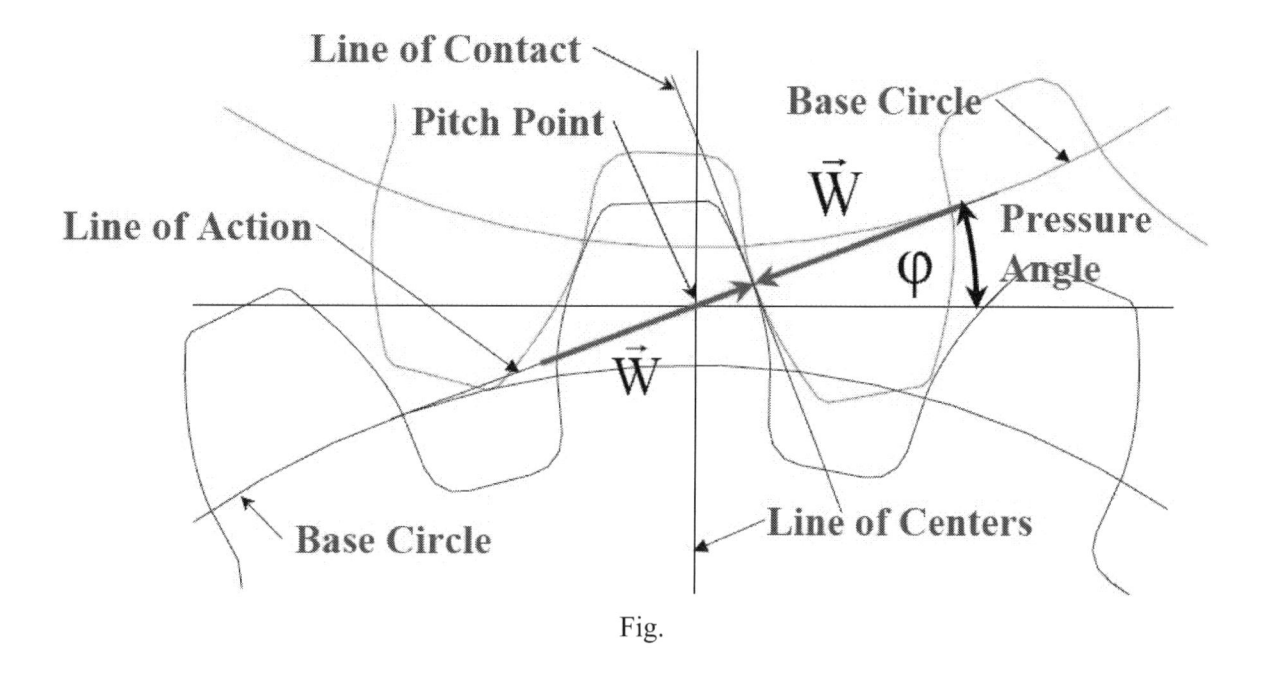

Fig.

Pitch Circle – Circle with origin at the gear center and passing through the pitch point.

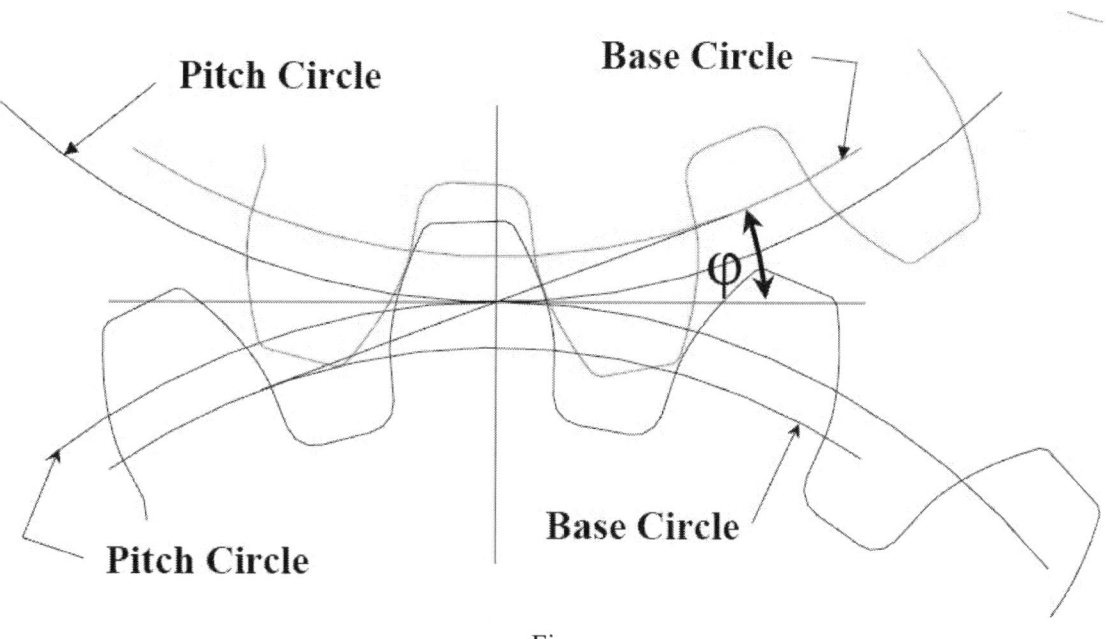

Fig.

Gear Train and Gear Design

Chapter 9

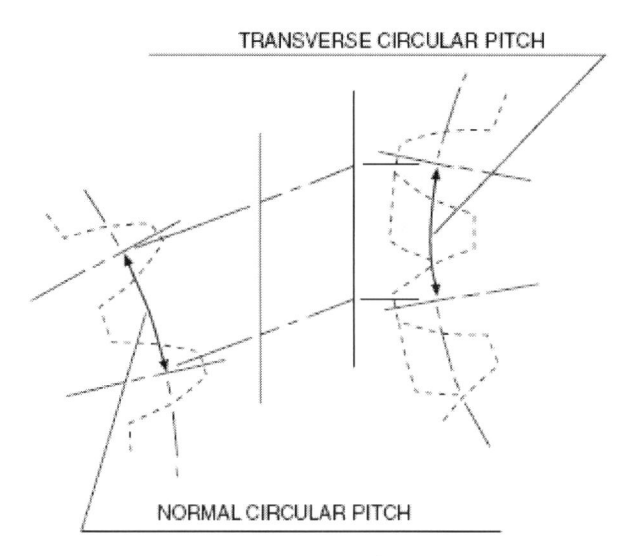

Fig.

Pressure angle – Angle between the line normal to the line of centers and the line of action.
- The pressure angle of a spur gear normally varies **from 14° to 20°**

- The value of pressure angle generally used for involute gears are **20°**

- Relationship Between Pitch and Base Circles

$$r_b = r\cos\phi$$

Fig.

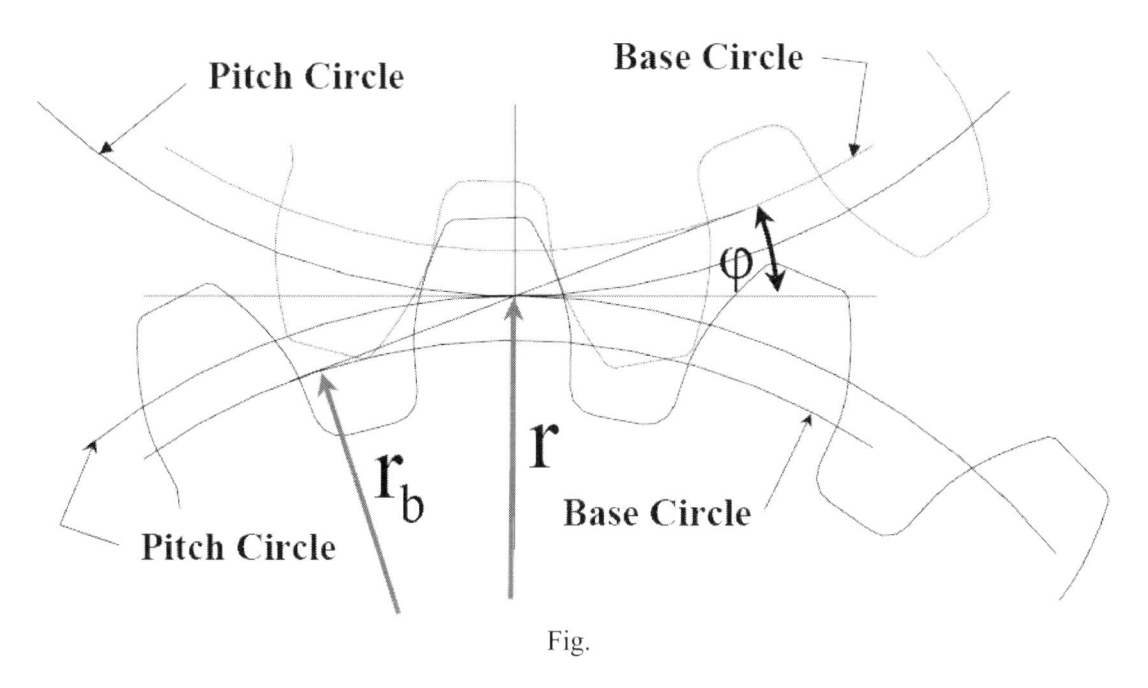

The following four systems of gear teeth are commonly used in practice.

| 1. | $14\frac{1}{2}^0$ | Composite system. |

Gear Train and Gear Design

Chapter 9

2.	$14\frac{1}{2}^0$	Full depth involute system.
3.	20^0	Full depth involute system
4	20^0	Stub involutes system.

The $14\frac{1}{2}^0$ *composite system* is used for general purpose gears. It is stronger but has no interchangeability. The tooth profile of this system has cycloidal curves at the top and bottom and involute curve at the middle portion. The teeth are produced by formed milling cutters or hobs. The tooth profile of the $14\frac{1}{2}^0$ *full depth involute system* was developed for use with gear hobs for spur and helical gears.

The tooth profile of the **20° *full depth involute system*** may be cut by hobs. The increase of the pressure angle from $14\frac{1}{2}^0$ to 20° results in a stronger tooth, because the tooth acting as a beam is wider at the base. The **20° *stub involute system*** has a **strong tooth** to take heavy loads.

Classification of Gears

Gears can be divided into several broad classifications.

1. Parallel axis gears:
(a) Spur gears
(b) Helical gears
(c) Internal gears.

2. Non-parallel, coplanar gears (intersecting axes):
(a) Bevel gears
(b) Face gears,
(c) Conical involute gearing.

3. Non-parallel, non- coplanar gears (nonintersecting axes):
(a) Crossed axis helical
(b) Cylindrical worm gearing
(c) Single enveloping worm gearing,
(d) Double enveloping worm gearing,
(e) Hypoid gears,
(f) Spiroid and helicon gearing,
(g) Face gears (off centre).

4. Special gear types:
(a) Square and rectangular gears,
(b) Elliptical gears.

RACK

RACKS are yet another type of spur gear. Unlike the basic spur gear, racks have their teeth cut into the surface of a straight bar instead of on the surface of a cylindrical blank.

Gear Train and Gear Design

Chapter 9

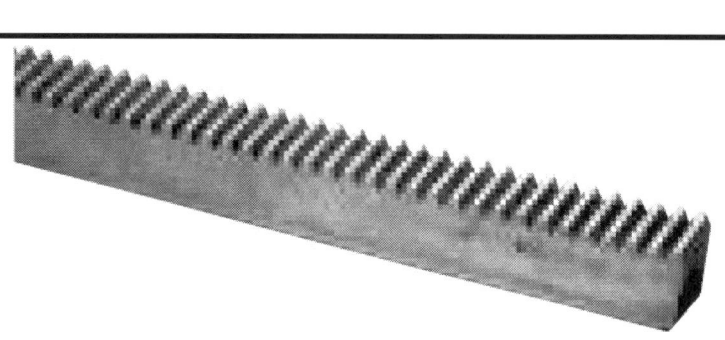

Fig. Rack

Helical gear

The helical gears may be of ***single helical type*** or ***double helical type.*** In case of single helical gears there is some axial thrust between the teeth, which is a disadvantage. In order to eliminate this axial thrust, double helical gears (*i.e.* **herringbone gears**) are used. It is equivalent to two single helical gears, in which equal and opposite thrusts are provided on each gear and the resulting axial **thrust is zero.**

Herringbone gears

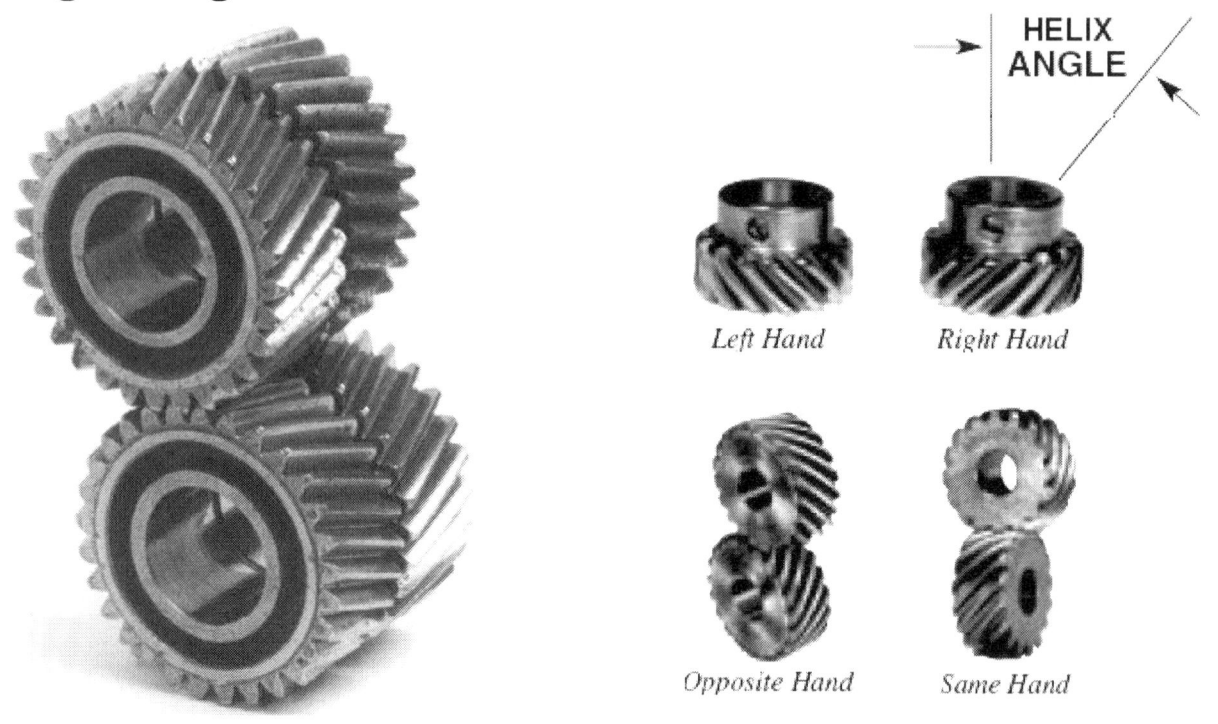

Figure-Herringbone gear

Gear Train and Gear Design

Chapter 9

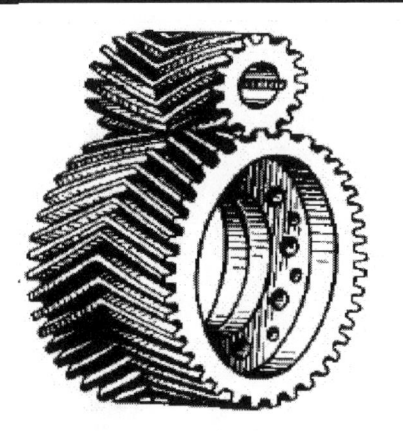

Figure- Herringbone gear

Figure- Crossed axis helical gears

- In spur gears, the contact between meshing teeth occurs along the entire face width of the tooth, resulting in a sudden application of the load which, in turn, results in impact conditions and generates noise.

Gear Train and Gear Design

Chapter 9

- In helical gears, the contact between meshing teeth begins with a point on the leading edge of the tooth and gradually extends along the diagonal line across the tooth. There is a gradual pick-up of load by the tooth, resulting in smooth engagement and silence operation.

Bevel Gears

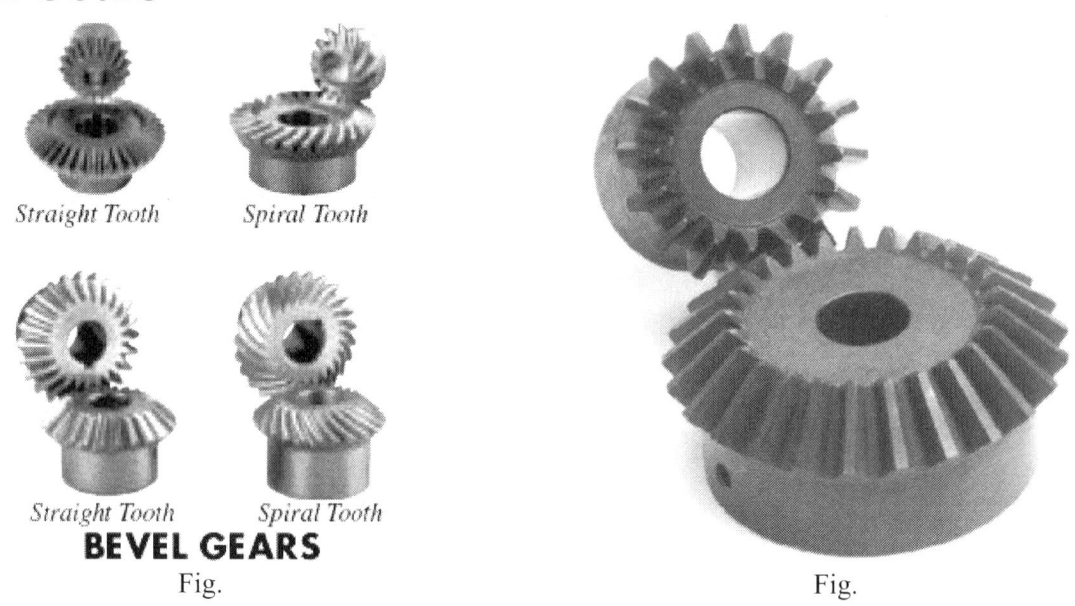

Straight Tooth *Spiral Tooth*

Straight Tooth *Spiral Tooth*
BEVEL GEARS
Fig.

Fig.

Worm Gear

FACE

Right Hand Worm

FACE

Worm Gear

90°

Worm and Gear Single Thread   *Worm and Gear Four Thread*
Fig.

Fig.

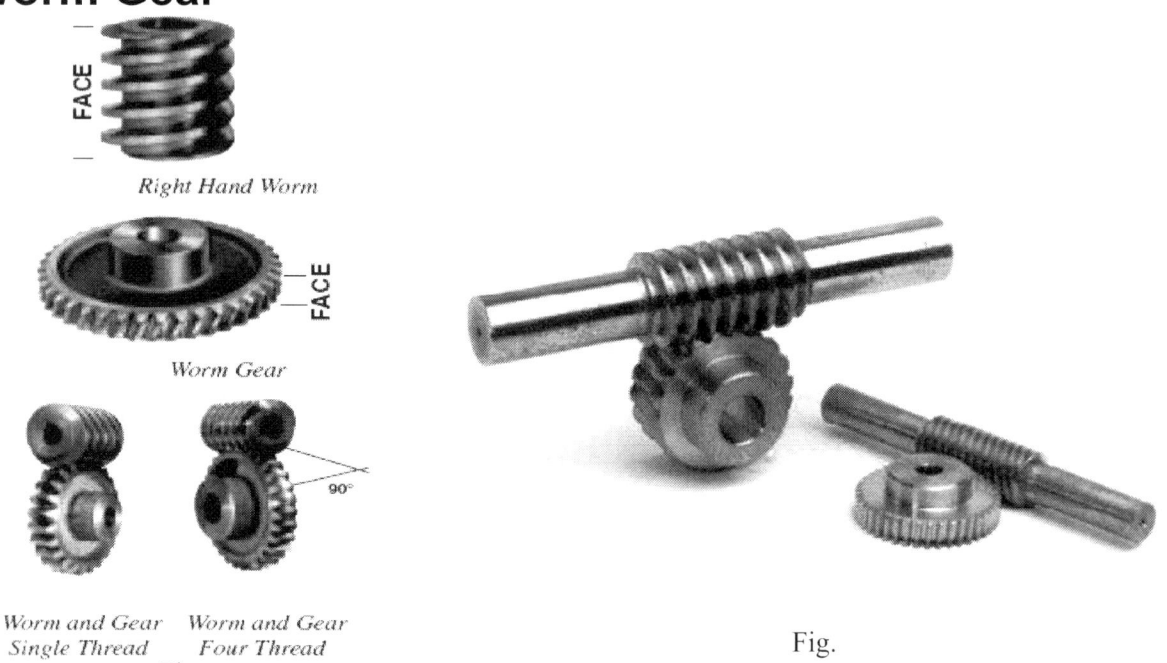

Hypoid Gears

Hypoid gears resemble bevel gears trand spiral bevel gears and are used on crossed-axis shafts. The distance between a hypoid pinion axis and the axis of a hypoid gear is called the *offset*. Hypoid pinions may have as few as five teeth in a high ratio set. Ratios can be obtained with hypoid gears that are not available with bevel gears. High ratios are easy to obtain with the hypoid gear system.

Gear Train and Gear Design

Chapter 9

Hypoid gears are matched to run together, just as zero or spiral bevel gear sets are matched. The geometry of hypoid teeth is defined by the various dimensions used to set up the machines to cut the teeth.

- Hypoid gears are similar in appearance to spiral-bevel gears. They differ from spiral-bevel gears in that the axis of the pinion is offset from the axis of the gear.

Fig.

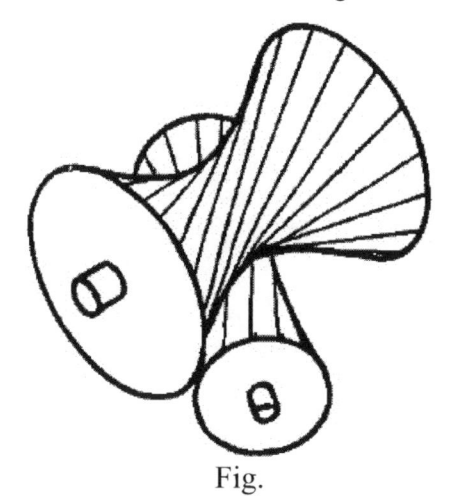

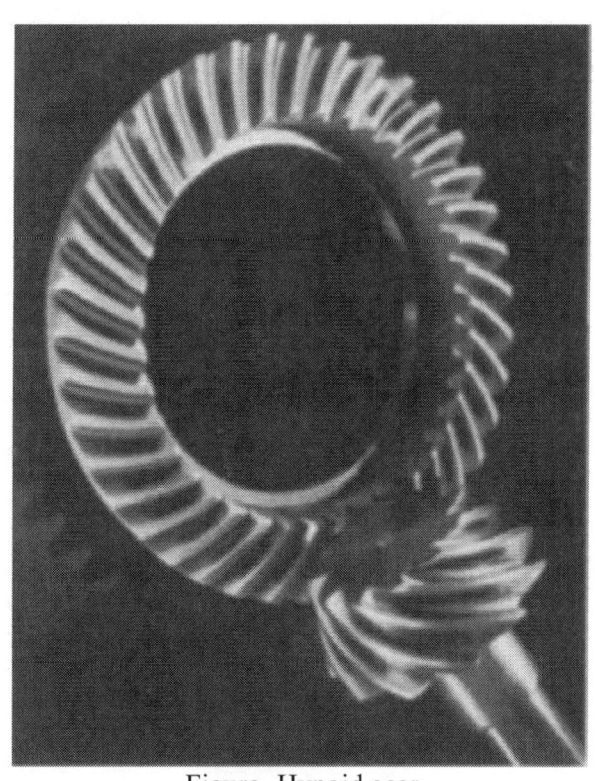

Figure- Hypoid gear

Gear Train and Gear Design

Chapter 9

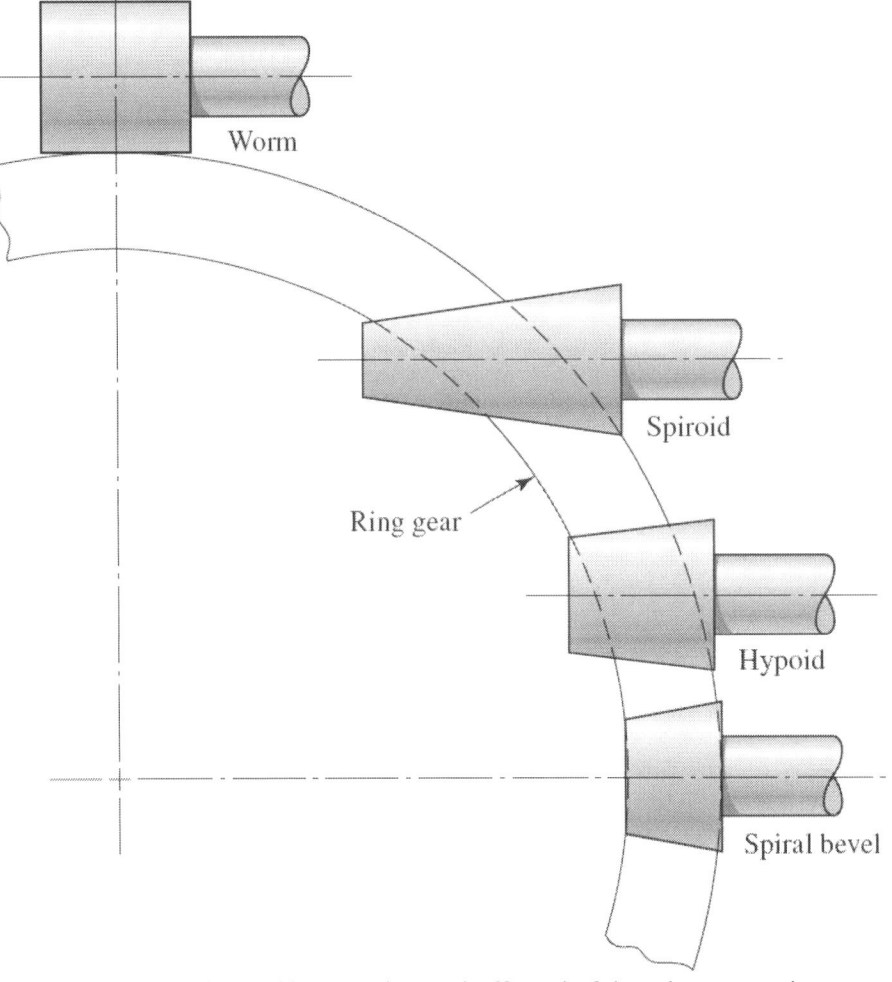

Figure- Comparison of intersecting and offset-shaft bevel-type gearings

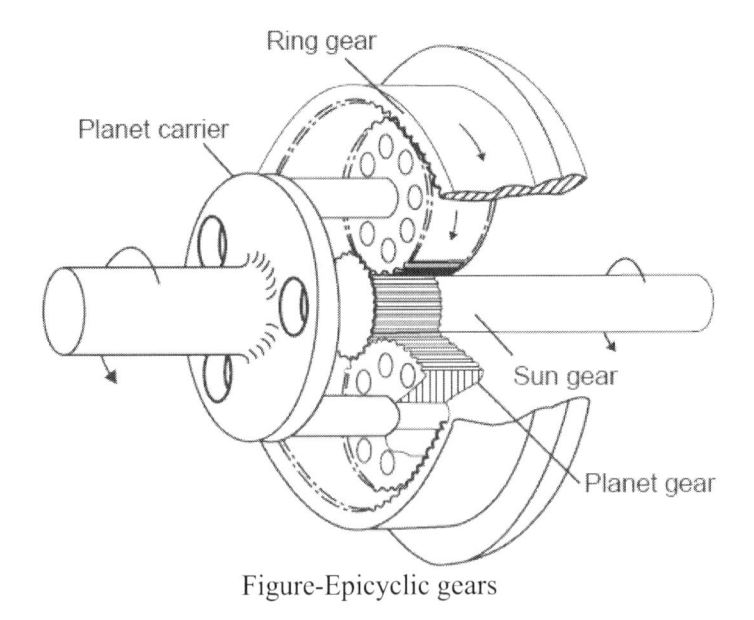

Figure-Epicyclic gears

Mitres gear

Gear Train and Gear Design

Chapter 9

Miter gears are identical to bevel gears except that in a miter gear set, both gears always have the same number of teeth. Their ratio, therefore, is **always 1 to 1**. As a result, miter gears are not used when an application calls for a change of speed.

- **When equal bevel gears (having equal teeth) connect two shafts whose axes are mutually perpendicular, then the bevel gears are known as *mitres*.**

Straight Tooth   *Spiral Tooth*

Figure- Miter gears

Minimum Number of Teeth on the Pinion in Order to Avoid Interference

The number of teeth on the pinion (T_p) in order to avoid interference may be obtained from the following relation:

$$T_p = \frac{2A_w}{G\left[\sqrt{1+\frac{1}{G}\left(\frac{1}{G}+2\right)\sin^2\phi}-1\right]}$$

Where A_W = Fraction by which the standard addendum for the wheel should be Multiplied, (generally $A_W = 1$)
 G = Gear ratio or velocity ratio = $T_G / T_P = D_G / D_P$,
 ϕ = Pressure angle or angle of obliquity.

- Minimum number of teeth for involute rack and pinion arrangement for pressure angle of **20°** is

$$T_{min} = \frac{2A_R}{\sin^2\theta} = \frac{2\times 1}{\sin^2 20°} = 17.1 \quad as > 17 \quad So,\ T_{min} = 18$$

- The minimum number of teeth on the pinion to operate without interference in standard full height involute teeth gear mechanism with 20° pressure angle is **18**.

- In **full depth** $14\frac{1}{2}°$ degree involute system, the smallest number of teeth in a pinion which meshes with rack without interference is **32**.

Gear Train and Gear Design

Chapter 9

Forms of teeth

Cycloidal teeth

A *cycloid* is the curve traced by a point on the circumference of a circle which rolls without slipping on a fixed straight line. When a circle rolls without slipping on the outside of a fixed circle, the curve traced by a point on the circumference of a circle is known as *epicycloid*. On the other hand, if a circle rolls without slipping on the inside of a fixed circle, then the curve traced by a point on the circumference of a circle is called *hypocycloid*.

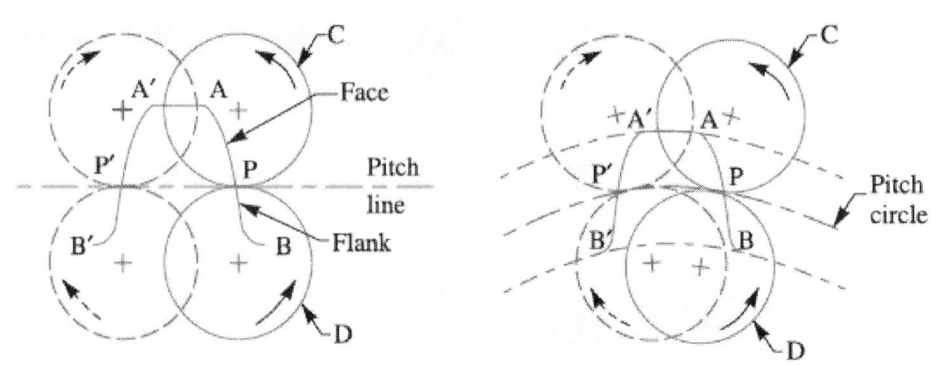

Fig. cycloidal teeth of a gear

Advantages of cycloidal gears

Following are the advantages of cycloidal gears:
1. Since the cycloidal teeth have wider flanks, therefore the cycloidal gears are stronger than the involute gears for the same pitch. Due to this reason, the cycloidal teeth are preferred especially for cast teeth.

2. In cycloidal gears, the contact takes place between a convex flank and concave surface, whereas in involute gears, the convex surfaces are in contact. This condition results in less wear in cycloidal gears as compared to involute gears. However the difference in wear is negligible.

3. In cycloidal gears, **the interference does not occur** at all. Though there are advantages of cycloidal gears but they are outweighed by the greater simplicity and flexibility of the involute gears.

Involute teeth

An involute of a circle is a plane curve generated by a point on a tangent, which rolls on the circle without slipping or by a point on a taut string which is unwrapped from a reel as shown in figure below. In connection with toothed wheels, the circle is known as base circle.

Gear Train and Gear Design

Chapter 9

Fig.

Figure-involute teeth

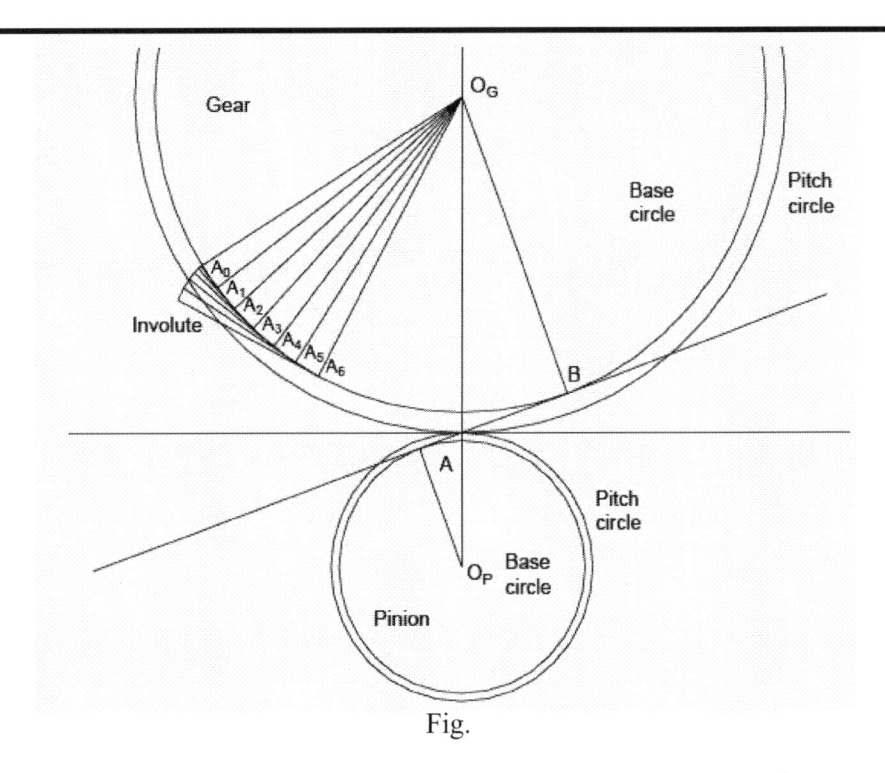

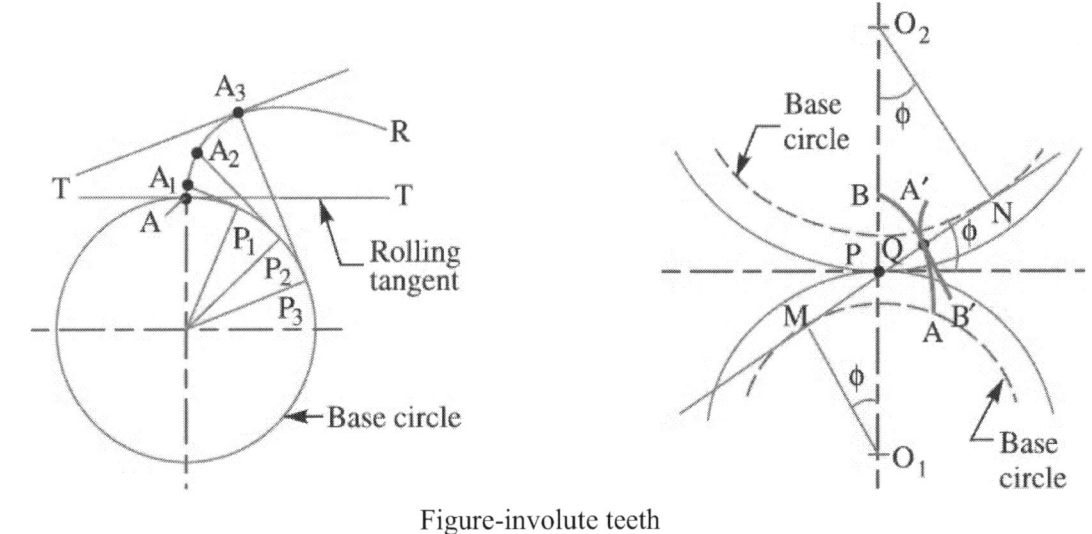

- The tooth profile most commonly used in gear drives for power transmission is an involute. It is due to easy manufacturing.

Advantages of involute gears

Following are the advantages of involute gears:

1. The most important advantage of the involute gears is that the centre distance for a pair of involute gears can be varied within limits without changing the velocity ratio. This is not true for cycloidal gears which require exact centre distance to be maintained.

2. In involute gears, the pressure angle, from the start of the engagement of teeth to the end of the engagement, remains **constant**. It is necessary for smooth running and less wear of gears. But in cycloidal gears, the pressure angle is maximum at the beginning of engagement, reduces to zero at pitch point, starts

Gear Train and Gear Design

Chapter 9

increasing and again becomes maximum at the end of engagement. This results in less smooth running of gears.

3. The face and flank of involute teeth are generated by a single curve whereas in cycloidal gears, double curves (*i.e.* epicycloids and hypocycloid) are required for the face and flank respectively.
Thus the involute teeth are easy to manufacture than cycloidal teeth. In involute system, the basic rack has straight teeth and the same can be cut with simple tools.

Note: The only disadvantage of the involute teeth is that **the interference occurs** with pinions having smaller number of teeth. This may be avoided by altering the heights of addendum and dedendum of the mating teeth or the angle of obliquity of the teeth.

Contact ratio

Note: The ratio of the length of arc of contact to the circular pitch is known as *contact ratio* i.e. number of pairs of teeth in contact.

$$\text{Contact ratio} = \frac{\text{length of arc of contact}}{\text{circular pitch}}$$

$$= \frac{\sqrt{R_{A^2} - R^2 \cos^2 \phi} + \sqrt{r_{A^2} - r^2 \cos^2 \phi} - (R+r)\sin\phi}{P_c(\cos\phi)}$$

Fig.

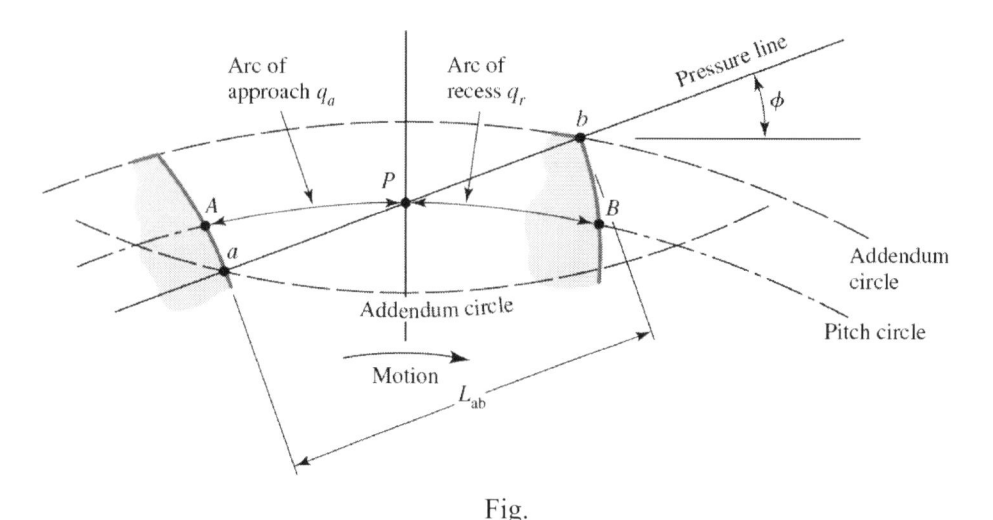

The zone of action of meshing gear teeth is shown in figure above. We recall that tooth
Contact begins and ends at the intersections of the two addendum circles with the pressure line. In figure above initial contact occurs at *a* and final contact at *b*. Tooth profiles drawn through these points intersect the pitch circle at *A* and *B*, respectively. As shown, the distance *AP* is called **the arc of approach** (q_a), and the distance *P B*, **the arc of recess** (q_r). The sum of these is the *arc of action* (q_t).

- The ratio of the length of arc of contact to the circular pitch is known as *contact ratio* i.e. number of pairs of teeth in contact. The contact ratio for gears is greater than one. **Contact ratio should be at least 1.25.** For maximum smoothness and quietness, the contact ratio should be between 1.50 and 2.00. High-speed applications should be designed with a face-contact ratio of 2.00 or higher for best results.

Gear Train and Gear Design

Chapter 9

Interference

- **The contact of portions of tooth profiles that are not conjugate is called *interference*.**

- Contact begins when the tip of the driven tooth contacts the flank of the driving tooth. In this case the flank of the driving tooth first makes contact with the driven tooth at point *A*, and this occurs *before* the involute portion of the driving tooth comes within range. In other words, contact is occurring below the base circle of gear 2 on the *noninvolute* portion of the flank. The actual effect is that the involute tip or face of the driven gear tends to dig out the noninvolute flank of the driver.

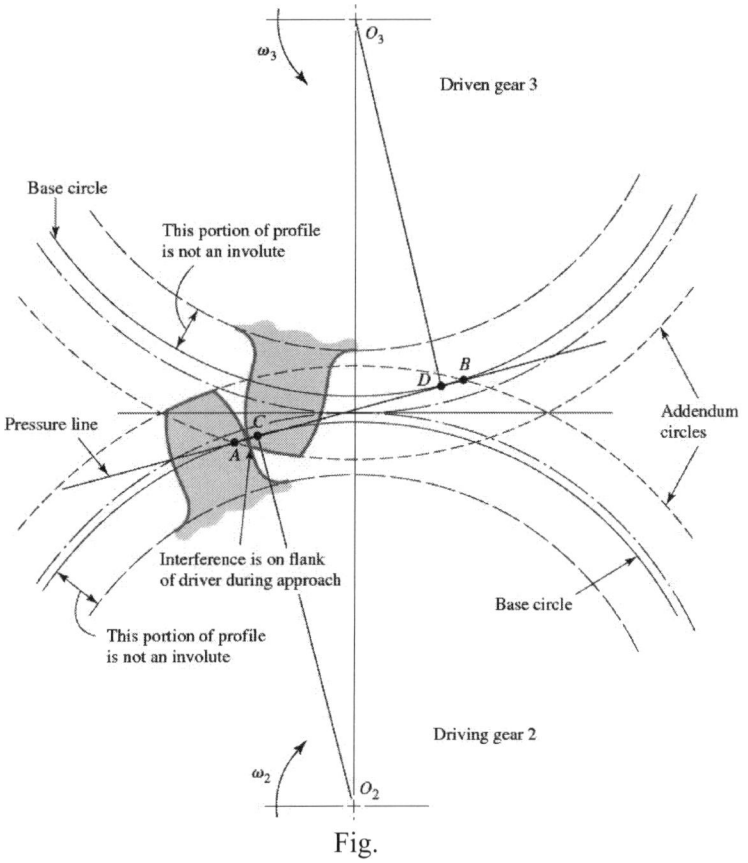

Fig.

- Interference can be eliminated by using more teeth on the pinion. However, if the pinion is to transmit a given amount of power, more teeth can be used only by increasing the pitch diameter.

- Interference can also be reduced by using a larger pressure angle. This results in a smaller base circle, so that more of the tooth profile becomes involute.

- The demand for smaller pinions with fewer teeth thus favors the use of a 25^0 pressure angle even though the frictional forces and bearing loads are increased and the contact ratio decreased.

- **There are several ways to avoid interfering:**
  i. Increase number of gear teeth
  ii. Modified involutes
  iii. Modified addendum
  iv. Increased centre distance.

Face Width

Gear Train and Gear Design

Chapter 9

Face width. It is the width of the gear tooth measured parallel to its axis.

Face width.
  We know that **face width,**
    $b = 10\ m$

Where, m is module.

Fig.

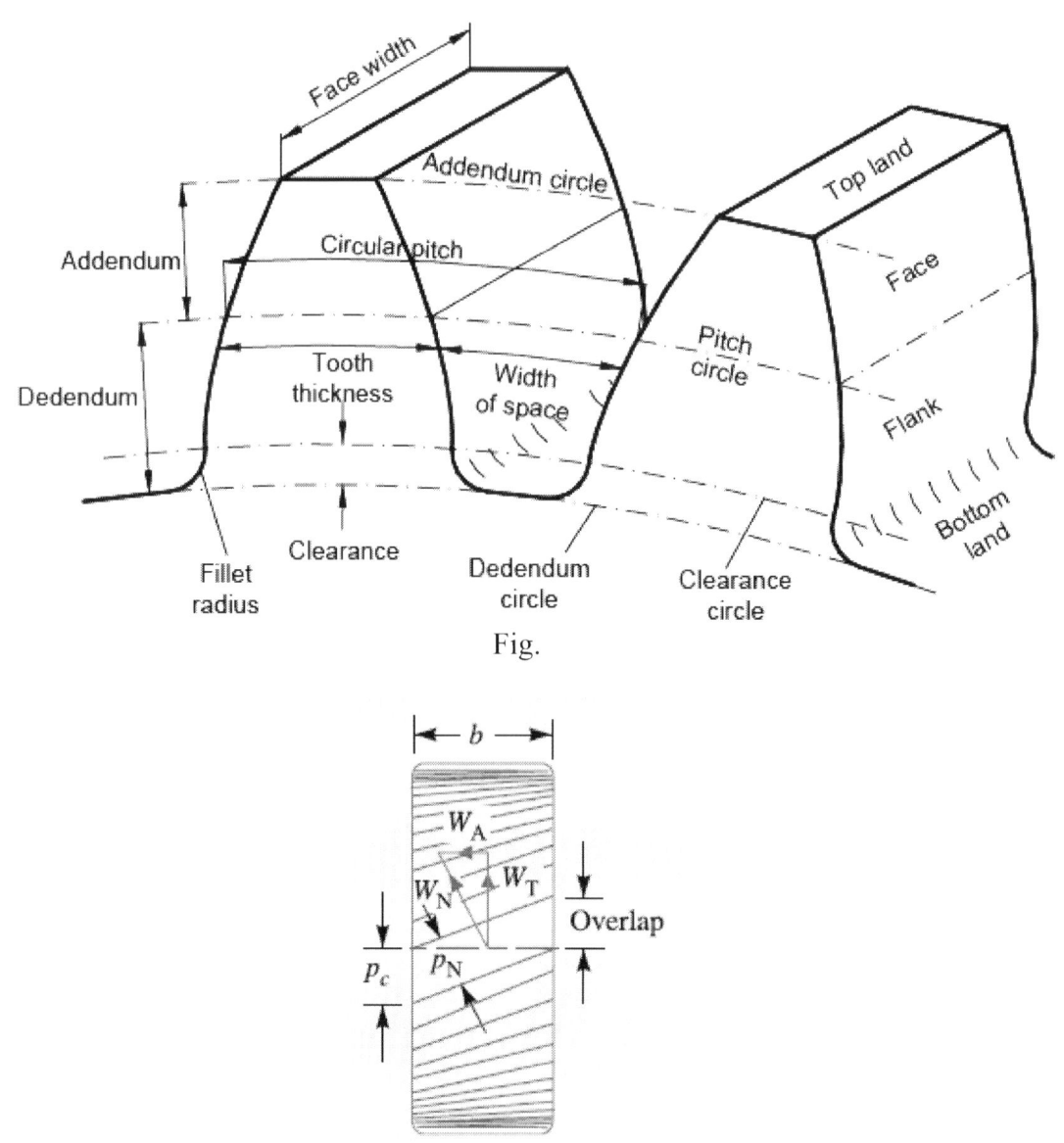

Fig. Face width of helical gear.

Beam Strength of Gear Tooth

The beam strength of gear teeth is determined from an equation **(known as Lewis equation)** and the load carrying ability of the toothed gears as determined by this equation gives satisfactory results. In the investigation, Lewis assumed that as the load is being transmitted from one gear to another, it is all given and taken by one tooth, because it is not always safe to assume that the load is distributed among several teeth, considering each tooth as a cantilever beam.

Notes: (*i*) The **Lewis equation** am applied only to the weaker of the two wheels (*i.e.* pinion or gear).

Gear Train and Gear Design

Chapter 9

(ii) When both the pinion and the gear are made of the same material, then pinion is the weaker.

(iii) When the pinion and the gear are made of different materials, then the product of $(\sigma_w \times y)$ or $(\sigma_o \times y)$ is the deciding factor. The Lewis equation is used to that wheel for which $(\sigma_w \times y)$ or $(\sigma_o \times y)$ is less.

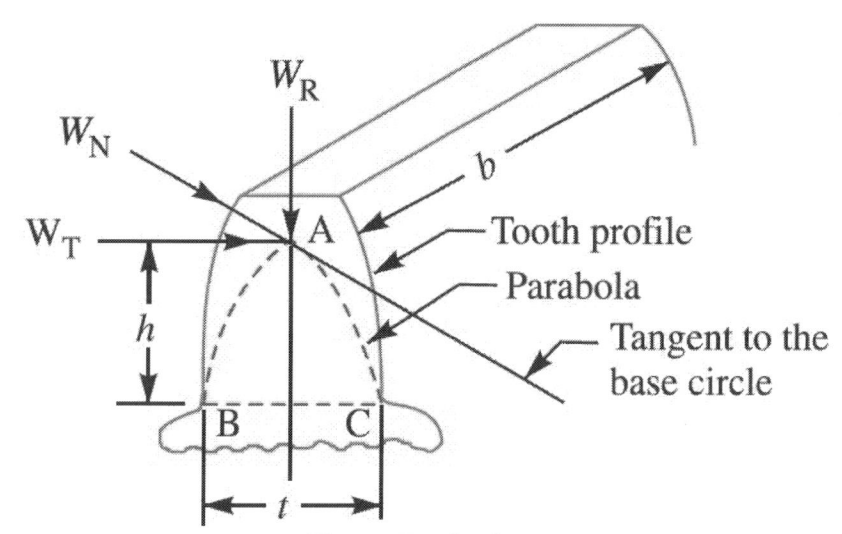

Figure- Tooth of a gear

The maximum value of the bending stress (or the permissible working stress):

$$\sigma_w = \frac{(W_T \times h)\, t/2}{b.t^3/12} = \frac{(W_T \times h) \times 6}{b.t^2}$$

Where
M = Maximum bending moment at the critical section $BC = W_T \times h$,
W_T = Tangential load acting at the tooth,
h = Length of the tooth,
y = Half the thickness of the tooth (t) at critical section $BC = t/2$,
I = Moment of inertia about the centre line of the tooth = $b.t^3/12$,
b = Width of gear face.

Lewis form factor or tooth form factor

$$W_T = \sigma_w \cdot b \cdot p_c \cdot y = \sigma_w \cdot b \cdot \pi m \cdot y$$

The quantity y is known as *Lewis form factor* or *tooth form factor* and W_T (which is the tangential load acting at the tooth) is called the *beam strength of the tooth.*

Lewis form factor or **tooth form factor**

$y = 0.124 - \dfrac{0.684}{T}$, for $14\frac{1}{2}^0$ composite and full depth involute system.

$= 0.154 - \dfrac{0.912}{T}$, for 20^0 full depth involute system.

$= 0.175 - \dfrac{0.841}{T}$, for 20^0 stub system.

Gear Train and Gear Design

Chapter 9

Example: A spur gear transmits 10 kW at a pitch line velocity of 10 m/s; driving gear has a diameter of 1.0 m. find the tangential force between the driver and the follower, and the transmitted torque respectively.

Solution: Power transmitted = Force × Velocity

$\Rightarrow \quad 10 \times 10^3 = \text{Force} \times 10$

$\Rightarrow \quad \text{Force} = \dfrac{10 \times 10^3}{10} = 1000 \text{ N/m}$

Torque Transmitted = Force × $\dfrac{\text{diameter}}{2}$

$= 1000 \times \dfrac{1}{2} = 1000 \times 0.5$

$= 500 \text{ N}-\text{m} = 0.5 \text{ kN}-\text{m}$

Wear Strength of Gear Tooth

Wear strength $(\sigma_\omega) = bQdpK$,

Where, $Q = \dfrac{2T_g}{T_g + T_p}$ for external gear

$= \dfrac{2T_g}{T_g - T_p}$ for internal gear

load - stress factor $(k) = \dfrac{\sigma_c^2 \sin\phi \cos\phi}{1.4}\left(\dfrac{1}{E_1} + \dfrac{1}{E_2}\right)$

$= 0.16\left(\dfrac{BHN}{100}\right)^2$

Gear Lubrication

All the major oil companies and lubrication specialty companies provide lubricants for gearing and other applications to meet a very broad range of operating conditions. General gear lubrication consists of high-quality machine oil when there are no temperature extremes or other adverse ambient conditions. Many of the automotive greases and oils are suitable for a broad range of gearing applications.

For adverse temperatures, environmental extremes, and high-pressure applications, consult the lubrication specialty companies or the major oil companies to meet your particular requirements or specifications.

The following points refer especially to spiral and hypoid bevel gears:

(a) Both spiral and hypoid bevel gears have combined rolling and sliding motion between the teeth, the rolling action being beneficial in maintaining a film of oil between the tooth mating surfaces.

(b) Due to the increased sliding velocity between the hypoid gear pair, a more complicated lubrication system may be necessary.

Gear Train and Gear Design

Chapter 9

Simple Gear train

A gear train is one or more pairs of gears operating together to transmit power. When two gears are in mesh, their pitch circles roll on each other without slippage.

If r_1 is pitch radius of gear 1; r_2 is pitch radius of gear 2; ω_1 is angular velocity of gear 1; and ω_2 is angular velocity of gear 2 then the pitch line velocity is given by

$$V = |r_1 \omega_1| = |r_2 \omega_2|$$

The velocity ratio is

$$\left|\frac{\omega_1}{\omega_2}\right| = \frac{r_2}{r_1}$$

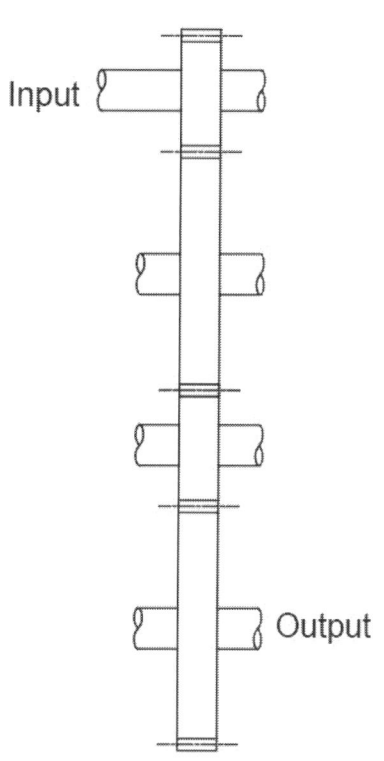

Figure- Simple gear train

Compound gear train

Gear Train and Gear Design

Chapter 9

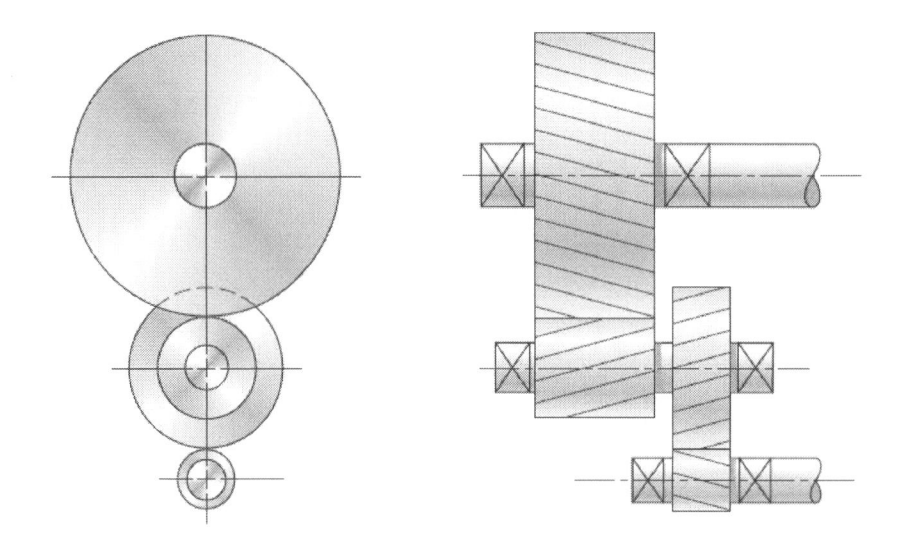

Figure -Compound gear train

The velocity ratio in the case of the compound train of wheels is equal to

$$= \frac{\text{Product of teeth on the followers}}{\text{Product of teeth on the drivers}}$$

The velocity ratio of the following gear train is

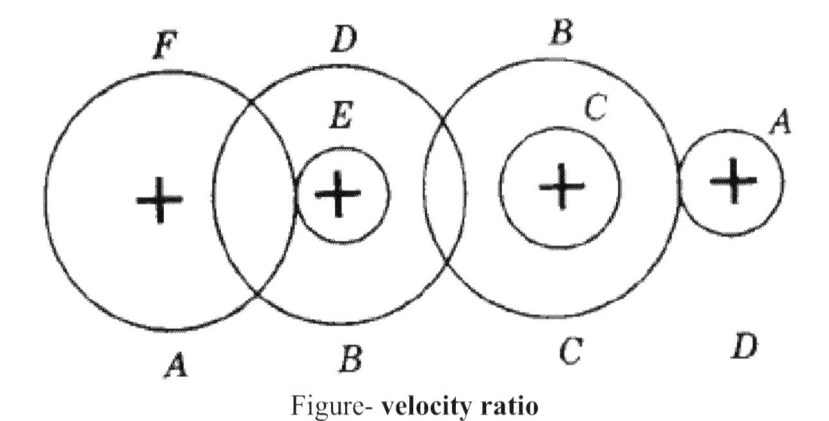

Figure- velocity ratio

$$\boxed{\frac{N_F}{N_A} = \frac{T_A \times T_C \times T_E}{T_B \times T_D \times T_F}}$$

Reverted gear train

Gear Train and Gear Design

Chapter 9

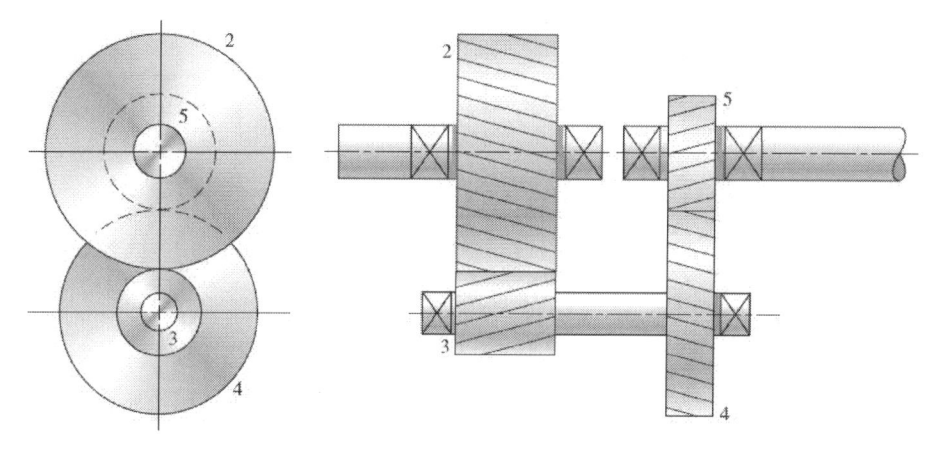

Figure- reverted gear train or a compound reverted gear train

It is sometimes desirable for the input shaft and the output shaft of a two-stage compound gear train to be in-line, as shown in Fig above. This configuration is called a *compound reverted gear train*. This requires the distances between the shafts to be the same for both stages of the train.

The distance constraint is

$$\frac{d_2}{2} + \frac{d_3}{2} = \frac{d_4}{2} + \frac{d_5}{2}$$

The diametric pitch relates the diameters and the numbers of teeth, $P = T/d$. Replacing All the diameters give

$$T_2/(2P) + T_3/(2P) = T_4/(2P) + T_5/(2P)$$

Assuming a constant diametral pitch in both stages, we have the geometry condition Stated in terms of numbers of teeth:

$$T_2 + T_3 = T_4 + T_5$$

This condition must be exactly satisfied, in addition to the previous ratio equations, to Provide for the in-line condition on the input and output shafts.

In the compound gear train shown in the figure, gears A and C have equal numbers of teeth and gears B and D have equal numbers of teeth.
From the figure $r_A + r_B = r_C + r_D$ or $T_A + T_B = T_C + T_D$ and as $N_B + N_C$ it must be $T_B = T_D$ & $T_A = T_C$
Or $\dfrac{N_B}{N_A} = \dfrac{N_D}{N_C}$ or $N_C = \sqrt{N_A N_D}$
[where $N_B = N_c$]

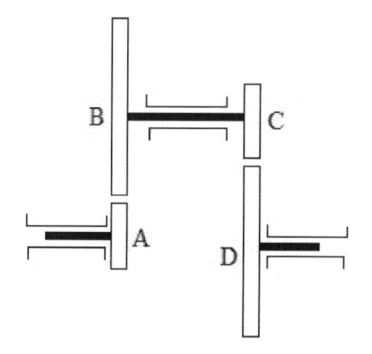

Epicyclic gear train

Consider the following Epicyclic gear trai

Gear Train and Gear Design

Chapter 9

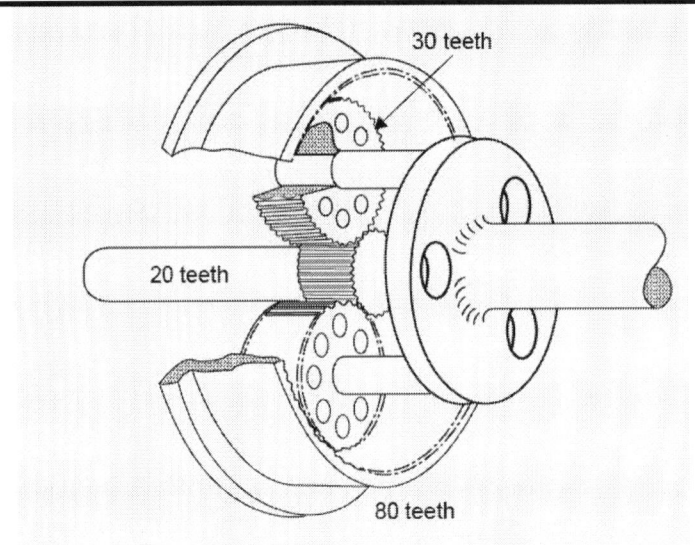

Figure- Epicyclic gear train

For the epicyclic gearbox illustrated in figure, determine the speed and direction of the final drive and also the speed and direction of the planetary gears. The teeth numbers of the sun, planets and ring gear are 20, 30 and 80, respectively. The speed and direction of the sun gear is 1000 rpm clockwise and the ring gear is held stationary.

Solution

$$n_{arm} = \frac{n_{sun}}{(80/20)+1} = \frac{-1000}{5} = -200 \; rpm$$

The speed of the final drive is 200 rpm clockwise. The reduction ratio for the gearbox is Given by $n_{sun}/n_{arm} = 1000/200 = 5$. To determine the speed of the planets use

The planets and sun are in mesh, so

$$\frac{n_{planet}/n_{arm}}{n_{sum}/n_{arm}} = -\frac{N_S}{N_P}$$

$$\frac{n_{planet} - n_{arm}}{n_{sum} - n_{arm}} = -\frac{N_S}{N_P}$$

$$\frac{n_{planet} - (-200)}{-1000 - (-200)} = -\frac{20}{30}$$

$$n_{planet} = -\frac{20}{30} \times (-800) - 200 = 333 \; rpm$$

The speed of rotation of the planetary gears is 333 rpm counter-clockwise.

Now make a table for **the epicyclic gear arrangement shown in the figure below.**

Gear Train and Gear Design

Chapter 9

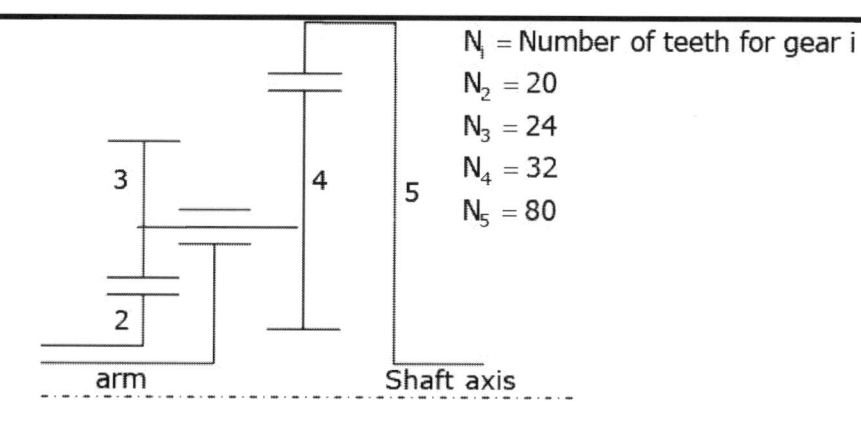

N_i = Number of teeth for gear i
$N_2 = 20$
$N_3 = 24$
$N_4 = 32$
$N_5 = 80$

	Arm	2	3	4	5
1.	0	+x	$-\dfrac{N_2}{N_3}x$	$-\dfrac{N_2}{N_3}x$	$-\dfrac{N_4}{N_5} \times \dfrac{N_2}{N_3}x$
2.	y	y	y	y	y
	y	x + y	$y - \dfrac{N_2}{N_3}x$		$y - \dfrac{N_4}{N_5} \times \dfrac{N_2}{N_3}x$

Formula List for Gears:

(a) Spur Gear

Name	Speed ratio	
1. Spur & Helical	6:1 to 10:1	for high speed helical. For high speed spur.
2. Bevel	1:1 to 3:1	
3. Worm	10:1 to 100:1	provided $\angle 100$ KW

SPUR GEAR

(i) Circular pitch $(p) = \dfrac{\pi d}{T}$

(ii) Diametral pitch $(P) = \dfrac{T}{d}$

(iii) $pP = \pi$

(iv) Module $(m) = \dfrac{d}{T} = \dfrac{1}{P}$ or $d = mT$

(v) Speed ratio $(G) = \dfrac{\omega_p}{\omega_g} = \dfrac{T_g}{T_p}$

(vi) centre-to-centre distance $= \dfrac{1}{2}(d_g + d_p)$
$= \dfrac{1}{2}m(T_g + T_p)$

(vii) Addendum $(h_a) = 1m$
$(h_f) = 1.25m$

Gear Train and Gear Design

Chapter 9

Clearance (C) = 0.25 m

(viii)
$$P_t = \frac{2T}{d}$$
$$P_r = P_t \tan\alpha$$
$$P_N = \frac{P_t}{\cos\alpha}$$

(ix) Minimum number of teeth to avoid interference

$$T_{min} = \frac{2A_w}{\sin^2\phi} \quad || \quad \text{For } 20° \text{ full depth } T = 18 \text{ to } 20$$

(x)

$$T_{min, pinion} = \frac{2 \times A_w}{G\left[\sqrt{1 + \frac{1}{G}\left(\frac{1}{G} + 2\right)\sin^2\phi} - 1\right]}$$

$$G = \textbf{Gear ratio} = \frac{Z_g}{Z_p} = \frac{\omega_p}{\omega_g}$$

A_p = fraction of addendum to module = $\dfrac{h_a}{m}$ for pinion

A_w = fraction of addendum to module = $\dfrac{h_f}{m}$ for gear (generally 1)

(xi) Face width 8m < b < 12m; usually b = 10m

(xii) Beam strength $\quad \sigma_b = mb\sigma_b Y \rightarrow$ **Lewis equation**

Where $\sigma_b = \dfrac{\sigma_{ult}}{3}$

Lewis form factor, $Y = \left(0.154 - \dfrac{0.912}{z}\right)$ **for 20° full depth gear.**

(xiii) Wear strength $(\sigma_\omega) = bQdpK$

Where $Q = \dfrac{2 T_g}{T_g + T_p}$ for external gear

$\quad\quad\quad = \dfrac{2 T_g}{T_g - T_p}$ for internal gear

load-stress factor $(k) = \dfrac{\sigma_c^2 \sin\phi\cos\phi}{1.4}\left(\dfrac{1}{E_1} + \dfrac{1}{E_2}\right)$

$\quad\quad\quad\quad = 0.16\left(\dfrac{BHN}{100}\right)^2$

(xiv)

Gear Train and Gear Design

Chapter 9

$$P_{eff} = \frac{C_s}{C_v} P_t \quad \text{where } C_v = \frac{3}{3+V}, \quad \text{for ordinary cut gear } v < 10 \text{ m/s}$$

$$= \frac{6}{6+v}, \quad \text{for hobbed generated } v > 20 \text{ m/s}$$

$$= \frac{5.6}{5.6+\sqrt{v}}, \quad \text{for precision gear } v > 20 \text{ m/s}$$

(xv) **Spott's equation,** $(P_{eff}) = C_s P_t + P_d$

where $P_d = \dfrac{e n_p T_p b r_1 r_2}{2530\sqrt{r_1^2 + r_2^2}}$ for steel pinion and steel gear

$e = 16.00 + 1.25\phi$ for grade 8

$\phi = m + 0.29\sqrt{d}$ for all.

Gear Train and Gear Design
Chapter 9

(b) Helical Gear

(1) $p_n = p\cos\alpha$ where p = transverse circular pitch.
          p_n = normal circular pitch.

(2) $P_n = \dfrac{P}{\cos\alpha}$ where P_n = normal diametral pitch.
          P = transverse diametral pitch.
          α = helix angle.

(3) $pP = \pi$

(4) $m_n = m\cos\alpha$ $\left[P = \dfrac{1}{m}\right]$ m = transverse module.

          m_n = normal module.

(5) $p_a = \dfrac{p}{\tan\alpha} = \dfrac{\pi m}{\tan\alpha}$; p_a = axial pitch.

(6) $\cos\alpha = \dfrac{\tan\phi_n}{\tan\phi}$ ϕ_n = normal pressure angle (usually 20°).

          ϕ = transverse pressure angle.

(7) $d = \dfrac{TP}{\pi} = zm = \dfrac{Tm_n}{\cos\alpha}$; d = pitch circle diameter.

(8) $a = \dfrac{m_n}{2\cos\alpha}\{T_1 + T_2\}$ → centre - to - centre distance.

(9) $T' = \dfrac{T}{\cos^3\alpha}$; $d' = \dfrac{d}{\cos^2\alpha}$

(10) An imaginary spur gear is considered with a pitch circle diameter of d' and module m_n is called „formative" or „virtual" spur gear.

(11) Helix angle α varies from **15 to 25°**.

(12) Preference value of normal modus (m_n) = 1, 1.25, 1.5, 2, 3, 4, 5, 6, 8

(13) Addendum $(h_a) = m_n$; dedendum $(h_f) = 1.25\, m_n$, clearance = $0.25\, m_n$

(14) $P_t = \dfrac{2M_t}{d} = P\cos\phi_n \cos\alpha$; $M_t = \dfrac{60 \times 10^6 \times (KW)}{2\pi N}$ N-mm

   $P_r = P_t\left(\dfrac{\tan\phi_n}{\cos\alpha}\right) = P\sin\alpha\phi_n$

   $P_a = P_t \tan\alpha = P\cos\phi_n \sin\alpha$

(15) Beam strength $S_b = m_n b \sigma_b Y'$

(16) Wear strength $S_w = \dfrac{bQd_pK}{\cos^2\alpha}$

(17)(17)

Gear Train and Gear Design
Chapter 9

$$P_{eff} = \frac{C_s P_t}{C_v} \;;\; C_v = \frac{5.6}{5.6+\sqrt{v}} \;;\; P_d = \frac{en_p T_p b r_1 r_2}{2530\sqrt{r_1^2 + r_2^2}}$$

$$P_{eff} = C_s P_t + P_d \cos\alpha_n \cos\psi$$

(c) Worm Gear

(i) Specified and designated by $T_1 / T_2 / q / m$

Where: q is diametric quotient $= \dfrac{d_1}{m}$

(ii) The threads of the worm have an involute helicoids profile.

(iii) Axial pitch (p_x) = distance between two consecutive teeth-measured along the axis of the worm.

(iv) The lead (l) = when the worm is rotated one revolution, a distance that a point on the helical profile will move.

(v) $\quad l = p_x \times T_1; \qquad d_z = mT_2$

(vi) Axial pitch of the worm = circular pitch of the worm wheel

$$P_x = \frac{\pi d_2}{T_2} = \pi m \qquad\qquad [ICS-04]$$

$$l = P_x T_1 = \pi m T_1$$

(vii) Lead angle $(\delta) = \tan^{-1}\left(\dfrac{T_1}{q}\right) = \tan^{-1}\left(\dfrac{l}{\pi d_1}\right)$

(viii) centre-to-centre distance $(a) = \dfrac{1}{2}(d_1 + d_2) = \dfrac{1}{2}m(T_1 + T_2)$

(ix) Preferred values of q: 8, 10, 12.5, 16, 20, 25

(x) Number of starts T_1 usually taken 1, 2, or 4

(xi)
addendum $(h_{a_1}) = m$	$h_{a_2} = m(2\cos\delta - 1)$
dedendum $(h_{f_1}) = (2.2\cos\delta - 1)m$	$h_{f_2} = m(1 + 0.2\cos\delta)$
clearance $(c) = 0.2m\cos\delta$	$c = 0.2m\cos\delta$

(xii) $F = 2m\sqrt{q+1}$ effective face width of the root of the worm wheel.

(xiii) $\delta = \sin^{-1}\left(\dfrac{\Box F}{d_{a_1} + 2C}\right); l_r = (d_{a_1} + 2C)\sin^{-1}\left(\dfrac{\Box F}{d_{a_1} + 2C}\right)$ = length of the root of the worm wheel teeth

Gear Train and Gear Design

Chapter 9

(xiv) $(P_1)_t = \dfrac{2m_t}{d_1}$; $(P_1)_a = (P_1)_t \dfrac{\cos\alpha\cos\delta - \mu\sin\delta}{\cos\alpha\sin\delta + \mu\cos\delta}$; $(P_1)_r = (P_1)_t \dfrac{\sin\alpha}{\cos\alpha\sin\delta + \mu\cos\delta}$

(xv) Efficiency $(\eta) = \dfrac{\text{Power output}}{\text{Power input}} = \dfrac{\cos\alpha - \mu\tan\delta}{\cos\alpha + \mu\tan\delta}$

(xvi) Rubbing velocity $(V_s) = \dfrac{\pi d_1 n_1}{60000 \cos\delta}$ m/s (remaining 4 cheak)

(xvii) Thermal consideration $H_g = 1000(1-\eta) \times (KW)$

$$H_d = K(t - t_0)A$$

$$KW = \dfrac{K(t-t_0)A}{1000(1-\eta)}$$

Gear Train and Gear Design

Chapter 9

Objective Questions (IES, IAS, GATE)

Previous 20-Years GATE Questions

Spur gear

GATE-1. Match the type of gears with their most appropriate description. [GATE-2008]

Type of gear	Description
P Helical	1. Axes non parallel and intersecting
Q Spiral	2. Axes parallel and teeth are inclined to the axis
R Hypoid	3. Axes parallel and teeth are parallel to the axis
S Rack and pinion	4. Axes are perpendicular and intersecting, and teeth are inclined to the axis
	5. Axes are perpendicular and used for large speed reduction
	6. Axes parallel and one of the gears has infinite radius

(a) P-2, Q-4, R-1, S-6   (c) P-2, Q-6, R-4, S-2
(b) P-1, Q-4, R-5, S-6   (d) P-6, Q-3, R-1, S-5

GATE-2. One tooth of a gear having 4 module and 32 teeth is shown in the figure. Assume that the gear tooth and the corresponding tooth space make equal intercepts on the pitch circumference. The dimensions $'a'$ and $'b'$, respectively, are closest to   [GATE-2008]

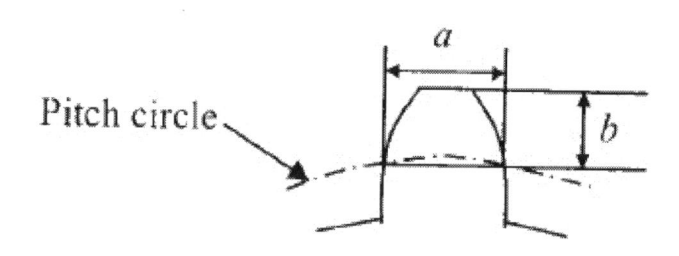

(a) 6.08 mm, 4 mm   (b) 6.48 mm, 4.2 mm
(c) 6.28 mm, 4.3 mm   (d) 6.28 mm, 4.1

Classification of Gears

GATE-3. Match the following   [GATE-2004]

Type of gears	Arrangement of shafts
P. Bevel gears	1. Non-parallel off-set shafts
Q. Worm gears	2. Non-parallel intersecting shafts
R. Herringbone gears	3. Non-parallel non-intersecting shafts
S. Hypoid gears	4. Parallel shafts

(a) P-4 Q-2 R-1 S-3   (b) P-2 Q-3 R-4 S-1
(c) P-3 Q-2 R-1 S-4   (d) P-1 Q-3 R-4 S-2

Pitch point

GATE-4. In spur gears, the circle on which the involute is generated is called the
(a) Pitch circle   (b) clearance circle   [GATE-1996]

Gear Train and Gear Design

Chapter 9

(c) Base circle (d) addendum circle

Minimum Number of Teeth

GATE-5. The minimum number of teeth on the pinion to operate without interference in standard full height involute teeth gear mechanism with 20° pressure angle is [GATE-2002]
(a) 14 (b) 12 (c) 18 (d) 32

Interference

GATE-6 Tooth interference in an external in volute spur gear pair can be reduced by
(a) Decreasing center distance between gear pair [GATE-2010]
(b) Decreasing module
(c) Decreasing pressure angle
(d) Increasing number of gear teeth

GATE-7. Interference in a pair of gears is avoided, if the addendum circles of both the gears intersect common tangent to the base circles within the points of tangency.
[GATE-1995]
(a) True (b) False
(c) Insufficient data (d) None of the above

GATE-8. Twenty degree full depth involute profiled 19-tooth pinion and 37-tooth gear are in mesh. If the module is 5 mm, the centre distance between the gear pair will be [GATE-2006]
(a) 140 mm (b) 150 mm
(c) 280 mm (d) 300 mm

Beam Strength of Gear Tooth

GATE-9. A spur gear has a module of 3 mm, number of teeth 16, a face width of 36 mm and a pressure angle of 20°. It is transmitting a power of 3 kW at 20 rev/s. Taking a velocity factor of 1.5, and a form factor of 0.3, the stress in the gear tooth is about [GATE-2008]
(a) 32 MPa (b) 46 MPa
(c) 58 MPa (d) 70 MPa

Statement for Linked Answer GATE-10 and GATE-11:
A 20° full depth involute spur pinion of 4 mm module and 21 teeth is to transmit 15 kW at 960 rpm. Its face width is 25 mm.

GATE-10. The tangential force transmitted (in N) is [GATE -2009]
(a) 3552 (b) 261 1 (c) 1776 (d) 1305

GATE-11. Given that the tooth geometry factor is 0.32 and the combined effect of dynamic load and allied factors intensifying the stress is 1.5; the minimum allowable stress (in MPa) for the gear material is [GATE -2009]
(a) 242.0 (b) 166.5 (c) 121.0 (d) 74.0

Simple Gear train

Note: - Common Data for GATE-12 & GATE-13.
A gear set has a opinion with 20 teeth and a gear with 40 teeth. The pinion runs at 0 rev/s and transmits a power of 20 kW. The teeth are on the 20° full –depth system and have module of 5 mm. The length of the line of action is 19 mm.

Gear Train and Gear Design

Chapter 9

GATE-12. The center distance for the above gear set in mm is [GATE-2007]
   (a) 140   (b) 150   (c) 160   (d) 170.

GATE-13 The contact ratio of the contacting tooth [GATE-2007]
   (a) 1.21   (b) 1.25   (c) 1.29   (d) 1.33

GATE-14. The resultant force on the contacting gear tooth in N is: [GATE-2007]
   (a) 77.23   (b) 212.20   (c) 225.80   (d) 289.43

Compound gear train

Data for GATE-15 & GATE-16 are given below. Solve the problems and choose correct answers.

A compacting machine shown in the figure below is used to create a desired thrust force by using a rack and pinion arrangement. The input gear is mounted on tile motor shaft. The gears have involute teeth of 2 mm module.

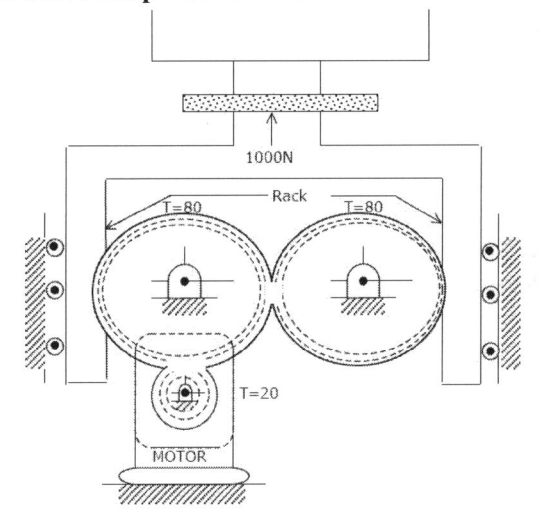

GATE-15. If the drive efficiency is 80%, then torque required on the input shaft to create 1000 N output thrust is [GATE-2004]
   (a) 20 Nm   (b) 25 Nm   (c) 32 Nm   (d) 50 Nm

GATE-16. If the pressure angle of the rack is 20°, then force acting along the line of action between the rack and the gear teeth is [GATE-2004]
   (a) 250 N   (b) 342 N   (c) 532 N   (d) 600 N

Reverted gear train

Data for GATE-17 & GATE-18 are given below. Solve the problems and choose correct answers.

The overall gear ratio in a 2 stage speed reduction gear box (with all spur gears) is 12. The input and output shafts of the gear box are collinear. The countershaft which is parallel to the input and output shafts has a gear (Z_2 teeth) and pinion (Z_3 = 15 teeth) to mesh with pinion (Z_1 = 16 teeth) on the input shaft and gear (Z_4 teeth) on the output shaft respectively. It was decided to use a gear ratio of 4 with 3 module in the first stage and 4 module in the second stage.

GATE-17. Z_2 and Z_4 are [GATE-2003]
   (a) 64 and 45   (b) 45 and 64   (c) 48 and 60   (d) 60 and 48

GATE-18. The centre distance in the second stage is [GATE-2003]
   (a) 90 mm   (b) 120 mm   (c) 160 mm   (d) 240mm

Gear Train and Gear Design

Chapter 9

Epicyclic gear train

GATE-19. For the epicyclic gear arrangement shown in the figure, $\omega_2 = 100$ rad/s clockwise (CW) and $\omega_{arm} = 80$ rad/s counter clockwise (CCW). The angular velocity ω_5, (in rad/s) is

[GATE-2010]

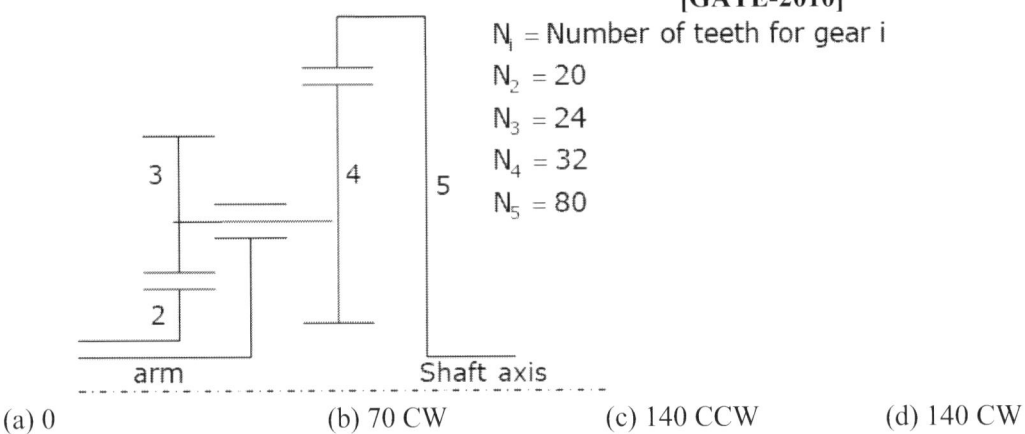

N_i = Number of teeth for gear i
$N_2 = 20$
$N_3 = 24$
$N_4 = 32$
$N_5 = 80$

(a) 0 (b) 70 CW (c) 140 CCW (d) 140 CW

GATE-20. An epicyclic gear train is shown schematically in the adjacent figure.

The sun gear 2 on the input shaft is a 20 teeth external gear. The planet gear 3 is a 40 teeth external gear. The ring gear 5 is a 100 teeth internal gear. The ring gear 5 is fixed and the gear 2 is rotating at 60 rpm (ccw = counter-clockwise and cw = clockwise).

The arm 4 attached to the output shaft will rotate at
(a) 10 rpm ccw
(b) 10 rpm ccw
(c) 12 rpm cw
(d) 12 rpm ccw

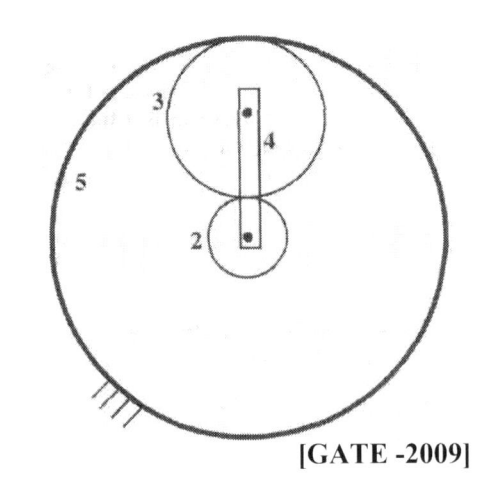

[GATE -2009]

GATE-21 The arm OA of an epicyclic gear train shown in figure revolves counter clockwise about O with an angular velocity of 4 rad/s. Both gears are of same size. Tire angular velocity of gear C, if the sun gear B is fixed, is

[GATE-1995]

(a) 4 rad / sec
(b) 8 rad / sec
(c) 10 rad / sec
(d) 12 rad / sec

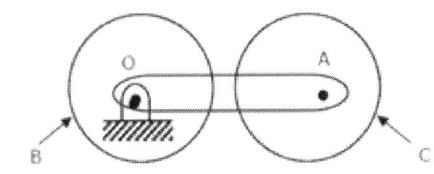

GATE-22. The sun gear in the figure is driven clockwise at 100 rpm. The ring gear is held stationary. For the number of teeth shown on the gears, the arm rotates at
(a) 0 rpm   (b) 20 rpm
(c) 33.33 rpm   (d) 66.67 rpm

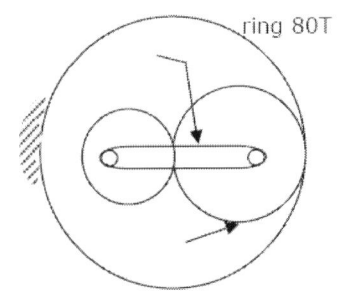

Gear Train and Gear Design

Chapter 9

[GATE-2001]

GATE-23. Two mating spur gears have 40 and 120 teeth respectively. The pinion rotates at 1200 rpm and transmits a torque of 20 Nm. The torque transmitted by the gear is [GATE-2004]
(a) 6.6 Nm (b) 20 Nm (c) 40 Nm (d) 60 Nm

Common Data for GATE-24, GATE-25:
A planetary gear train has four gears and one carrier. Angular velocities of the gears are $\omega_1, \omega_2, \omega_3,$ and ω_4 respectively. The carrier rotates with angular velocity ω_5,

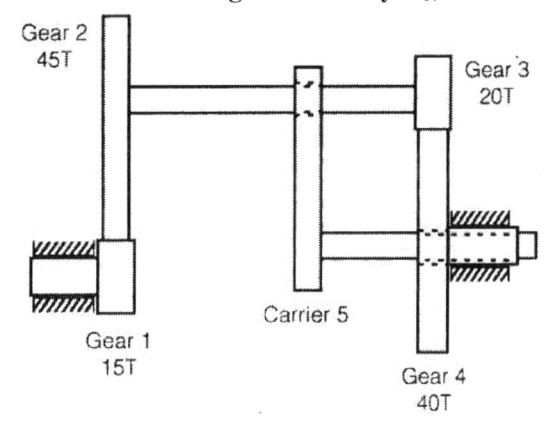

Gear 2 45T, Gear 3 20T, Gear 1 15T, Carrier 5, Gear 4 40T

GATE-24. What is the relation between the angular velocities of Gear 1 and Gear 4? [GATE-2006]

GATE-25. For (ω_1 = 60 rpm clockwise (cw) when looked from the left, what is the angular velocity of the carrier and its direction so that Gear 4 rotates in counter clockwise (ccw) direction at twice the angular velocity of Gear 1 when looked from the left? [GATE-2006]
(a) 130 rpm, cw (b) 223 rpm, ccw
(c) 256 rpm, cw (d) 156 rpm, ccw

Worm Gears

GATE-26. Large speed reductions (greater than 20) in one stage of a gear train are possible through [GATE-2002]
(a) Spur gearing (b) Worm gearing (c) Bevel gearing (d) Helical gearing

GATE-27. A 1.5 kW motor is running at 1440 rev/min. It is to be connected to a stirrer running at 36 rev/min. The gearing arrangement suitable for this application is [GATE-2000]
(a) Differential gear (b) helical gear
(c) Spur gear (d) worm gear

GATE-28. To make a worm drive reversible, it is necessary to increase [GATE-1997]
(a) centre distance (b) worm diameter factor
(c) Number of starts (d) reduction ratio

Gear Train and Gear Design

Chapter 9

Previous 20-Years IES Questions

Spur gear

IES-1. The velocity ratio between pinion and gear in a gear drive is 2.3, the module of teeth is 2.0 mm and sum of number of teeth on pinion and gear is 99. What is the centre distance between pinion and the gear? [IES 2007]

(a) 49.5 mm (b) 99 mm (c) 148.5 mm (d) 198 mm

IES-2. Consider the following statements: [IES-2001]
When two gears are meshing, the clearance is given by the
1. Difference between dedendum of one gear and addendum of the mating gear.
2. Difference between total and the working depth of a gear tooth.
3. Distance between the bottom land of one gear and the top land of the mating gear.
4. Difference between the radii of the base circle and the dedendum circle.
Which of these statements are correct?
(a) 1, 2 and 3 (b) 2, 3 and 4 (c) 1, 3 and 4 (d) 1, 2 and 4

IES-3. The working surface above the pitch surface of the gear tooth is termed as [IES-1998]
(a) Addendum (b) dedendum (c) flank (d) face

IES-4. Match the following $14\frac{1}{2}^o$ composite system gears

[IES-1992]

List I	List II
A. Dedendum	1. $\dfrac{2}{pd}$
B. Clearance	2. $\dfrac{0.157}{pd}$
C. Working depth	3. $\dfrac{1.157}{pd}$
D. Addendum	4. $\dfrac{1}{pd}$

Code: A B C D   A B C D
(a)   1 2 3 4   (b) 4 3 2 1
(c)   3 2 1 4   (d) 3 1 2 4

IES-5. Match List I with List II and select the correct answer using the codes given below the lists: [IES-1993]

List I (Standard tooth/arms)	List II (Advantages or disadvantages)
A. 20° and 25° systems	1. Results in lower loads on bearing
B. 14.5° stub-tooth system	2. Broadest at the base and strongest in bending
C. 25° Full depth system	3. Obsolete
D. 20° Full depth system	4. Standards for new applications

Code: A B C D   A B C D
(a)   4 3 2 1   (b) 3 1 2 4
(c)   3 2 1 4   (d) 4 2 3 1

IES-6. Assertion (A): When one body drives another by direct contact, their contact points must have equal components of velocity normal to the surfaces at the point of contact.

Gear Train and Gear Design

Chapter 9

Reason (R): Two points in the same body must have the same component of velocity relative to the third body, in the direction of the line joining the two points. [IES-1993]
(a) Both A and R are individually true and R is the correct explanation of A
(b) Both A and R are individually true but R is **not** the correct explanation of A
(c) A is true but R is false
(d) A is false but R is true

Classification of Gears

IES-7. Match List I with List II and select the correct answer [IES-1996]

List I
A. Helical gears
B. Herring bone gears
C. Worm gears
D. Hypoid Gears

List II
1. Non-interchangeable
2. Zero axial thrust
3. Quiet motion
4. Extreme speed reduction

Codes:
	A	B	C	D		A	B	C	D
(a)	1	2	3	4	(b)	3	2	1	4
(c)	3	1	4	2	(d)	3	2	4	1

IES-8. Match List-l (Type of Gears) with List-II (Characteristics) and select the correct answer using the code given below the Lists: [IES-2006]

List-I
A. Helical gearing
B. Herringbone gearing
C. Worm gearing
D. Hypoid gearing

List -II
1. Zero axial thrust
2. Non-inter-changeable
3. Skew shafts
4. Parallel shafts

	A	B	C	D		A	B	C	D
(a)	4	1	3	2	(b)	3	2	4	1
(c)	4	2	3	1	(d)	3	1	4	2

IES-9. Match List I with List II and select the correct answer using the code given below the Lists:

List I [IES 2007]
A. Worm gear
B. Spur gear
C. Herringbone gear
D. Spring level gear

List II
1. Imposes no thrust load on the shaft
2. To transmit power between two non-intersecting shafts which are perpendicular to each other
3. To transmit power when the shafts are parallel
4. To transmit power when the shafts are at right angles to one another

Code:
	A	B	C	D		A	B	C	D
(a)	1	2	3	4	(b)	2	3	1	4
(c)	1	2	4	3	(d)	2	3	4	1

IES-10. Match List I (Type of Gear/Gear Train) with List II (Different Usage and Drive) and select the correct answer using the code given below the Lists:

List I [IES-2005]
A Epicyclic gear train
B. Bevel Gear
C. Worm-worm Gear
D. Herringbone Gear

List II
1. Reduces end thrust
2. Low gear ratio
3. Drives non-parallel nonintersecting shafts
4. Drives non-parallel intersecting shafts
5. High gear ratio

	A	B	C	D		A	B	C	D
(a)	5	4	3	1	(b)	2	3	4	5
(c)	5	3	4	1	(d)	2	4	3	5

IES-11. Which type of gear is used for shaft axes having an offset? [IES-2004]

Gear Train and Gear Design

Chapter 9

(a) Mitre gears   (b) Spiral bevel gears
(c) Hypoid gears   (d) Zerol gears

IES-12. The gears employed for connecting two non-intersecting and non-parallel, i.e., non-coplanar shafts are [IES-2003; 2005]
(a) Bevel gears   (b) Spiral gears   (c) Helical gears   (d) Mitre gears

IES-13. When two shafts are neither parallel nor intersecting, power can be transmitted by using [IES-1998]
(a) A pair of spur gears   (b) a pair of helical gears
(c) An Oldham's coupling   (d) a pair of spiral gears

IES-14. In a single reduction, a large velocity ratio is required. The best transmission is [IES-1999]
(a) Spur gear drive   (b) helical gear drive
(c) Bevel gear drive   (d) worm gear drive

IES-15. Which one of the following pairs is not correctly matched? [IES-1995]
(a) Positive drive......Belt drive
(b) High velocity ratio......Worm gearing
(c) To connect non-parallel and non- intersecting shafts......Spiral gearing.
(d) Diminished noise and smooth operation......Helical gears.

Mitres gear

IES-16. Mitre gears [IES-1992]
(a) spur-gears with gear ratio 1: 1
(b) Skew gears connecting non-parallel and nonintersecting shafts
(c) Bevel gears transmitting power at more than or less than 90°
(d) Bevel gears in which the angle between the axes is 90° and the speed ratio of the gears is 1: 1

IES-17. Match List-I (Gears) with List-II (Configurations) and select the correct answer using the codes given below the Lists: [IES-2003]

List-I (Gears)
A Spur
B. Bevel
C. Helical
D. Mitre

List-II (Configurations)
1. Connecting two non-parallel or intersecting but coplanar shafts
2. Connecting two parallel and coplanar shafts with teeth parallel to the axis of the gear wheel
3. Connecting two parallel and coplanar shafts with teeth inclined to the axis of the gear wheel
4. Connecting two shafts whose axes are mutually perpendicular to each other

Codes:
	A	B	C	D		A	B	C	D
(a)	2	4	3	1	(b)	3	1	2	4
(c)	2	1	3	4	(d)	3	4	2	1

Pitch point

IES-18. Gearing contact is which one of the following? [IES 2007]
(a) Sliding contact   (b) Sliding contact, only rolling at pitch point
(c) Rolling contact   (d) Rolling and sliding at each point of contact

IES-19. When two spur gears having involute profiles on, their teeth engage, the line of action is tangential to the [IES-2003]
(a) Pitch circles   (b) Dedendum circles

Gear Train and Gear Design

Chapter 9

(c) Addendum circles (d) Base circles

Pressure angle

IES-20. What is the value of pressure angle generally used for involute gears?
[IES-2006]
(a) 35° (b) 30° (c) 25° (d) 20°

IES-21. Consider the following, modifications regarding avoiding the interference between gears: [IES-2003]
1. The centre distance between meshing gears be increased
2. Addendum of the gear be modified
3. Teeth should be undercut slightly at the root
4. Pressure angle should be increased
5. Circular pitch be increased

Which of these are effective in avoiding interference?
(a) 1, 2 and 3 (b) 2, 3, 4 and 5 (c) 1, 4 and 5 (d) 3, 4 and 5

IES-22. An external gear with 60 teeth meshes with a pinion of 20 teeth, module being 6 mm. What is the centre distance in mm? [IES-2009]
(a) 120 (b) 180 (c) 240 (d) 300

IES-23. **Assertion (A):** An involute rack with 20° pressure angle meshes with a pinion of 14.5° pressure angle. [IES-2002]
Reason (R): Such a matching is impossible.
(a) Both A and R are individually true and R is the correct explanation of A
(b) Both A and R are individually true but R is **not** the correct explanation of A
(c) A is true but R is false
(d) A is false but R is true

IES-24. Compared to gears with 20^0 pressure angle involute full depth teeth, those with 20^0 pressure angle and stub teeth have [IES 2007]
1. Smaller addendum. 2. Smaller dedendum.
3. Smaller tooth thickness. 4. Greater bending strength.
Which of the statements given above are correct?
(a) 1, 2 and 3 (b) 1, 2 and 4
(c) 1, 3 and 4 (d) 2, 3 and 4

IES-25. Consider the following statements: [IES-1999]

A pinion of $14\frac{1}{2}^o$ pressure angle and 48 involute teeth has a pitch circle diameter of 28.8 cm. It has
1. Module of 6 mm 2. Circular pitch of 18 mm
3. Addendum of 6 mm 4. Diametral pitch of $\frac{11}{113}$

Which of these statements are correct?
(a) 2 and 3 (b) 1 and 3 (c) 1 and 4 (d) 2 and 4

IES-26. Which of the following statements are correct? [IES-1996]
1. For constant velocity ratio transmission between two gears, the common normal at the point of contact must always pass through a fixed point on the line joining the centres of rotation of the gears.
2. For involute gears the pressure angle changes with change in centre distance between gears.

Gear Train and Gear Design

Chapter 9

3. The velocity ratio of compound gear train depends upon the number of teeth of the input and output gears only.
4. Epicyclic gear trains involve rotation of at least one gear axis about some other gear axis.
(a) 1, 2 and 3 (b) 1, 3 and 4 (c) 1, 2 and 4 (d) 2, 3 and 4

IES-27. Which one of the following is true for involute gears? [IES-1995]
(a) Interference is inherently absent
(b) Variation in centre distance of shafts increases radial force
(c) A convex flank is always in contact with concave flank
(d) Pressure angle is constant throughout the teeth engagement.

IES-28. In involute gears the pressure angle is [IES-1993]
(a) Dependent on the size of teeth (b) dependent on the size of gears
(c) Always constant (d) always variable

Minimum Number of Teeth

IES-29. Which one of the following statements is correct? [IES-2004]
Certain minimum number of teeth on the involute pinion is necessary in order to
(a) Provide an economical design (b) avoid Interference
(c) Reduce noise in operation (d) overcome fatigue failure of the teeth

IES-30. A certain minimum number of teeth is to be kept for a gear wheel
(a) So that the gear is of a good size [IES-1999]
(b) For better durability
(c) To avoid interference and undercutting
(d) For better strength

IES-31. In full depth $14\frac{1}{2}^o$ degree involute system, the smallest number of teeth in a pinion which meshes with rack with out interference is [IES-1992]
(a) 12 (b) 16 (c) 25 (d) 32

IES-33. Match List I with List II and select the correct answer using the codes given below the lists:
List I (Terminology) List II (Relevant terms) [IES-1995]
A. Interference 1. Arc of approach, arc of recess, circular pitch
B. Dynamic load on tooth 2. Lewis equation
C. Static load 3. Minimum number of teeth on pinion
D. Contract ratio 4. Inaccuracies in tooth profile
Codes: A B C D   A B C D
(a)   3 4 1 2 (b) 1 2 3 4
(c)   4 3 2 1 (d) 3 4 2 1

IES-34 Assertion (A): When a pair of spur gears of the same material is in mesh, the design is based on pinion. [IES-2002; 1993]
Reason (R): For a pair of gears of the same material in mesh, the 'strength factor' of the pinion is less than that of the gear.
(a) Both A and R are individually true and R is the correct explanation of A
(b) Both A and R are individually true but R is **not** the correct explanation of A
(c) A is true but R is false
(d) A is false but R is true

Gear Train and Gear Design

Chapter 9

Cycloidal teeth

IES-35. The curve traced by a point on the circumference of a circle which rolls along the inside of affixed circle, is known as [IES-1992]
(a) Epicycloid (b) hypocycloid
(c) Cardiod (d) involute

IES-36

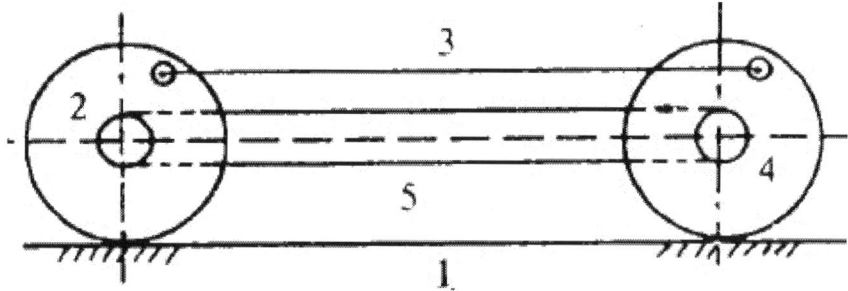

In the mechanism shown above, link 3 has [IES-2004]
(a) Curvilinear translation and all points in it trace out identical cycloids
(b) Curvilinear translation and all points in it trace out identical involutes
(c) Linear translation & all points in it trace out identical helices
(d) Linear translation & all points in it trace out identical ellipses

IES-37. A thin circular disc is rolling with a uniform linear speed, along a straight path on a plane surface. [IES-1994]
Consider the following statements in this regard:
1. All points on the disc have the same velocity.
2. The centre of the disc has zero acceleration.
3. The centre of the disc has centrifugal acceleration.
4. The point on the disc making contact with the plane surface has zero acceleration of these statements
(a) 1 and 4 are correct (b) 3 and 4 are correct
(c) 3 alone is correct (d) 2 alone is correct.

Involute teeth

IES-38. In the case of an involute toothed gear, involute starts from [IES-1997]
(a) Addendum circle (b) dedendum circle
(c) Pitch circle (d) base circle

IES-39. Consider the following statements: [IES-2006]
1. A stub tooth has a working depth larger than that of a full-depth tooth.
2. The path of contact for involute gears is an arc of a circle.
Which of the statements given above is/are correct?
(a) Only 1 (b) Only 2 (c) Both 1 and 2 (d) Neither 1 nor 2

IES-40. Consider the following statements regarding the choice of conjugate teeth for the profile of mating gears: [IES-1999]
1. They will transmit the desired motion
2. They are difficult to manufacture.
3. Standardisation is not possible
4. The cost of production is low.
Which of these statements are correct?
(a) 1, 2 and 3 (b) 1, 2 and 4 (c) 2, 3 and 4 (d) 1, 3 and 4

IES-41. Which one of the following is correct? [IES-2008]

Gear Train and Gear Design

Chapter 9

When two teeth profiles of gears are conjugate, the sliding velocity between them
(a) Is always zero, all through the path of contact?
(b) Is zero, at certain points along the path of contact?
(c) Is never zero anywhere on the path of contact?
(d) Can be made zero by proper selection of profiles

Contact ratio

IES-42. Which one of the following is the correct statement? [IES 2007]
In meshing gears with involute gears teeth, the contact begins at the intersection of the

(a) Line of action and the addendum circle of the driven gear
(b) Line of action and the pitch circle of the driven gear
(c) Dedendum circle of the driver gear and the addendum circle of the driven gear
(d) Addendum circle of the driver gear and the pitch circle of the driven gear

IES-43. Common contact ratio of a pair of spur pinion and gear is [IES-2008]
(a) Less than 1·0　　　(b) equal to 1
(c) Between 2 and 3　　(d) greater than 3

Interference

IES-44. Interference between an involute gear and a pinion can be reduced by which of the following?
1. Increasing the pressure angle of the teeth in the pair, the number of teeth remaining the same.
2. Decreasing the addendum of the gear teeth and increasing the same for the pinion teeth by the corresponding amount. [IES-2008]
Select the correct answer using the code given below:
(a) 1 only　　(b) 2 only　　(c) Both 1 and 2　　(d) Neither 1 nor 2

IES-45. In gears, interference takes place when [IES-1993]
(a) The tip of a tooth of a mating gear digs into the portion between base and root circles
(b) Gears do not move smoothly in the absence of lubrication
(c) Pitch of the gear is not same
(d) gear teeth are undercut

IES-46. An involute pinion and gear are in mesh. If both have the same size of addendum, then there will be an interference between the [IES-1996]
(a) Tip of the gear tooth and flank of pinion.
(b) Tip of the pinion and flank of gear.
(c) Flanks of both gear and pinion.
(d) Tips of both gear and pinion.

IES-47. Interference between the teeth of two meshing involute gears can be reduced or eliminated by [IES 2007]
1. Increasing the addendum of the gear teeth and correspondingly reducing the addendum of the pinion.
2. Reducing the pressure angle of the teeth of the meshing gears.
3. Increasing the centre distance
Which of the statements given above is/are correct?
(a) 1 and 2　　(b) 2 and 3
(c) 1 only　　(d) 3 only

IES-48. Consider the following statements: [IES-2002]

Gear Train and Gear Design

Chapter 9

A 20° stub tooth system is generally preferred in spur gears as it results in
1. Stronger teeth
2. Lesser number of teeth on the pinion
3. Lesser changes of surface fatigue failure
4. Reduction of interference

Which of the above statements are correct?
(a) 1, 2 and 4   (b) 3 and 4   (c) 1 and 3   (d) 1, 2, 3 and 4

IES-49. Match List-I with List-II and select the correct answer using the codes given below the lists: [IES-2001]

List-I	List-II
A. Undercutting	1. Beam strength
B. Addendum	2. Interference
C. Lewis equation	3. Large speed reduction
D. Worm and wheel	4. Intersecting axes
	5. Module

Codes: A B C D   A B C D
   (a)   2 5 1 3   (b) 1 5 4 3
   (c)   1 3 4 5   (d) 2 3 1 5

IES-50. Which one of the following pairs is correctly matched? [IES-1999]
(a) Governors ....Interference
(b) Gears ....Hunting
(c) Klein's construction. ...Acceleration of piston
(d) Cam ....Pinion

IES-51. Consider the following characteristics: [IES-1998]
1. Small interference   2. Strong tooth.
3. Low production cost   4. Gear with small number of teeth.

Those characteristics which are applicable to stub 20° involute system would include
(a) 1 alone   (b) 2, 3 and 4   (c) 1, 2 and 3   (d) 1, 2, 3 and 4

IES-52. The motion transmitted between the teeth of two spur gears in mesh is generally [IES-1999]

(a) Sliding   (b) rolling
(c) Rotary   (d) party sliding and partly rolling

Beam Strength of Gear Tooth

IES-53. In heavy-duty gear drives, proper heat treatment of gears is necessary in order to: [IES-2006]

(a) Avoid interference
(b) Prevent noisy operation
(c) Minimize wear of gear teeth
(d) Provide resistance against impact loading on gear teeth

IES-54. Consider the following statements pertaining to the basic Lewis equation for the strength design of spur gear teeth: [IES-2005]
1. Single pair of teeth participates in power transmission at any instant.
2. The tooth is considered as a cantilever beam of uniform strength.
3. Loading on the teeth is static in nature.
4. Lewis equation takes into account the inaccuracies of the tooth profile.
5. Meshing teeth come in contract suddenly.

Gear Train and Gear Design

Chapter 9

Which of the statements given above are correct?
(a) 1, 3, 4 and 5 (b) 1, 2, 3 and 4 (c) 1, 2 and 3 (d) 2, 4 and 5

IES-55. **Assertion (A):** The Lewis equation for design of gear tooth predicts the static load capacity of a cantilever beam of uniform strength.
Reason (R): According to law of gears interchangeability is possible only when gears have same pressure angle and same module. [IES-2008]
(a) Both A and R are true and R is the correct explanation of A
(b) Both A and R are true but R is NOT the correct explanation of A
(c) A is true but R is false
(d) A is false but R is true

IES-56. In the formulation of Lewis equation for toothed gearing, it is assumed that tangential tooth load F_t, acts on the [IES-1998]
(a) Pitch point (b) tip of the tooth
(c) Root of the tooth (d) whole face of the tooth

IES-57. **Assertion (A):** The Lewis equation for gear tooth with involute profile predicts the static load capacity of cantilever beam of uniform strength. [IES-1994]
Reason (R): For a pair of meshing gears with involute tooth profile, the pressure angle and module must be the same to satisfy the condition of inter-changeability.
(a) Both A and R are individually true and R is the correct explanation of A
(b) Both A and R are individually true but R is **not** the correct explanation of A
(c) A is true but R is false
(d) A is false but R is true

IES-58. The dynamic load on a gear is due to [IES-2002]
1. Inaccuracies of tooth spacing
2. Irregularities in tooth profile
3. Deflection of the teeth under load
4. Type of service (i.e. intermittent, one shift per day, continuous per day).
Which of the above statements are correct?
(a) 1, 2 and 3 (b) 2, 3 and 4 (c) 1, 3 and 4 (d) 1, 2 and 4

IES-59. Consider the following statements:
The form factor of a spur gear tooth depends upon the [IES-1996]
1. Number of teeth 2. Pressure angle
3. Addendum modification coefficient 4. Circular pitch
Of these correct statements are
(a) 1 and 3 (b) 2 and 4 (c) 1, 2 and 3 (d) 1 and 4

IES-60. **Assertion (A):** If the helix angle of a helical gear is increased, the load carrying capacity of the tooth increases. [IES-1996]
Reason (R): The form factor of a helical gear increases with the increasing in the helix angle.
(a) Both A and R are individually true and R is the correct explanation of A
(b) Both A and R are individually true but R is **not** the correct explanation of A
(c) A is true but R is false
(d) A is false but R is true

IES-61. Match List I with List II and select the correct answer using the codes given below the Lists:
List I   List II   [IES-2000]
A. Unwin's formula   1. Bearings
B. Wahl factor   2. Rivets
C. Reynolds's equation   3. Gears

Gear Train and Gear Design

Chapter 9

	D. Lewis form factor				4. Springs			
Code:	A	B	C	D	A	B	C	D
(a)	1	4	2	3	(b) 2	3	1	4
(c)	1	3	2	4	(d) 2	4	1	3

IES-62. A spur gear transmits 10 kW at a pitch line velocity of 10 m/s; driving gear has a diameter of 1.0 m. Find the tangential force between the driver and the follower, and the transmitted torque respectively. **[IES-2009]**
(a) 1 kN and 0.5 kN-m
(b) 10 kN and 5 kN-m
(c) 0.5 kN and 0.25 kN-m
(d) 1 kN and 1 kN-m

Wear Strength of Gear Tooth

IES-63. The limiting wear load of spur gear is proportional to (where E_p = Young's modulus of pinion material; E_g = Young's modulus of gear material) **[IES-1997]**

(a) $(E_p + E_g)^{-1}$
(b) $\left(\dfrac{E_p + E_g}{E_p E_g}\right)$
(c) $\left(1 + \dfrac{E_p}{E_g}\right)$
(d) $\left(1 + \dfrac{E_g}{E_p}\right)$

Gear Lubrication

IES-64. Match List I (Types of gear failure) with List II (Reasons) and select the correct answer using the codes given below the Lists **[IES-2004]**

List I
A. Scoring
B. Pitting
C. Scuffing
D. Plastic flow

List II
1. Oil film breakage
2. Yielding of surface under heavy loads
3. Cyclic loads causing high surface stress
4. Insufficient lubrication

	A	B	C	D		A	B	C	D
(a)	2	1	3	4	(b)	2	3	4	1
(c)	4	3	1	2	(d)	4	1	3	2

Simple Gear train

IES-65. In a simple gear train, if the number of idler gears is odd, then the direction or motion of driven gear will **[IES-2001]**
(a) Be same as that of the driving gear
(b) Be opposite to that of the driving gear
(c) Depend upon the number of teeth on the driving gear
(d) Depend upon the total number of teeth on all gears of the train

IES-66. The gear train usually employed in clocks is a **[IES-1995]**
(a) Reverted gear train
(b) simple gear train
(c) Sun and planet gear
(d) differential gear.

Gear Train and Gear Design

Chapter 9

IES-67. In the figure shown above, if the speed of the input shaft of the spur gear train is 2400 rpm and the speed of the output shaft is 100 rpm, what is the module of the gear 4?
(a) 1.2   (b) 1.4
(c) 2   (d) 2.5

[IES-2005]

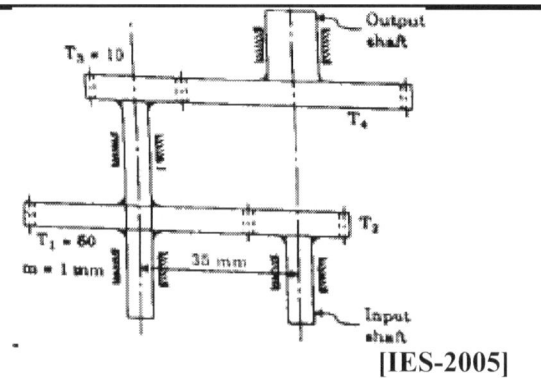

IES-68 In a machine tool gear box, the smallest and largest spindles are 100 rpm and 1120 rpm respectively. If there are 8 speeds in all, the fourth speed will be [IES-2002]
(a) 400 rpm   (b) 280 rpm   (c) 800 rpm   (d) 535 rpm

IES-69. A fixed gear having 200 teeth is in mesh with another gear having 50 teeth. The two gears are connected by an arm. The number of turns made by the smaller gear for one revolution of arm about the centre of the bigger gear is [IES-1996]
(a) $\frac{2}{4}$ (b) 3   (c) 4   (d) 5

Compound gear train

IES-70 The velocity ratio in the case of the compound train of wheels is equal to [IES-2000]

(a) $\dfrac{\text{No. of teeth on first driver}}{\text{No. of teeth on last follower}}$

(b) $\dfrac{\text{No. of teeth on last follower}}{\text{No. of teeth on first driver}}$

(c)(c) $\dfrac{\text{Product of teeth on the drivers}}{\text{Product of teeth on the followers}}$

(d) $\dfrac{\text{Product of teeth on the followers}}{\text{Product of teeth on the drivers}}$

IES-71. Consider the gear train shown in the given figure and table of gears and their number of teeth.

Gear: A   B   C   D   E   F
No of teeth: 20   50   25   75   26   65

[IES-1999]

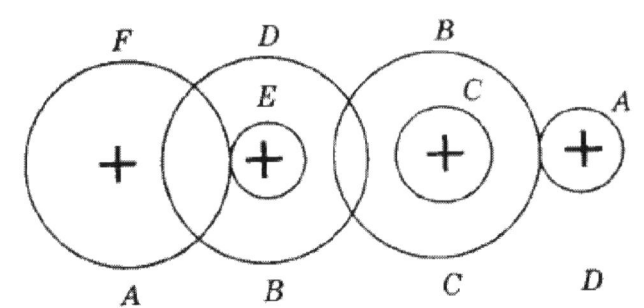

Gears BC and DE are moulded on parallel shaft rotating together. If the speed of A is 975 r.p.m., the speed of F will be

Gear Train and Gear Design

Chapter 9

IES-72. A compound train consisting of spur, bevel and spiral gears are shown in the given figure along with the teeth numbers marked against the wheels. Over-all speed ratio of the train is
(a) 8

(b) 2

(c) $\frac{1}{2}$

(d) $\frac{1}{8}$

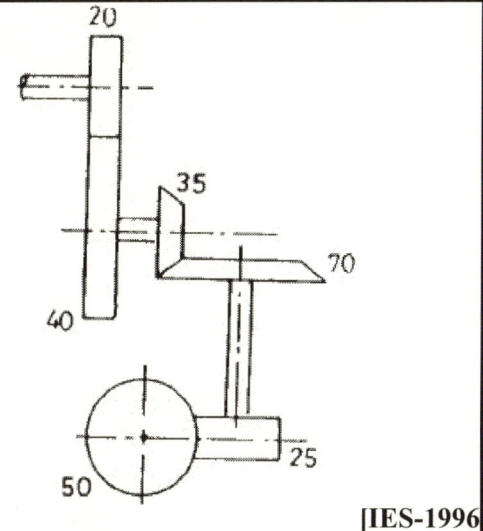

[IES-1996]

IES-73. In the compound gear train shown in the above figure, gears A and C have equal numbers of teeth and gears B and D have equal numbers of teeth. When A rotates at 800 rpm, D rotates at 200 rpm. The rotational speed of compound gears BC would then be
(a) 300 rpm
(b) 400rpm
(c) 500 rpm
(d) 600rpm

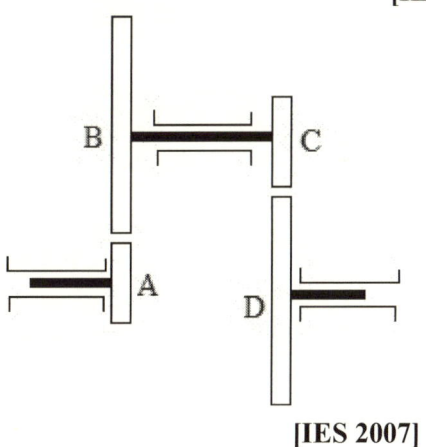

[IES 2007]

Reverted gear train

IES-74. Consider the following statements in case of reverted gear train: [IES-2002]
1. The direction of rotation of the first and the last gear is the same.
2. The direction of rotation of the first and the last gear is opposite.
3. The first and the last gears are on the same shaft.
4. The first and the last gears are on separate but co-axial shafts.
Which of these statements is/are correct?
(a) 1 and 3   (b) 2 and 3   (c) 2 and 4   (d) 1 and 4

IES-75. A reverted gear train is one in which the output shaft and input shaft
(a) Rotate in opposite directions   (b) are co-axial   [IES-1997]
(c) Are at right angles to each other   (d) are at an angle to each other

IES-76. In a reverted gear train, two gears P and Q are meshing, Q - R is a compound gear, and R and S are meshing. The modules of P and R are 4 mm and 5 mm respectively. The numbers of teeth in P, Q and R are 20, 40 and 25 respectively. The number of teeth in S is
[IES-2003]
(a) 23   (b) 35   (c) 50   (d) 53

IES-77. Two shafts A and B, in the same straight line are geared together through an intermediate parallel shaft. The parameters relating to the gears and pinions are given in the table:
[IES-2003]

Item	Speed	Teeth	PCD	Module
Driving wheel A	N_A	T_A	D_A	m

Gear Train and Gear Design

Chapter 9

Driven wheel B	N_B	T_B	D_B	m
Driven wheel C on the intermediate shaft	N_C	T_C	D_C	m
Driving wheel D on the intermediate shaft, in mesh with B	N_D	T_D	D_D	m

(a) $\dfrac{N_A}{N_B} = \dfrac{T_C}{T_A} \times \dfrac{T_B}{T_D}$

(b) $\dfrac{N_A}{N_B} = \dfrac{T_A}{T_C} \times \dfrac{T_D}{T_B}$

(c) $D_A + D_C = D_B + D_D$

(d) $T_A + T_C = T_B + T_D$

IES-78. A gear having 100 teeth is fixed and another gear having 25 teeth revolves around it, centre lines of both the gears being jointed by an arm. How many revolutions will be made by the gear of 25 teeth for one revolution of arm?
[IES-2009]
(a) 3   (b) 4   (c) 5   (d) 6

Epicyclic gear train

IES-79. If the annular wheel of an epicyclic gear train has 100 teeth and the planet wheel has 20 teeth, the number of teeth on the sun wheel is   [IES-2003]
(a) 80   (b) 60   (c) 40   (d) 20

IES-80. In the epicyclic gear train shown in the given figure, A is fixed. A has 100 teeth and B has 20 teeth. If the arm C makes three revolutions, the number of revolutions made by B will be
(a) 12
(b) 15
(c) 18
(d) 24

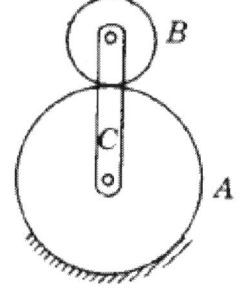

[IES-1997]

IES-81. An epicyclic gear train has 3 shafts A, B and C, A is an input shaft running at 100 rpm clockwise. B is an output shaft running at 250 rpm clockwise. Torque on A is 50 kNm (clockwise). C is a fixed shaft. The torque to fix C
(a) Is 20 kNm anticlockwise   [IES-2002]
(b) is 30 kNm anticlockwise
(c) Is 30 kNm clockwise
(d) Cannot be determined as the data is insufficient

Gear Train and Gear Design

Chapter 9

IES-82. A single epicyclic gear train is shown in the given figure. Wheel A is stationary. If the number of teeth on A and B are 120 and 45 respectively, then when B rotates about its own axis at 100 rpm, the speed of C would be

(a) 20 rpm
(b) $27\frac{3}{11}$ rpm
(c) $19\frac{7}{11}$ rpm
(d) 100 rpm

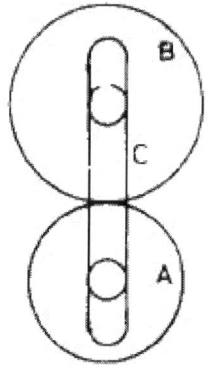

[IES-1994]

Terminology of Helical Gears

IES-83. If α = helix angle, and p_c = circular pitch; then which one of the following correctly expresses the axial pitch of a helical gear? [IES 2007]

(a) $p_c \cos\alpha$ (b) $\dfrac{p_c}{\cos\alpha}$ (c) $\dfrac{p_c}{\tan\alpha}$ (d) $p_c \sin\alpha$

IES-84. A helical gear has the active face width equal to b, pitch p and helix angle α. What should be the minimum value of b in order that contact IS maintained across the entire active face of the gear? [IES-2004]

(a) $p\cos\alpha$ (b) $p\sec\alpha$ (c) $p\tan\alpha$ (d) $p\cot\alpha$

IES-85. **Assertion (A):** Helical gears are used for transmitting motion and power between intersecting shafts, whereas straight bevel gears arc used for transmitting motion and power between two shafts intersecting each other at 90°. [IES-2000]
Reason (R): In helical gears teeth are inclined to axis of the shaft and arc in the form or a helix. Where as in bevel gears, teeth arc tapered both in thickness and height form one end to the other.
(a) Both A and R are individually true and R is the correct explanation of A
(b) Both A and R are individually true but R is **not** the correct explanation of A
(c) A is true but R is false
(d) A is false but R is true

IES-86. **Assertion (A):** Shafts supporting helical gears must have only deep groove ball-bearings.
Reason (R): Helical gears produce axial thrusts. [IES-1999]
(a) Both A and R are individually true and R is the correct explanation of A
(b) Both A and R are individually true but R is **not** the correct explanation of A
(c) A is true but R is false
(d) A is false but R is true

IES-87. **Assertion (A):** Crossed helical gears for skew shafts are not used to transmit heavy loads.
Reason (R) The gears have a point contact, and hence are not considered strong.
(a) Both A and R are individually true and R is the correct explanation of A
(b) Both A and R are individually true but R is **not** the correct explanation of A
(c) A is true but R is false
(d) A is false but R is true

[IES-1995]

Gear Train and Gear Design

Chapter 9

Bevel Gears

IES-88. In a differential mechanism, two equal sized bevel wheels A and B are keyed to the two halves of the rear axle of a motor car. The car follows a curved path. Which one of the following statements is correct? [IES-2004]
The wheels A and B will revolve at different speeds and the casing will revolve at a speed which is equal to the
(a) Difference of speeds of A and B
(b) Arithmetic mean of speeds of A and B
(c) Geometric mean of speeds of A and B
(d) Harmonic mean of speeds of A and B

Worm Gears

IES-89. **Assertion (A):** Tapered roller bearings must be used in heavy duty worm gear speed reducers.
Reason (R): Tapered roller bearings are suitable for large radial as well as axial loads.
(a) Both A and R are individually true and R is the correct explanation of A
(b) Both A and R are individually true but R is **not** the correct explanation of A
(c) A is true but R is false
(d) A is false but R is true [IES-2005]

IES-90. Consider the following statements in respect of worm gears: [IES-2005]
1. They are used for very high speed reductions.
2. The velocity ratio does not depend on the helix angle of the worm.
3. The axes of worm and gear are generally perpendicular and non-intersecting.
Which of the statements given above are correct?
(a) 1 and 2   (b) 1 and 3   (c) 2 and 3   (d) 1, 2 and 3

IES-91. For a speed ratio of 100 smallest gear box is obtained by using which of the following?
(a) A pair of spur gears [IES-2008]
(b) A pair of bevel and a pair of spur gears in compound gear train
(c) A pair of helical and a pair of spur gears in compound gear train
(d) A pair of helical and a pair of worm gears in compound gear train

IES-92. Consider the following statements regarding improvement of efficiency of worm gear drive: [IES-2004]
1. Efficiency can be improved by increasing the spiral angle of worm thread to 45° or more
2. Efficiency can be improved by adopting proper lubrication
3. Efficiency can be improved by adopting worm diameter as small as practicable to reduce sliding between worm-threads and wheel teeth
4. Efficiency can be improved by adopting convex tooth profile both for worm and wheel
Which of the statements given above are correct?
(a) 1, 2 and 3   (b) 1, 2 and 4   (c) 2, 3 and 4   (d) 1, 3 and 4

IES-93. The lead angle of a worm is 22.5 deg. Its helix angle will be [IES-1994]
(a) 22.5 deg.   (b) 45 deg.   (c) 67.5 deg.   (d) 90°C.

Previous 20-Years IAS Questions

Spur gear

IAS-1. Match List I (Terms) with List II (Definition) and select the correct answer using the codes given below the lists: [IAS-2001]
List I   List II

Gear Train and Gear Design

Chapter 9

    A. Module   1. Radial distance of a tooth from the pitch circle to the top of the tooth
    B. Addendum   2. Radial distance of a tooth from the pitch circle to the bottom of the tooth
    C. Circular pitch   3. Distance on the circumference of the pitch circle from a point of one tooth to the corresponding point on the next tooth
                                                       4. Ratio of a pitch circle diameter in mm to the number of teeth

Codes: A B C   A B C
(a)   4   1   3   (b)   4   2   3
(c)   3   1   2   (d)   3   2   4

IAS-2   Consider the following specifications of gears A, B, C and D: [IAS-2001]

Gears	A	B	C	D
Number of teeth	20	60	20	60
Pressure angle	$14\frac{1}{2}°$	$14\frac{1}{2}°$	20°	$14\frac{1}{2}°$
Module	1	3	3	1
Material	Steel	Brass	Brass	Steel

Which of these gears form the pair of spur gears to achieve a gear ratio of 3?
(a) A and B   (b) A and D   (c) B and C   (d) C and D

IAS-3.   If the number of teeth on the wheel rotating at 300 r.p.m. is 90, then the number of teeth on the mating pinion rotating at 1500 r.p. m. is   [IAS-2000]
(a) 15   (b) 18   (c) 20   (d) 60

IAS-4.   A rack is a gear of   [IAS-1998]
(a) Infinite diameter   (b) infinite module
(c) zero pressure angle   (d) large pitch

Classification of Gears

IAS-5.   **Assertion (A):** While transmitting power between two parallel shafts, the noise generated by a pair of helical gears is less than that of an equivalent pair of spur gears.   [IAS-2000]
Reason(R): A pair of helical gears has fewer teeth in contact as compared to an equivalent pair of spur gears.
(a) Both A and R are individually true and R is the correct explanation of A
(b) Both A and R are individually true but R is **not** the correct explanation of A
(c) A is true but R is false
(d) A is false but R is true

Pitch point

IAS-6.   An imaginary circle which by pure rolling action, gives the same motion as the actual gear, and is called   [IAS-2000]
(a) Addendum circle   (b) pitch circle
(c) Dedendum circle   (d) base circle

Pressure angle

IAS-7.   The pressure angle of a spur gear normally varies from   [IAS-2000]
(a) 14° to 20°   (b) 20° to 25°   (c) 30° to 36°   (d) 40° to 50°

Gear Train and Gear Design

Chapter 9

Minimum Number of Teeth

IAS-8. Minimum number of teeth for involute rack and pinion arrangement for pressure angle of 20° is [IAS-2001]
(a) 18 (b) 20 (c) 30 (d) 34

Cycloidal teeth

IAS-9. The tooth profile most commonly used in gear drives for power transmission is
(a) A cycloid (b) An involute (c) An ellipse (d) A parabola [IAS-1996]

Contact ratio

IAS-10. Which one of the following statements is correct? [IAS-2007]
(a) Increasing the addendum results in a larger value of contact ratio
(b) Decreasing the addendum results in a larger value of contact ratio
(c) Addendum has no effect on contact ratio
(d) Both addendum and base circle diameter have effect on contact ratio

IAS-11. The velocity of sliding of meshing gear teeth is [IAS-2002]

(a) $(\omega_1 \times \omega_2)x$ (b) $\left(\dfrac{\omega_1}{\omega_2}\right)x$ (c) $(\omega_1 + \omega_2)x$ (d) $\dfrac{(\omega_1 + \omega_2)}{x}$

(Where ω_1 and ω_2 = angular velocities of meshing gears
x = distance between point of contact and the pitch point)

Interference

IAS-12. For spur with gear ratio greater than one, the interference is most likely to occur near the [IAS-1997]
(a) Pitch point (b) point of beginning of contact
(c) Point of end of contact (d) root of the tooth

IAS-13. How can interference in involute gears be avoided? [IAS-2007]
(a) Varying the centre distance by changing the pressure angle only
(b) Using modified involute or composite system only
(c) Increasing the addendum of small wheel and reducing it for the larger wheel only
(d) Any of the above

IAS-14. Which one of the following statements in respect of involute profiles for gear teeth is not correct? [IAS-2003]
(a) Interference occurs in involute profiles,
(b) Involute tooth form is sensitive to change in centre distance between the base circles.
(c) Basic rack for involute profile has straight line form
(d) Pitch circle diameters of two mating involute gears are directly proportional to the base circle diameter

IAS-15. Assertion (A): In the case of spur gears, the mating teeth execute pure rolling motion with respect to each other from the commencement of engagement to its termination. [IAS-2003]
Reason (R): The involute profiles of the mating teeth are conjugate profiles which obey the law of gearing.
(a) Both A and R are individually true and R is the correct explanation of A
(b) Both A and R are individually true but R is **not** the correct explanation of A
(c) A is true but R is false
(d) A is false but R is true

Gear Train and Gear Design

Chapter 9

IAS-16. **Assertion (A):** Gears with involute tooth profile transmit constant velocity ratios between shafts connected by them. **[IAS-1997]**
Reason (R): For involute gears, the common normal at the point of contact between pairs of teeth always passes through the pictch point.
(a) Both A and R are individually true and R is the correct explanation of A
(b) Both A and R are individually true but R is **not** the correct explanation of A
(c) A is true but R is false
(d) A is false but R is true

Compound gear train

IAS-17. There are six gears A, B, C, D, E, F in a compound train. The numbers of teeth in the gears are 20, 60, 30, 80, 25 and 75 respectively. The ratio of the angular speeds of the driven (F) to the driver (A) of the drive is
(a) $\dfrac{1}{24}$ (b) $\dfrac{1}{8}$ (d) ~~12~~ **[IAS-1995]**
(c) 15

Epicyclic gear train

IAS-18. A fixed gear having 100 teeth meshes with another gear having 25 teeth, the centre lines of both the gears being joined by an arm so as to form an epicyclic gear train. The number of rotations made by the smaller gear for one rotation of the arm is **[IAS-1995]**
(a) 3 (b) 4 (b) 5 (d) 6

IAS-19. For an epicyclic gear train, the input torque = 100 Nm. RPM of the input gear is 1000 (clockwise), while that of the output gear is 50 RPM (anticlockwise). What is the magnitude of the holding torque for the gear train? **[IAS-2007]**
(a) Zero (b) 500 Nm (c) 2100 Nm (d) None of the above

IAS-20. In the figure shown, the sun wheel has 48 teeth and the planet has 24 teeth. If the sun wheel is fixed, what is the angular velocity ratio between the internal wheel and arm?
(a) 3.0

(b) 1.5

(c) 2.0

(d) 4.0

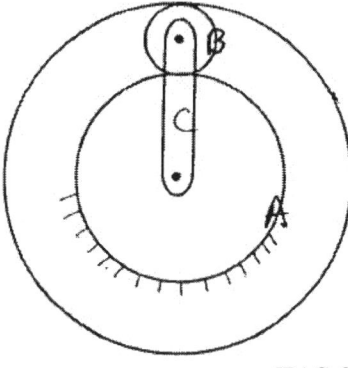

[IAS-2004]

IAS-21. 100 kW power is supplied to the machine through a gear box which uses an epicyclic gear train. The power is supplied at 100 rad/s. The speed of the output shaft of the gear box is 10 rad/s in a sense opposite to the input speed. What is the holding torque on the fixed gear of the train? **[IAS-2004]**
(a) 8 kNm (b) 9 kNm (c) 10 kNm (d) 11 kNm

Gear Train and Gear Design

Chapter 9

IAS-22. In the epicyclic gear train shown in the figure, $T_A = 40$, $T_B = 20$. For three revolutions of the arm, the gear B will rotate through

(a) 6 revolutions

(b) 2.5 revolutions

(c) 3 revolutions

(d) 9 revolutions

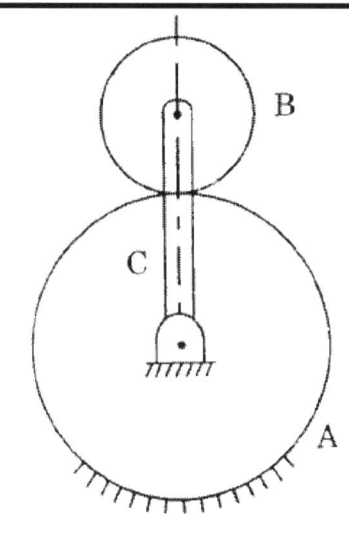

[IAS-2003]

Bevel Gears

IAS-23. **Assertion (A):** Spiral bevel gears designed to be used with an offset in their shafts are called „hypoid gears" [IAS-2004]
Reason (R): The pitch surfaces of such gears are hyperboloids of revolution.
(a) Both A and R are individually true and R is the correct explanation of A
(b) Both A and R are individually true but R is **not** the correct explanation of A
(c) A is true but R is false
(d) A is false but R is true

Worm Gears

IAS-24. If reduction ratio of about 50 is required in a gear drive, then the most appropriate gearing would be [IAS-1999]
(a) spur gears (b) bevel gears
(c) Double helical gears (d) worm and worm wheel

IAS-25. Speed reduction in a gear box is achieved using a worm and worm wheel. The worm wheel has 30 teeth and a pitch diameter of 210 mm. If the pressure angle of the worm is 20°, what is the axial pitch of the worm?
(a) 7 mm (b) 22 mm [IAS-2004]
(c) 14 mm (d) 63 mm

IAS-26. A speed reducer unit consists of a double-threaded worm of pitch = 11 mm and a worm wheel of pitch diameter = 84 mm. The ratio of the output torque to the input to ratio is
(a) 7·6 (b) 12 (c) 24 (d) 42 [IAS-2002]

IAS-27. The maximum efficiency for spiral gears in mesh is given by (Where (θ = shaft angle and ϕ, = friction angle) [IAS-1998]

(a) $\dfrac{1+\cos(\theta-\phi)}{1+\cos(\theta+\phi)}$

(b) $\dfrac{1+\cos(\theta+\phi)}{1+\cos(\theta-\phi)}$

(c) $\dfrac{1-\cos(\theta-\phi)}{1+\cos(\theta+\phi)}$

(d) $\dfrac{1-\cos(\theta+\phi)}{1+\cos(\theta-\phi)}$

IAS-28. **Assertion (A):** A pair of gears forms a rolling pair. [IAS-1996]
Reason (R): The gear drive is a positive drive.

Gear Train and Gear Design

Chapter 9

(a) Both A and R are individually true and R is the correct explanation of A
(b) Both A and R are individually true but R is **not** the correct explanation of A
(c) A is true but R is false
(d) A is false but R is true

Answers with Explanation (Objective)

Previous 20-Years GATE Answers

GATE-1. Ans. (a)
GATE-2. Ans. (a)
GATE-3. Ans. (b)
GATE-4. Ans. (a)
GATE-5. Ans. (c)
GATE-6. Ans. (d)

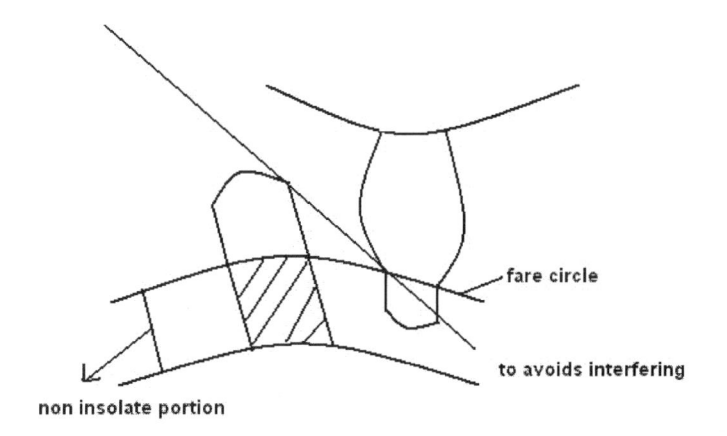

There are several ways to avoid interfering:
i. Increase number of gear teeth
ii. Modified involutes
iii. Modified addendum
iv. Increased centre distance

GATE-7. Ans. (a)

GATE-8. Ans. (a) Centre distance $= \dfrac{D_1 + D_2}{2} = \dfrac{mT_1 + mT_2}{2} = \dfrac{5(19+37)}{2} = 140$ mm

GATE-9. Ans. (c)
GATE-10. Ans. (a)
GATE-11. Ans. (b)
GATE-12. Ans. (b)
GATE-13. Ans. (c)
GATE-14. Ans. (c)
GATE-15. Ans. (b)

Gear Train and Gear Design

Chapter 9

Given : Module m = 2, $\dfrac{D}{T} = 2$

∴ $\quad D = 80 \times 2 = 160$ mm

$2F = 1000$, or $F = 500$ N

Let T_1 be the torque applied by motor.

T_2 be the torque applied by gear.

∴ Power transmission = 80%

Now, $T_1 \omega_1 = \dfrac{2T_2 \times \omega_1}{0.8}$

or $\quad T_1 = \dfrac{2 \times F \times (D/2)}{0.8} \times \dfrac{\omega_1}{\omega_2}$

$= 2 \times 500 \times \dfrac{0.16}{2} \times \dfrac{1}{0.8} \times \dfrac{1}{4}$

$= 25$ N–m.

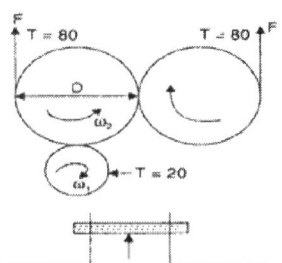

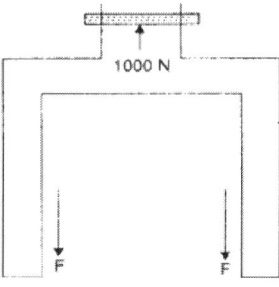

GATE-16. Ans. (c)

$$P \cos \phi = F$$

∴ Force acting along the line of action,

$$P = \dfrac{F}{\cos \phi}$$

$$= \dfrac{500}{\cos 20°}$$

$$= 532 N$$

GATE-17. Ans. (a)

Given, $\dfrac{N_1}{N_2} = 12$, $\dfrac{N_1}{N_2} = 4 = \dfrac{D_2}{D_1}$

$m_1 = 3$, $m_2 = 4$

Now, $\quad \dfrac{D_1}{Z_1} = \dfrac{D_2}{Z_2}$

$\Rightarrow \quad \dfrac{Z_1}{Z_2} = \dfrac{D_1}{D_2} = \dfrac{N_2}{N_1} = \dfrac{1}{4}$

$\Rightarrow \quad Z_2 = Z_1 \times 4 = 64$

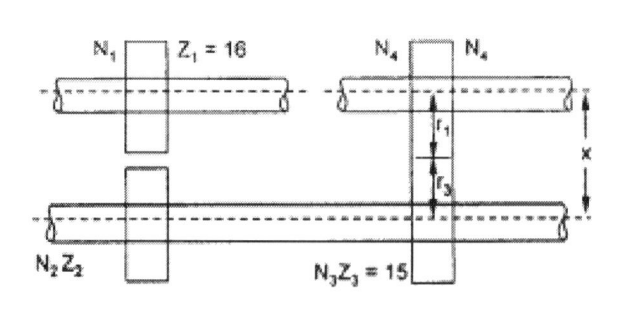

$\Rightarrow \quad 12 = \dfrac{D_4}{D_3}$

$\Rightarrow \quad \dfrac{D_4}{D_3} = 3$

Also, $\quad \dfrac{Z_3}{Z_4} = \dfrac{D_3}{D_4}$

$\Rightarrow \quad Z_4 = Z_3 \dfrac{D_4}{D_3} = Z_3 \times 3 = 15 \times 3$

$= 45$

GATE-18. Ans. (b)

Gear Train and Gear Design

Chapter 9

Now, $\quad x = r_4 + r_3 = \dfrac{D_4 + D_3}{2}$

But $\quad \dfrac{D_4}{Z_4} = \dfrac{D_3}{Z_3} = 4$

$\Rightarrow \quad D_4 = 180, D_3 = 60$

$\therefore \quad x = \dfrac{180 + 60}{2} = 120\,mm$

GATE-19. Ans. (c)

	Arm	2	3	4	5
1.	0	$+x$	$-\dfrac{N_2}{N_3}x$	$-\dfrac{N_2}{N_3}x$	$-\dfrac{N_4}{N_5} \times \dfrac{N_2}{N_3}x$
2.	y	y	y	y	y
	y	$x+y$	$y - \dfrac{N_2}{N_3}x$		$y - \dfrac{N_4}{N_5} \times \dfrac{N_2}{N_3}x$

$x + y = 100 \text{ (cw)}$

$y = -80 \text{ (ccw)}$

$Speed\ of\ Gear(W_5) = -80 - \dfrac{32}{80} \times \dfrac{20}{24} \times 180 = -140 = 140 \text{ (ccw)}$

GATE-20. Ans. (a)

GATE-21. Ans. (b)

Explanation

	Arm A	B	C
Fix arm A Give one rotation to B	0	1	-1
Multiply by x	0	$+x$	$-x$
Add y	y	$X+y$	$y-x$

B is fixed, therefore $\quad x + y = 0$

$\quad y = \text{rad/sec(ccw)}$

$\Rightarrow \quad x = -4 \text{ rad/sec(cw)}$

Angular velocity of gear $\quad C = y - x = 4 - (-4) = 8 \text{ rad/s}$

GATE-22. Ans. (b)

Arm	Sun	Planet	Ring
+1	+1	+1	+1
0	$\dfrac{80}{30} \times \dfrac{30}{20}$	$-\dfrac{80}{30}$	-1
1	5	$-\dfrac{5}{3}$	0

For 5 Revolutions Of Sun, Arm rotates by 1

$\therefore$ for 100 revolutions of Sun, Arm rotates by $\dfrac{100}{5} = 20$

GATE-23. Ans. (d)

Gear Train and Gear Design

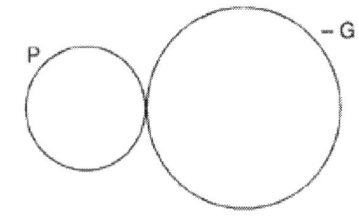

Chapter 9

We know $\dfrac{N_P}{N_G} = \dfrac{T_G}{T_P}$

where, N_P = speed of pinion, N_G = speed of gear wheel

T_G = number of teeth of gear,

T_P = number of teeth of pinion

∴ $\dfrac{1200}{N_G} = \dfrac{120}{40}$

or $N_G = 400$ r.p.m

Since power transmitted by both gear will be equal

i.e. $T_P \omega_P = T_G \omega_G$

where, T_P = torque transmitted by pinion, T_G = torque transmitted by gear wheel

∴ $\dfrac{20 \times 2\pi \times 1200}{60} = \dfrac{T_G \times 2\pi \times 400}{60}$

∴ torque transmitted by gear, $T_G = 60$ N.m.

GATE-24. Ans. (a)

$\dfrac{\omega_1 - \omega_5}{\omega_2 - \omega_5} = 3$ (with respect to arm 5 or carrier 5)

$\dfrac{\omega_3 - \omega_5}{\omega_4 - \omega_5} = 2$ (with respect to carrier 5)

As, $\omega_3 = \omega_2$

∴ $\dfrac{\omega_1 - \omega_5}{\omega_4 - \omega_5} = 6$

GATE-25. Ans. (d)

$\omega_1 = 60$ rpm (Clockwise)

$\omega_4 = 120$ rpm (Counter clock wise)

$\dfrac{60 - \omega_5}{-120 - \omega_5} = 6$

∴ $\omega_5 = -156$ i.e. counter clockwise

GATE-26. Ans. (b)
GATE-27. Ans. (d) speed reduction = 1440/36 = 40
GATE-28. Ans. (c)

Previous 20-Years IES Answers

IES-1. Ans. (b) Centre distance = $\dfrac{D_1 + D_2}{2} = \dfrac{mT_1 + mT_2}{2} = \dfrac{m}{2}(T_1 + T_2) = \dfrac{2}{2} \times 99 = 99$ mm

Gear Train and Gear Design

Chapter 9

IES-2. Ans. (a)
IES-3. Ans. (d)
IES-4. Ans. (c)
IES-5. Ans. (a)
IES-6. Ans. (a)
IES-7. Ans. (d)
IES-8. Ans. (a)
IES-9. Ans. (b)
IES-10. Ans. (a)
IES-11. .Ans. (c)
IES-12. Ans. (b)
IES-13. Ans. (d)
IES-14. Ans. (a)
IES-15. Ans. (a)
IES-16. Ans. (d)
IES-17. Ans. (c)
IES-18. Ans. (b) When pair of teeth touch at the pitch point, they have for the instant pure rolling action. At any other position they have the sliding action.
IES-19. Ans. (d)
IES-20. Ans. (d)
IES-21. Ans. (b)
IES-22. Ans. (c)

$$\text{Centre distance in mm} = \frac{m}{2}(T_1 + T_2)$$
$$= \frac{6}{2}(60 + 20)$$
$$= 240 \text{ mm}$$

IES-23. Ans. (d)
IES-24. Ans. (b)
IES-25. Ans. (b) Module $= \frac{d}{T} = \frac{288}{48} = 6$ mm

Circular pitch $= \frac{\pi d}{T} = \pi \times 6 = 18.84$ mm ; addendum = 1 module = 6 mm

diametral pitch $= \frac{T}{d} = \frac{1}{6}$

Circular pitch = - = 1t X 6 = 18.84 mm

IES-26. Ans. (c)
IES-27. Ans. (d) For involute gears, the pressure angle is constant throughout the teeth engagement.
IES-28. Ans. (c) The pressure angle is always constant in involute gears.
IES-29. Ans. (b)
IES-30. Ans. (c)
IES-31. Ans. (d)
IES-33. Ans. (d)
IES-34. Ans. (a)
IES-35. Ans. (b)
IES-36. Ans. (a)
IES-37. Ans. (d)
IES-38. Ans. (b)
IES-39. Ans. (d) 1. A stub tooth has a working depth lower than that of a full-depth tooth.
2. The path of contact for involute gears is a line.
IES-40. Ans. (a) Cost of production of conjugate teeth, being difficult to manufacture is high.

Gear Train and Gear Design

Chapter 9

IES-41. Ans. (a)
IES-42. Ans. (a)
IES-43. Ans. (c) The ratio of the length of arc of contact to the circular pitch is known as **contact ratio** i.e. number of pairs of teeth in contact. The contact ratio for gears is greater than one. Contact ratio should be at least 1.25. For maximum smoothness and quietness, the contact ratio should be between 1.50 and 2.00. High-speed applications should be designed with a face-contact ratio of 2.00 or higher for best results.
IES-44. Ans. (c)
IES-45. Ans. (a) In gears, interference takes place when the tip of a tooth of a mating gear digs into the portion between base .and root circle.
IES-46. Ans. (a)
IES-47. Ans. (d)
IES-48. Ans. (a)
IES-49. Ans. (a)
IES-50. Ans. (c)
IES-51. Ans. (b) Involute system is very interference prone.
IES-52. Ans. (b)
IES-53. Ans. (c)
IES-54. Ans. (c)
IES-55. Ans. (b) The beam strength of gear teeth is determined from an equation (known as Lewis equation) and the load carrying ability of the toothed gears as determined by this equation gives satisfactory results. In the investigation, Lewis assumed that as the load is being transmitted from one gear to another, it is all given and taken by one tooth, because it is not always safe to assume that the load is distributed among several teeth.

Notes: (*i*) The Lewis equation is applied only to the weaker of the two wheels (*i.e.* pinion or gear).
(**ii**) When both the pinion and the gear are made of the same material, then pinion is the weaker.
(**iii**) When the pinion and the gear are made of different materials, then the product of $(\sigma_w \times y)$ *or* $(\sigma_o \times y)$ is the deciding factor. The Lewis equation is used to that wheel for which $(\sigma_w \times y)$ *or* $(\sigma_o \times y)$ is less.

IES-56. Ans. (b)
IES-57. Ans. (c) For a pair of meshing gears with involute tooth profile, the pressure angle and module must be the same to satisfy the condition of inter-changeability it is not correct. Due to law of gearing.
IES-58. Ans. (a)
IES-59. Ans. (c)
IES-60. Ans. (a)
IES-61. Ans. (d)
IES-62. Ans. (a) Power transmitted = Force × Velocity
$$\Rightarrow 10 \times 10^3 = \text{Force} \times 10$$
$$\Rightarrow \text{Force} = \frac{10 \times 10^3}{10} = 1000 \text{ N/m}$$
$$\text{Torque Transmitted} = \text{Force} \times \frac{\text{diameter}}{2}$$
$$= 1000 \times \frac{1}{2} = 1000 \times 0.5$$
$$= 500 \text{ N} - \text{m} = 0.5 \text{ kN} - \text{m}$$

IES-63. Ans. (b)
IES-64. Ans. (b)
IES-65. Ans. (a)
IES-66. Ans. (a)
IES-67. Ans. (b)

Gear Train and Gear Design

Chapter 9

$$\frac{mT_2 + mT_1}{2} = 35$$

or $T_2 = 10$

$$N_1 = -N_i \times \frac{T_2}{T_1} = N_3$$

$$N_4 = \frac{-N_3 T_3}{T_4} = +N_i \times \frac{T_2}{T_1} \times \frac{T_3}{T_4} \text{ or } 100 = 2400 \times \frac{10}{60} \times \frac{10}{T_4} \text{ or } T_4 = 40$$

$$\frac{m'T_3 + m'T_4}{2} = 35 \text{ or } m' = \frac{70}{(40+10)} = 1.4$$

IES-68. Ans. (b)

IES-69. Ans. (d) $1 + 200/50 = 1 + 4 = 5$

IES-70. Ans. (d)

IES-71. Ans. (b) Speed ratio $\dfrac{N_F}{N_A} = \dfrac{T_A \times T_C \times T_E}{T_B \times T_D \times T_F} = \dfrac{20 \times 25 \times 26}{50 \times 75 \times 65} = \dfrac{4}{75}$ or $N_F = 975 \times \dfrac{4}{75} = 52$ rpm

IES-72. Ans. (a)

Elements of higher pair like follower in cam
is under the action of gravity or spring force.

$$\text{Train value} = \frac{\text{speed of lost driven or follower}}{\text{speed of the first gear}}$$

$$\text{Train value} = \frac{\text{product of no. of teeth no the drives}}{\text{product of no. of teeth on the drives}}$$

$$\frac{\text{speed of the first drive}}{\text{speed of the last driven or follower}}$$

IES-73. Ans. (b) From the figure $r_A + r_B = r_C + r_D$ or $T_A + T_B = T_C + T_D$ and as $N_B + N_C$ it must be $T_B = T_D$ & $T_A = T_C$

Or $\dfrac{N_B}{N_A} = \dfrac{N_D}{N_C}$ or $N_C = \sqrt{N_A N_D} = \sqrt{800 \times 200} = 400$ rpm $[\because N_B = N_c]$

IES-74. Ans. (d)

IES-75. Ans. (b)

IES-76. Ans. (a)

Summation of radius will be constant.

$R_P + R_Q = R_R + R_G$

or $D_P + D_Q = D_R + D_S$

or $m_1(T_P + T_Q) = m_2(T_R + T_S)$

or $4(20 + 40) = 5(25 + T_S)$

or $T_S = 23$

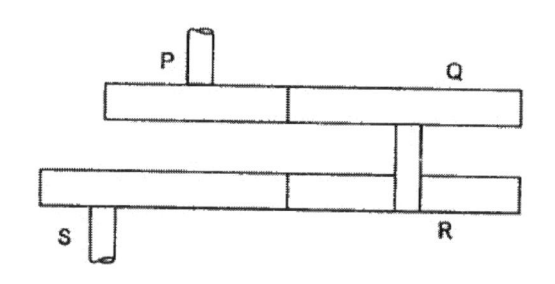

IES-77. Ans. (b)

(i) $D_A + D_C = D_B + D_D$

(ii) $mT_A + mT_C = mT_B + mT_D$

(iii) $\dfrac{N_A}{N_B} = \dfrac{N_A}{N_C} \times \dfrac{N_C}{N_B} = \dfrac{T_C}{T_A} \times \dfrac{T_B}{N_D}$

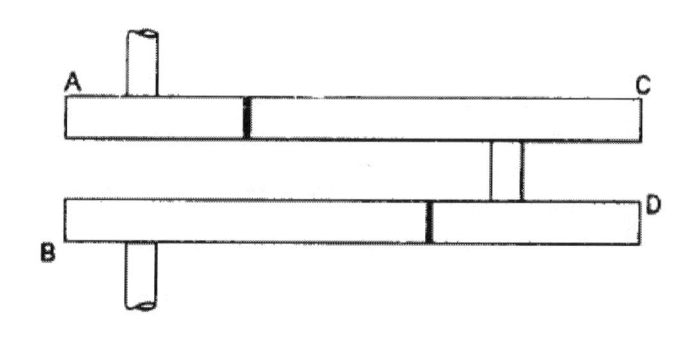

Gear Train and Gear Design

Chapter 9

IES-78. Ans. (c)

Arm	N_A	N_B
0	+1	$\dfrac{-100}{25}$

Multiplying through out by x

| 0 | +x | $\dfrac{-100}{25}x$ |
| y | y + x | y − 4x |

Given that y + x = 0 ∴ x = -y = -1

($\because$ y = 1)

∴ N_B = y − 4x = 5

IES-79. Ans. (b) From geometry

$2d_p + d_s = d_A$

or $2T_p + T_s = T_A$

or $T_s = T_A - 2T_p = 100 - 2 \times 20 = 60$

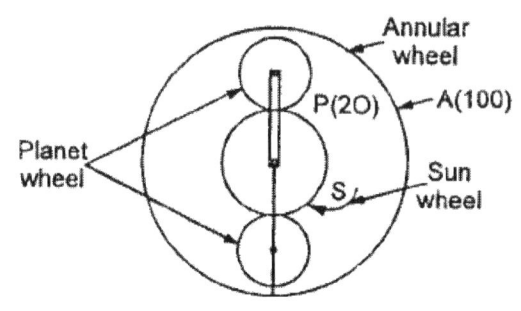

IES-80. Ans. (c) For 1 revolution of C,

$N_B = 1 + \dfrac{T_A}{T_B} = 1 + \dfrac{100}{20} = 6$ ∴ for 3 revolution, $N_D = 6 \times 3 = 18$

IES-81. Ans. (b)

Now $\omega_1 M_1 - \omega_2 M_2 = 0$

∴ $M_2 = \dfrac{100 \times 50}{250} = 20$ KNm(anticlockwise)

and $M_1 + M_2 + M_3 = 0$

50 − 20 + M_3 = 0

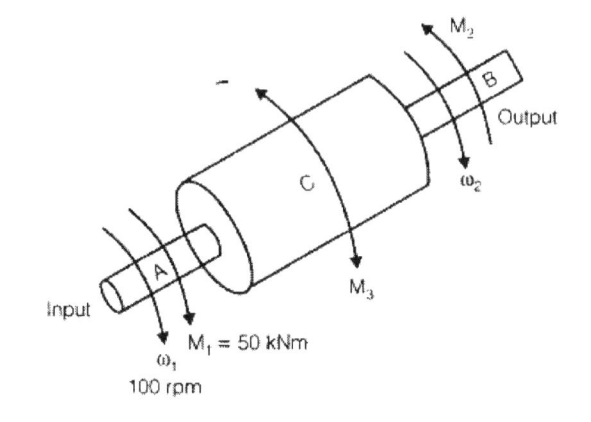

∴ $M_3 = -30$ kNm(clockwise) = 30 kNm(anticlockwise)

IES-82. Ans. (c)

IES-83. Ans. (c)

IES-84. Ans. (d) $b \geq \dfrac{P}{\tan \alpha}$

IES-85. Ans. (d)

IES-86. Ans. (a)

IES-87. Ans. (b)

IES-88. Ans. (d)

IES-89. Ans. (a)

IES-90. Ans. (d)

IES-91. Ans. (d)

IES-92. Ans. (a)

Gear Train and Gear Design

Chapter 9

$$\text{Gear } \eta_{wormgear} = \frac{\tan \lambda}{\tan(\phi_v + \lambda)}$$

$\tan \phi_v = \pi_v$

$\tan \lambda = \dfrac{z_w \cdot m}{d_W}$

The face of wormgear is made concave to envelope the worm.

IES-93. Ans. (c) α = Pressure angle ≅ lead angle; α + β = 90°; β = helix angle = 90° - 22.5° = 67.5°

Previous 20-Years IAS Answers

IAS-1. Ans. (a)

IAS-2. Ans. (b) For a gear pair i) module must be same
(ii) Pressure angle must be same.

IAS-3. Ans. (b) Peripheral velocity (πDN) = constant. $\pi D_1 N_1 = \pi D_2 N_2$ and D = mT

or $\pi m T_1 N_1 = \pi m T_1 N_1$ or $T_2 = T_1 \times \dfrac{N_1}{N_2} = 90 \times \dfrac{300}{1500} = 18$

Or you may say speed ratio, $\dfrac{N_1}{N_2} = \dfrac{T_2}{T_1}$

IAS-4. Ans. (a)

IAS-5. Ans. (c) In spur gears, the contact between meshing teeth occurs along the entire face width of the tooth, resulting in a sudden application of the load which, in turn, results in impact conditions and generates noise.

In helical gears, the contact between meshing teeth begins with a point on the leading edge of the tooth and gradually extends along the diagonal line across the tooth. There is a gradual pick-up of load by the tooth, resulting in smooth engagement and silence operation.

IAS-6. Ans. (b)

IAS-7. Ans. (a)

IAS-8. Ans. (a) $T_{min} = \dfrac{2h_f}{\sin^2 \theta} = \dfrac{2 \times 1}{\sin^2 20^o} = 17.1 \quad as > 17 \quad So\, T_{min} = 18$

IAS-9. Ans. (b) It is due to easy manufacturing.

IAS-10. Ans. (d) contact ratio $= \dfrac{length\ of\ arc\ of\ contant}{circular\ pitch}$

$= \dfrac{\sqrt{R_A^2 - R^2 \cos^2 \theta} + \sqrt{r_A^2 - r^2 \cos^2 \theta} - (R+r)\sin\theta}{P_c(\cos\theta)}$

IAS-11. Ans. (c)

IAS-12. Ans. (d)

IAS-13. Ans. (d)

IAS-14. Ans. (b)

IAS-15. Ans. (a)

IAS-16. Ans. (a)

IAS-17. Ans. (a) The ratio of angular speeds of F to A $= \dfrac{T_A \cdot T_C \cdot T_E}{T_B \cdot T_D \cdot T_F} = \dfrac{20 \times 30 \times 25}{60 \times 80 \times 75} = \dfrac{1}{24}$

IAS-18. Ans. (c) Revolution of 25 teeth gear $= 1 + \dfrac{T_{100}}{T_{25}}$ (for one rotation of arm) $= 1 + \dfrac{100}{25} = 5$

Gear Train and Gear Design

Chapter 9

IAS-19. Ans. (c) $T_i + T_o + T_{arm} = 0$ and $T_i\omega_i + T_o\omega_o + T_{arm}\omega_{arm} = 0$

Gives, $T_{arm} = T_i\left(\dfrac{\omega_i}{\omega_o} - 1\right) = T_i\left(\dfrac{N_i}{N_o} - 1\right) = 100 \times \left(\dfrac{-1000}{50} - 1\right) = -2100\ Nm$

IAS-20. Ans. (a) $\dfrac{N_B - N_C}{N_A - N_C} = -\dfrac{T_A}{T_B} \quad \because N_A = 0$

$\dfrac{N_B - N_C}{-N_C} = -\dfrac{48}{24}\quad or\ -\dfrac{N_B}{N_C}+1 = -2 \quad or\ \dfrac{N_B}{N_C} = 2+1 = 3$

IAS-21. Ans. (b) $T_1 + T_2 + T_3 = 0$

$T_1W_1 + T_2W_2 + T_3W_3 = 0$

$W_3 = 0$

$T_1W_1 = 100\,kW,\ W_1 = 100\,rad/s$

$\therefore T_1 = 1\,kNm$

Or $T_2 = -\dfrac{T_1W_1}{W_2} = \dfrac{-100}{(10)} = -10\,kNm$

$T_3 = -T_2 - T_1 = -(-10) - 1 = 9\,kNm$

IAS-22. Ans. (d)

IAS-23. Ans. (a)

IAS-24. Ans. (d)

IAS-25. Ans. (b) $m = \dfrac{210}{30} = 7$ and $P_x = \pi m = \dfrac{22}{7} \times 7 = 22\,mm$

Axial pitch = circular pitch of the worm wheel = πm

IAS-26. Ans. (a) $\dfrac{\text{Output torque}}{\text{Input torque}} = \dfrac{\text{pitch diameter of worm wheel}}{\text{pitch of worm}} = \dfrac{84}{11} = 7.6$

IAS-27. Ans. (b)

IAS-28. Ans. (d) In rolling pair one link rolls over another fixed link.

Miscellaneous

Chapter 10

10. Miscellaneous

Objective Questions (IES, IAS, GATE)

IES-1. He mass moment of inertia of the two rotors in a two rotor system is 100 kg m² and 10 kg m². The length of the shaft of uniform diameter between the rotors is 110 cm. The distance of node from the rotor of lower moment of inertia is   [IES-2002]
(a) 80 cm   (b) 90 cm   (c) 100 cm   (d) 110 cm

IES-2. Consider a harmonic motion x = 1.25 sin (5t −π/6) cm. Match List-I with List-II and select the correct answer using the .codes given below the lists:
List I	List II	[IES-2001]
A. Amplitude (cm)	1. 5/2 π	
B. Frequency (cycle/s)	2. 1.25	
C. Phase angle (rad)	3. 1/5	
D. Time period (s)	4. π /6	

Codes: A B C D   A B C D
(a)   4 1 2 3   (b) 2 3 4 1
(c)   4 3 2 1   (d) 2 1 4 3

IES-3. The pitching of a ship in the ocean is an oscillatory periodic motion. A ship is pitching 6° above and 6° below with a period of 20s from its horizontal plane. Consider the following statements in this regard:
1. The motion has a frequency of oscillation (i.e. pitching) of 3 cycles/minute
2. The motion has an angular frequency of 3.14 rad/s.
3. The angular velocity of precession of ship's rotor is π²/300 rad/s.
4. The amplitude of pitching is π/30 rad.
Which of these statements are correct?   [IES-2000]
(a) 1 and 2   (b) 1, 2 and 4   (c) 2, 3 and 4   (d) 1, 3 and 4

IES-4. Two geared shafts A and B having moments of inertia I_a and I_b and angular acceleration α_a and α_b respectively are meshed together. B rotates at G times the speed of A. If the gearing efficiency of the two shafts in η, then in order to accelerate B, the torque which must be applied to A will be
(a) $I_a \alpha_a + G^2 I_b \alpha_b / \eta$ (b) $G^2 I_a \alpha_a / \eta$ [IES-1998]
(c) $G^2 I_b \alpha_a / \eta$ (d) $G^2 I_b \alpha_a / \eta$

IES-5. In S.H.M., with respect to the displacement vector, the positions of Velocity vector and Acceleration vector will be respectively   [IES-1998]
(a) 180° and 90°   (b) 90° and 180°   (c) 0° and 90°   (d) 90° and 0°

IES-6. Two links OA and OB are connected by a pin joint at 'O'. The link OA turns with angular velocity ω₁ radians per second in the clockwise direction and the link OB turns with angular velocity ω₂ radians per second in the

Miscellaneous

Chapter 10

anticlockwise direction. If the radius of the pin at 'O' is 'r', then the rubbing velocity at the pin joint 'O' will be [IES-1998]

(a) $\omega_1 \omega_2 r$ (b) $(\omega_1 - \omega_2)r$ (c) $(\omega_1 + \omega_2)r$ (d) $(\omega_1 - \omega_2)2r$

IES-7. A torsional system with discs of moment of inertia I_1 and I_2 shown in the given figure, is gear driven such that the ratio of the speed of shaft B to shaft A is 'n'. Neglecting the inertia of gears, the equivalent inertia of disc on B at the speed of shaft A is equal to

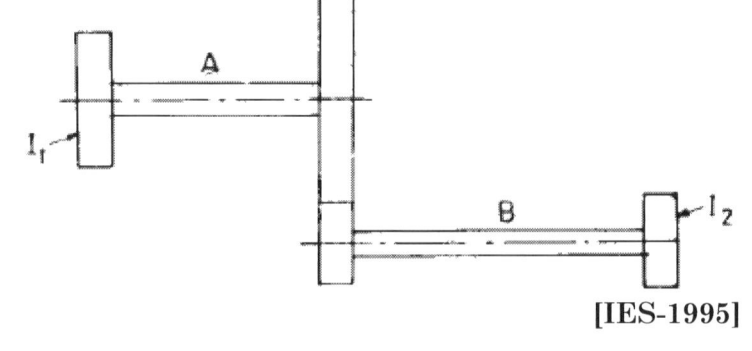

[IES-1995]

(a) nI_2 (b) n^2I_2 (c) I_2/n^2 (d) I_2/n

IES-8. In the figure shown crank AB is 15 cm long and is rotating at 10 rad/s. C is vertically above A. CA equals 24 cm. C is a swivel trunnion through which BD (40 cm) slides. If ABCD becomes a vertical line during its motion, the angular velocity of the swivel trunnion at that instant will be
(a) Zero
(b) (100/25) rad/s
(c) (100/15) rad/s
(d) (100/10) rad/s

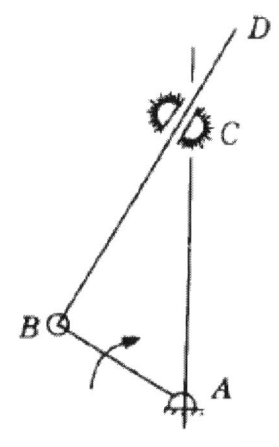

[IES-1997]

Previous 20-Years IES Answers

IES-1. Ans. (c)
IES-2. Ans. (d)
    Amplitude → 1.25
    Frequency → $\dfrac{5}{2\pi}$
    Phase angle → $\dfrac{\pi}{6}$
    Time period → $\dfrac{1}{5}$
IES-3. Ans. (d)

Miscellaneous

Chapter 10

$$\theta = 6° = \frac{6° \times \pi}{180}$$

$T_p = 20 \text{ sec}$

$$\therefore \omega = \frac{2\pi\theta}{T_p} = \frac{2\pi \times 6 \times \pi}{180 \times 20} = \frac{\pi^2}{300} \, r/s$$

$$\text{amplitude} = \frac{6\pi}{180} = \frac{\pi}{30} \, \text{rad}$$

IES-4. Ans. (a)
IES-5. Ans. (b)
IES-6. Ans. (c)
IES-7. Ans. (b) $I'_B = I_B (\text{on B}) \times \left(\frac{|\omega_B|}{\omega_A}\right)^2 = n^2 I_2$

IES-8. Ans. (a)

Made in the USA
Columbia, SC
13 July 2024